Analytical Applications of Spectroscopy II

INTERNATIONAL CONFERENCE ON SPECTROSCOPY ACROSS THE SPECTRUM

1576313

Analytical Applications of Spectroscopy II

BLDSC BOOK·NET
T 04 / 4337
REDISTRIBUTED MATERIAL

Edited by

A.M.C. Davies
Norwich Near-Infrared Consultancy

and

C.S. Creaser
School of Chemical Sciences, University of East Anglia

UNIVERSITY OF GREENWICH
1 - AUG 1996
LIBRARY

The Proceedings of the Second International Conference on Spectroscopy Across the Spectrum: Techniques and Applications of Analytical Spectroscopy, held 9–12th July, 1990 in Hatfield, England.

A catalogue record for this book is available from the British Library.

ISBN 0-85186-403-1

© The Royal Society of Chemistry, 1991

All Rights Reserved
No part of this book may be reproduced or transmitted in any form or by any means—graphic, electronic, including photocopying, recording, taping, or information storage and retrieval systems—without written permission from the Royal Society of Chemistry

Published by The Royal Society of Chemistry,
Thomas Graham House, The Science Park, Cambridge CB4 4WF

Printed and bound in Great Britain by Bookcraft (Bath) Ltd.

LI

Th

WC

This

9409779 MBCSBN

Preface

The soft hyphen

This second volume of 'Analytical Applications of Spectroscopy' includes papers presented at the *'Spectroscopy Across the Sepctrum'* conference held at Hertford, UK, in July 1990. This meeting was similar in format to the first conference held in 1987, in that a number of spectroscopists gathered to present reviews of their specialist areas within the context of themes common to all spectroscopies. We have been encouraged by the reception of the first volume, and by the growing number of spectroscopists who have adopted a multidisciplinary approach to spectroscopic analysis, to edit the papers from the recent meeting into a volume under the same title.

In the preface to the first volume, we referred to one of Thomas Hirschfeld's many contributions — *the hy-phen-ated technique*. That is, the combination of spectroscopic, or spectroscopic and chromatographic, techniques designated by a hyphen, such as GC–FTIR and MS–MS. These combined (hyphenated) techniques were prominent in 'Analytical Applications of Spectroscopy, Volume I'. This volume reflects the change of emphasis which was apparent at the second conference. In 1987, Hirschfeld's hyphen represented the hardware link for the transport of sample between techniques. In 1990, two of the distinctive themes were chemometrics and the combination of spectroscopy and microscopy. In spectroscopy/microscopy, **radiation** links the two techniques by passing through the microscope to the spectrometer; while in chemometrics it is software which provides the means to combine **information** from different techniques. By analogy with instrumental combined techniques, in which **sample** is passed from one instrument to another in a hyphenated system, these links can be represented by a *'soft hyphen'*. While the difference may seem trivial, it is important because of the unlikely combinations it brings to mind. This is particularly evident in contributions involving near infrared spectroscopy but, as Professor Grasselli's introduction makes clear, the development of combined techniques which do not require the mechanical transport of sample are of increasing importance to the future of analytical spectroscopy.

This volume has been organized, like the first volume, according to the principal spectroscopic technique discussed in each paper. However, for many papers, it was necessary to make a rather arbitrary decision on the most appropriate section. As in the first volume, we encourage readers to consult the contents and subject index pages in order to locate all that is relevant to their interests.

We would like to express our gratitude to the authors whose contributions are included in this volume, to Dorothy Gee for re-typing several of the papers, to Megan Davies and Claire Creaser for their help in the preparation of the index, and to Catherine Lyall of RSC Publications.

Norwich
January 1991

A.M.C. Davies
C.S. Creaser

Contents

Dedication

This volume is dedicated to the memory of Professor Harry Willis without whom there would be no *Spectroscopy Across the Spectrum.*

Section 1

INTRODUCTION

OPERATION SUPER SLEUTH: APPLICATIONS OF SPECTROSCOPY TO PROBLEMS IN SCIENCE

J. G. Grasselli

OHIO UNIVERSITY, ATHENS, OH 45701, USA

1 INTRODUCTION

The interaction of light with matter allows us to analyze materials by spectroscopic methods. Such methods, utilizing X-rays, ultraviolet, visible, infrared and radio parts of the electromagnetic spectrum, are widely used by industry, academia, forensic and government laboratories for "sleuthing" and providing unique and rapid identifications of samples, no matter what size or shape they may be. Typical examinations from the industrial lab may be identification of valve deposits, stains on metal, contaminants in polymers; or structural analyses of organic as well as inorganic materials; or quantitative analysis of trace amounts of toxic or hazardous materials in air, water or waste samples from the environment. Samples come from operating plants, technical service, or research.

Spectroscopy is but one of the many important tools of analytical science. Science, in general, has been instrumental in advancing the well-being of mankind and analytical sciences have been the facilitating factor which made that possible. This century, and particularly the period since World War II, has been a time of absolutely breathtaking technological change. In a famous quote from *Future Shock*, Alvin Toffler highlighted the degree to which the time period for change in our society has been shortened. He pointed out that "if the last 50,000 years of man's existence were divided into lifetimes of approximately 62 years each, there have been about 800 such lifetimes Of these 800, fully 650 were spent in caves.

Only during the last 70 lifetimes has it been possible to communicate effectively from one lifetime to another -- as writing made it possible to do. Only during the last four has it been possible to measure time with any precision. Only in the last two has anyone anywhere used an electric motor, and the overwhelming majority of all material goods we use in daily life today have been developed within the present, the 800th lifetime."

There are some interesting ways of measuring how quickly things become different. Television, space travel, digital computers, biotechnology, advanced composite materials, organ transplants, SSTs and the elimination of most infectious diseases have occurred in the last 50 years. By the time a child born today graduates from college, the amount of knowledge in the world will have doubled. Also, just imagine -- 80% of all the recognised scientists who have ever lived are alive today.

It has been said that we know only what we can measure. If that is true, it forms the most solid basis for my belief that analytical science, with its capability to delineate structure and composition of materials, measure kinetics and mechanisms of reactions in a rigourous and reproducible way, is a fundamental catalyst for the technological gains we have made. It is nearly impossible to believe that the transistor, the microchip, the laser, our sophisticated catalysts, the pacemaker or all the advances in biotechnology and our knowledge of diseases could have arisen without this capability.

We can prove this point in more detail by looking at several aspects of the technical progress in analytical science over the last 35 years (Fig. 1). Our detection limits have gone from micrograms to the subpicogram level. A millimeter sized sample presented a challenge

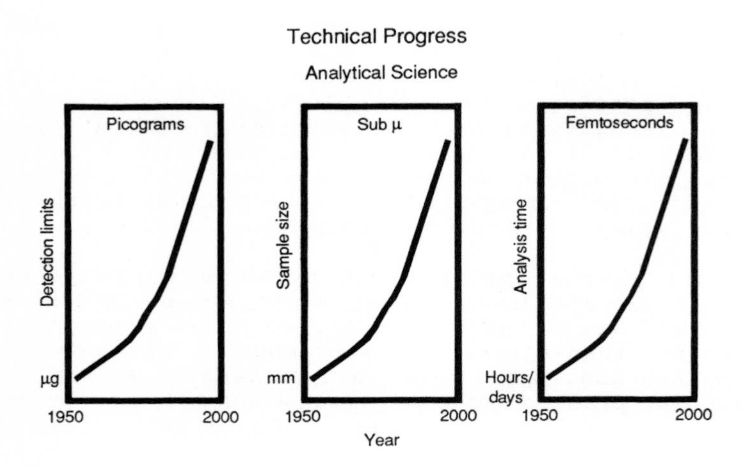

Figure 1

in 1950; but today we look at submicron, in fact, at single atoms resolved in analytical electron microscopes in an almost routine manner. Analysis time has changed from hours or sometimes even days for a complete qualitative and quantitative analysis on a sample to modern capabilities which include real-time spectroscopy on in situ experiments which give us data on molecular dynamics in the femtosecond time regime. Marie Curie once said "nothing in life is to be feared; it is only to be understood." Analytical science today is indeed the catalyst and the key to our understanding.

However, if the analytical scientist is to be effective in solving the problems of society, there is another dimension to his/her rôle which must be recognised. The successful analytical chemist must be adept at a number of techniques and must be at the forefront of chemical knowledge. The function of the analytical chemist is to provide information of sufficiently valid nature, that is, of the requisite statistical significance so that meaningful decisions can be made about materials or problems. The analytical chemist is essentially a "problem solver" and to do the job most effectively he/she should use a complete analytical approach to solving the problems. This analytical approach is defined by the following steps:

- Defining the problem correctly
- Ensuring that the samples available are representative of the problem
- Interacting with the client to obtain his/her knowledge of the problem and to define the boundary of timeliness and accuracy required
- Developing an analytical plan involving an evaluation of the sequence and best methods to be employed
- Completing the work using the highest level of expertise and good chemical knowledge
- Communicating answers, not data, including the precision and reliability on all numbers and specifying cautions or constraints on the use of the data
- Interpreting the information and results in a clear, consistent, and meaningful report that clearly addresses the problem.

In addition, the analyst must balance the constraints between time, cost and accuracy which are key elements in every "real world" problem. A good answer, on time, is much better than a perfect answer that is too late. This disciplined and logical approach should be the basis for the ongoing practice of analytical chemistry.

2. DISCUSSION

Today's problems in science which present challenges and opportunities to our analytical "super sleuths", are often grouped as the ABC's of technology: Advanced materials, Biotechnology, and Computers (encompassing information science, in general). To illustrate the power of modern instrumentation, let's look at a few "case histories" of problem solving in these areas using vibrational spectroscopy as our "super sleuth".

Advanced Materials

In addition to ceramics, composites and catalysts, polymers continue to dominate the materials market in the U.S. in both volume of sales and in applications. During the development of the plastic bottle for the soft drink beverage market, copolymers of acrylonitrile/styrene were under study because of the excellent barrier properties which such polymers exhibited. Bottles blown from this polymer were already in the pilot plant stage when a production problem occurred. The bottles were marked with "fish-eyes" - small globs of gelatinous material which marred the physical appearance of the bottle and also deteriorated its properties. The identification of the "fish-eyes" presented a formidable analytical problem. Surface scraping and reflection methods of sampling for FT-IR spectroscopic analysis were not successful. The bottle wall was about 4 mls thick but, even so, the tremendous advantage of FT-IR for obtaining information on energy-limited samples allowed us to obtain a transmission spectrum through the wall. A sample mask was used to obtain the spectrum of the clear portion of the bottle and subtract it from the spectrum obtained from the gel or "fish eye". (A microscope

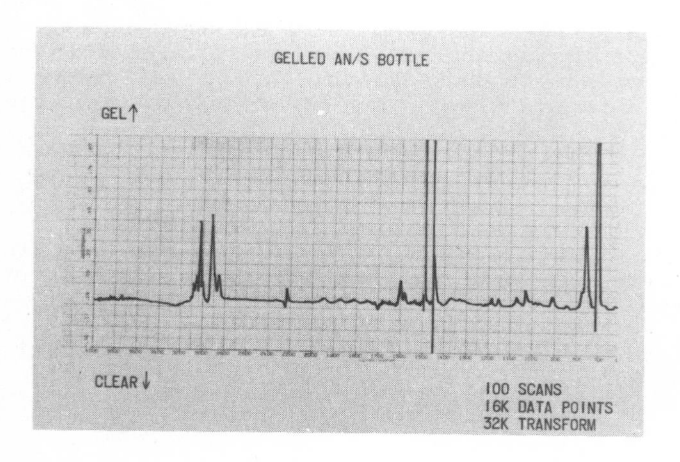

Figure 2

accessory would also be useful for this kind of sample). The spectrum shown in Fig. 2 is the resulting absorbance spectrum from the subtraction. It was easily identified as polystyrene. The problem could now be traced to either improper mixing in the polymerization reactor, or to an improper charge of styrene. Instead of a regular alternating polymer of acrylonitrile and styrene, the polymer had blocks of polystyrene, which melts at a different temperature than AN/S, resulting in the "fish-eyes" in the blown bottle.

Biotechnology

The dramatic emergence of biotechnology in the last two decades holds the promise of developing into at least a $60 billion industry by the turn of the century. Exciting research in diagnostics, genetics (both plant and animal) and in biologically catalyzed reactions will transform entire industries. In particular, the field of molecular biology is benefiting greatly from advances in instrumentation in both infrared and Raman spectroscopies, which can now give structural and compositional information on dilute samples in water and on highly fluorescent materials. The "FT" revolution in both techniques has been responsible for these new capabilities. FT-Raman in particular is making dramatic advances in helping to elucidate the structure of amino acids, proteins, peptides, polypeptides and polysaccharides.

Raman and infrared spectroscopy provide complementary information on molecules, and both are necessary to yield a complete vibrational analysis. To illustrate the power of the combined techniques. Fig. 3 shows the FT-IR and FT-Raman spectra of Trypotophan. This

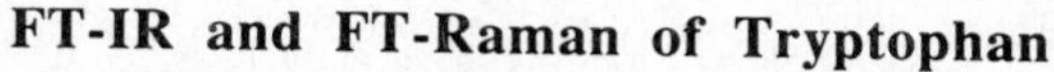

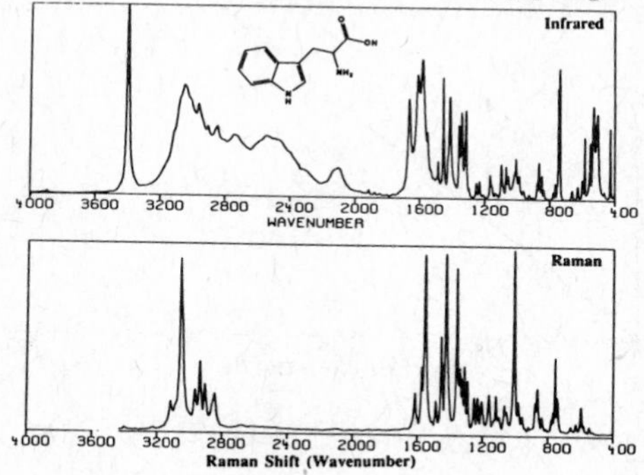

Figure 3

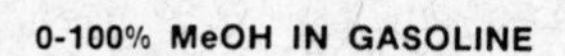

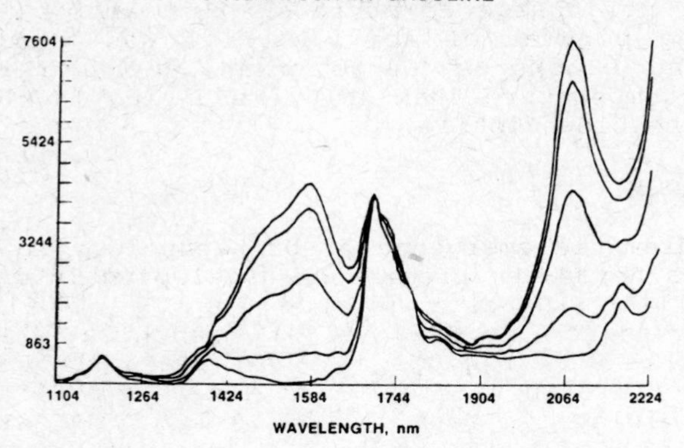

Figure 4

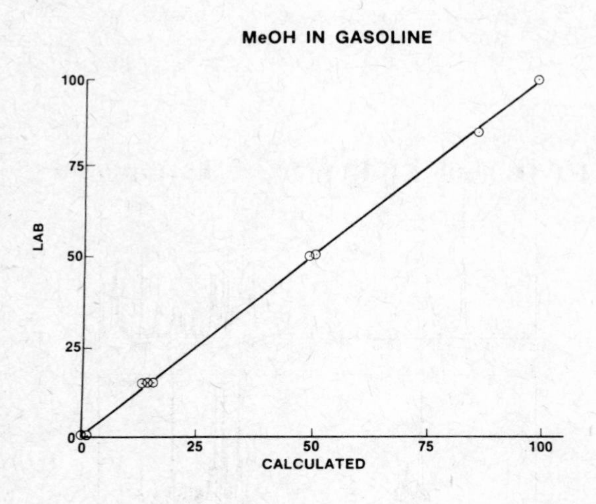

Figure 5

amino acid has an indole group which consists of a pyrrole ring and a benzene ring. The pyrrole ring has a "free" NH group which has a very strong stretching mode at 3400 cm^{-1} in the infrared spectrum. The highly polar amide and carboxylic acid groups are easily identified in the infrared spectrum.

Computers (information science)

Computers have become pervasive in every aspect of our professional and private lives. In analytical science they have introduced another set of ABC's - Analysis Becomes Computerized. From use on our simplest pH meters to the most sophisticated Scanning Electron Microscopes, the microprocessor, PC or minicomputers are throughout the laboratories. Their impact is on instrument operation and control, data manipulation, automatic searching through reference collections for identification, sample tracking and lab information systems, networking and process and product quality control.

The issue of industrial competitiveness in a global economy has been seen by many to reduce to the concept that "The cost of quality may be expensive, but the cost on nonquality is nonexistence". Therefore, perhaps the fastest growing area of analytical science today is that of process and product quality control. The trend in process instrumentation is from off-line to at-line to on-line and now to in-line. The reasons are: to avoid sampling, to improve response time, and to enhance safety. Near infrared spectroscopy, utilizing fibre optic sampling probes and correlation spectroscopy, has become the instrument of choice in many applications. One of these is shown in Figs. 4 and 5. Fig. 4 shows the near IR curves for a series of gasolines containing 0-100% methanol. By using the entire spectral range between 1264 and 2224 nm as a measure of the concentration of methanol, the correlation between the measured near IR lab results and the known values of methanol is shown in Fig. 5. This rapid and accurate method of analysis is the preferred method for determining oxygenates in gasoline, and serves as only one example of the power of NIR for quality control.

3 RESULTS AND DISCUSSION

The instrument revolution in which we currently find ourselves must be among the most exciting things ever to have happened in science. The problems of science are more complex and more challenging than ever before. But our ability to address those problems has also increased exponentially.

The frontiers of excitement in analytical science today are in our ability to observe and understand:

- Dynamic measurements under real-world conditions
- Surfaces and interfaces
- Atomic resolution and micro structures

To do this we will use the:

- Ultimate in separations, speciation and trace analysis
- Expert systems

The future could not be brighter for every "Operation Super Sleuth" applied to problems in science.

Section 2

VIBRATIONAL SPECTROSCOPY

FOURIER TRANSFORM RAMAN SPECTROSCOPY

Bruce Chase

CENTRAL RESEARCH AND DEVELOPMENT, DUPONT COMPANY, WILMINGTON, DELAWARE, USA

Since the advent of available visible lasers in the late 1960's, great expectations have been expressed for Raman spectroscopy, especially in the industrial laboratory. The complementary nature to infrared spectrometry coupled with the ease of sampling should have resulted in Raman spectrometry becoming a widely used technique for qualitative analysis. Unfortunately, the very low efficiency of the Raman scattering process, one part in 10^8, allows other radiative processes to dominate. Specifically, fluorescence, arising either from the sample itself or from impurities contained in the sample is often many times more intense than the Raman signal of interest. There have been many approaches targeted towards reducing this background fluorescence, ranging from time resolved detection in the nanosecond time domain to UV excited resonance Raman measurements. Many of the approaches proved successful for specific samples, but none provided a universal solution. It was recognized very early on that the optimal approach for fluorescence minimization would be to reduce the energy of the incident photons to below the threshold for excitation of the electronic excited state which is fluorescent. This would involve the use of red or near infrared lasers.

Unfortunately, moving the excitation laser towards the red end of the spectrum has several associated problems. The Raman scattering cross-section for non resonant excitation follows a ν^4 law. Shifting the laser from 514.5 nm to 1064 nm results in a 16 fold loss in Raman intensity. Additionally, the shot noise limited detectors available for visible excitation are not usable beyond 1000 nm. It was recognised in 1964 by Chantry and co-workers[1] that interferometry might provide a solution to this overall loss in sensitivity by making the near infrared excited Raman measurement with a multiplexing spectrometer. Unfortunately this work was not pursued until the late 1980's when the use of interferometers with near infrared excitation from a Nd/YAG laser was shown to be quite successful in the elimination of background fluorescence[2-4]. Currently, the performance of these FT-

Raman instruments has been increased to the point where rapid, reliable operation is easily attainable and a wide variety of samples can be examined.

The basic layout of an FT-Raman instrument based on a Bomem DA3.02 interferometer is shown in figure 1. As has been pointed out by several workers[5], these Raman experiments can also be readily accomplished with the bench top model interferometer. The output from a laser, usually Nd/YAG operating at 1064nm is focused onto a sample held at the focus of a collection optic. In this case 180° back scattering is observed with a 90° off axis ellipsoid.

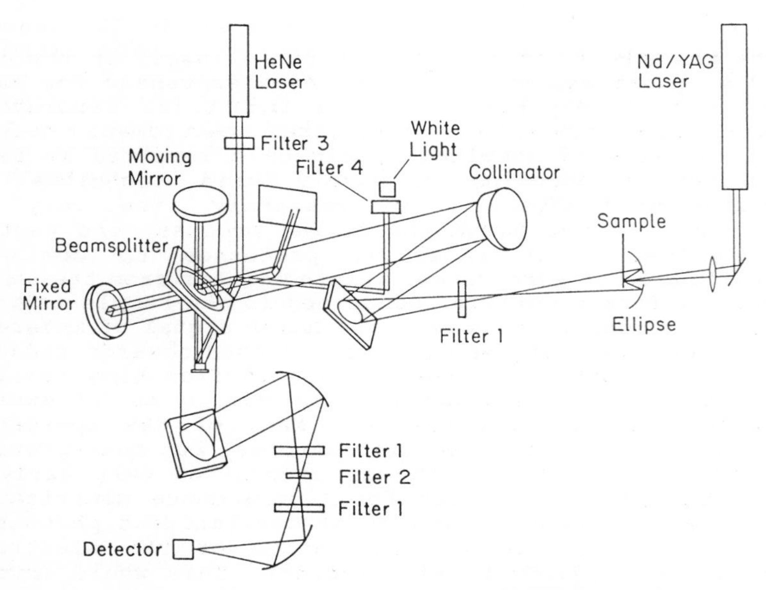

Figure 1. Instrumental layout for FT-Raman using a Bomem DA3.02.

Equivalently 90° scattering and/or lens collection can be employed. The scattered radiation is collected and focused onto the entrance aperture of the interferometer. A quartz or a CaF2 beamsplitter can be employed for the near infrared region and the detector is usually either an InGaAs element or a Ge element. Typical NEP's are in the range of 10^{-14} to 10^{-15} watts/Hz. Care must be taken to filter out unwanted optical signals such as the Raleigh line, non lasing lines from the internal HeNe laser and near infrared radiation from the internal white light reference source if present.

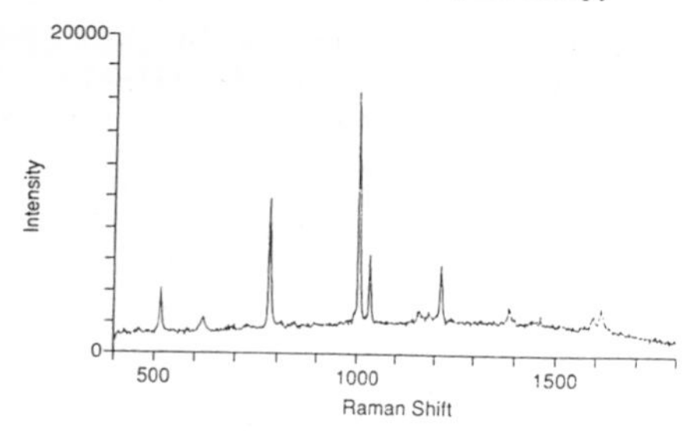

Figure 2a. Dispersive Raman spectrum of toluene, 90 mwatts 514.5 nm 4cm^{-1}.

The performance of FT-Raman with respect to fluorescence minimization has been well documented by many different research groups and will not be repeated here. The current performance in comparison to conventional scanning instruments is shown in figure 2a and 2b. The upper trace of toluene was recorded on a double monochrometer system equipped with photon counting and using 90 milliwatts of 514.5um radiation.

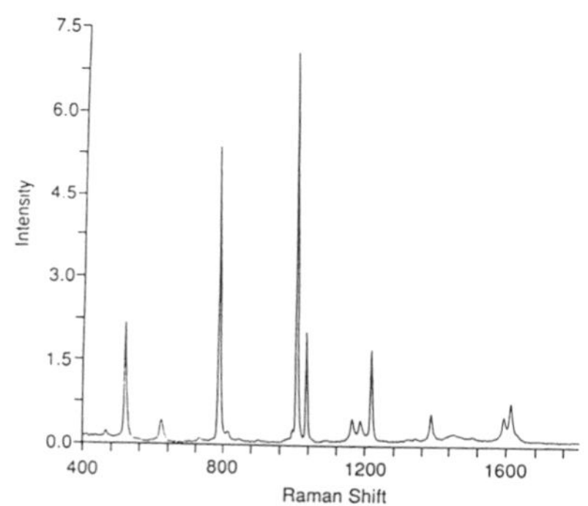

Figure 2b. FT-Raman spectrum of toluene, same conditions as 2a except 90 mwatts 1064nm.

Figure 2b shows the FT-Raman spectrum obtained at the same spectral resolution after apodization and using 90 milliwatts of 1064nm radiation. The total measurement time was equivalent to the time required to scan the full (0-3600 cm^{-1}) spectrum on the visible based system. In this case the FT-Raman instrument performance is superior and is probably due to the higher quantum efficiency of the near infrared detector and the high transmission and efficiency of the interferometer when compared to a double grating monochrometer. Recent communications with other research groups[6] have indicated that the performance of this visible instrument is not at the state of the art and significant improvements can be obtained. However, the performance of FT-NIR Raman is still comparable to that of a visible based instrument.

One of the potential advantages of interferometers is the potential for high frequency precision available with a

laser referenced instrument. Figure 3a shows the spectrum of 1% benzene in toluene. Figure 3b shows the result after subtracting a reference spectrum of toluene.

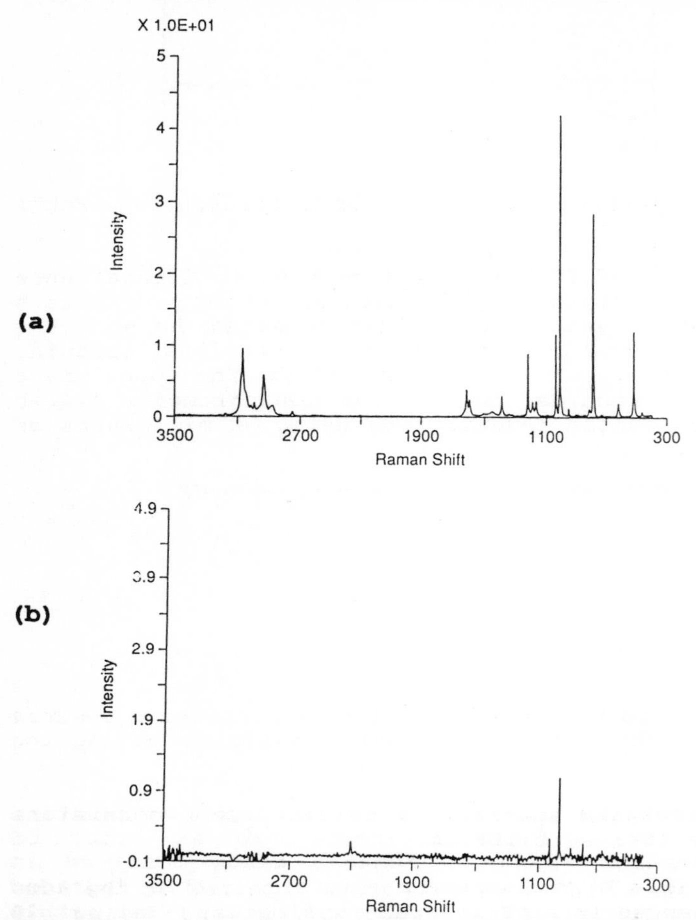

Figure 3. (a) Ft-Raman spectrum 1% benzene-d6/toluene 300 mwatts. (b) Result after subtraction of toluene spectrum from (a).

The benzene modes at 990 cm^{-1} and 2400 cm^{-1} are clearly seen. The degree to which high scale expansion subtractions can be done is determined by the magnitude of artifacts arising from loss of frequency registration.

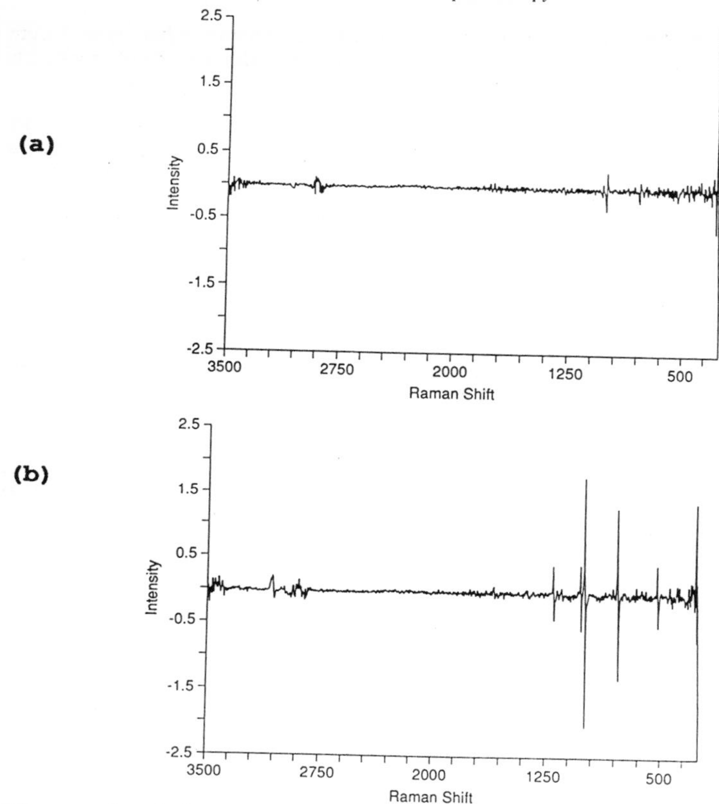

Figure 4. (a) One to one subtraction of successive spectra of toluene. **(b)** Same as 4a, after shifting collection optics.

Figure 4a shows the subtraction result from two separate toluene spectra. These artifacts are an order of magnitude smaller than the benzene bands observed in figure 3b. This high frequency precision can be degraded by subtle changes in experimental conditions. When sample focus and collection mirror position were changed slightly and a new spectrum recorded, the frequency base is changed by the slight change in optical path through the instrument. Figure 4b shows the subtraction result using spectra of toluene taken before and after changing the collection optic position. Clearly, one must take care to ensure reproducible position of sample, collection optics and focusing optics to ensure reliable spectral subtractions.

Though much of the experimentation in FT-Raman is similar to visible based systems, there are some critical differences. Since the chance of sample (or impurity) absorption is reduced in the near infrared, higher power levels can be used with less chance of thermal or photochemical degradation. Spectra have been obtained using several watts of power with no indication of sample degradation for certain samples. It is wise, however, to limit incident powers to several hundred milliwatts or less.

Another difference relates to the allowable spot size at the illuminated sample. If one uses f/1 collection optics and f/4 matching optics (for either an interferometer or a grating instrument), a 4x magnification of the illuminated sample area results at the interferometer (or spectrometer) entrance aperture. For a typical grating instrument the slit widths are normally 200 microns for 4 cm^{-1} resolution. Thus for a sample spot size greater than 50 microns, overfilling of the entrance slits will occur. For our interferometer, the effective entrance aperture is 4mm when using a 1mm detector and operating at 1cm^{-1} or lower resolution. Thus a sample spot size of 1mm is allowed. Since the Raman intensity is proportional to power while thermal degradation processes grow with power density, higher power levels can be used in a one mm spot maintaining low power densities. In this way the etendue of the interferometer can be put to an advantage in FT-Raman spectrometry.

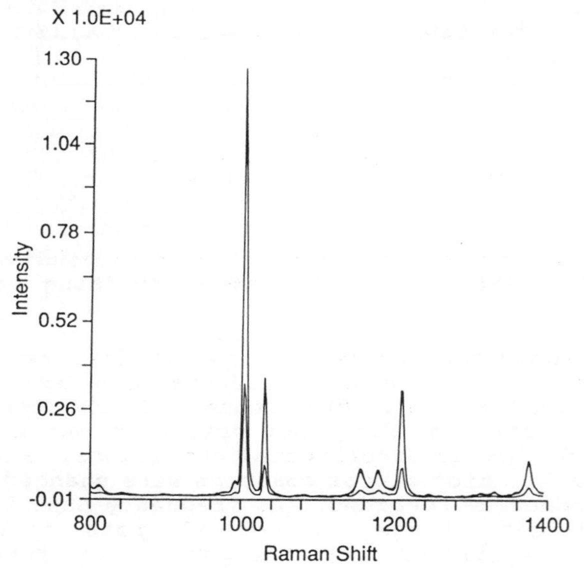

Figure 5. FT-Raman spectra of toluene taken with 0.5mm, 1.0mm, and 2.0mm laser spot size.

Figure 5 illustrates this effect with spectra of anthracene recorded with 0.5mm spot size, 1mm spot size and 2mm spot size. The loss in observed Raman intensity occurs only at a spot size greater than 1mm.

There have now been published many examples of how well FT-Raman works, including an entire journal issue[7] .

Equally important is to realize where FT-Raman fails or at least what sort of artifacts might occur. Some of these are quite easy to recognize and to correct. Incomplete filtering of the near-infrared output of the helium neon reference laser will result in sharp strong neon emission lines throughout the Raman spectrum. Most Nd/YAG lasers have additional non lasing emission lines found near 400cm^{-1}, 2000cm^{-1} and 3000cm^{-1} states shift. Proper filtering of the output of the laser will eliminate these. Sample heating caused by the focused laser beam can produce intensity in the near infrared (the exponential tail of the black body curve). This can be recognized by a large increase in background near 3000cm^{1} Stokes shift. Some of the current instruments available offer a Rayleigh line filter based on a chevron design. This allows measurement of both the Stokes and anti-Stokes lines, and thus an approximate sample temperature can be calculated. These artifacts are all rather well understood and easily recognized.

However, in Fourier spectroscopy there is the possibility of artifact lines appearing at a frequency when there is no radiation of that frequency present in the spectrometer. These artifact lines can arise from aliasing. Figure 6a shows the spectrum of benzene in the overtone region recorded in the visible with a multichannel spectrometer. Figure 6b shows the same region for benzene recorded with an FT-Raman instrument. Note the appearance of an artifact line at 1966 cm^{-1}. This arises from the strong benzene line at 992 cm^{-1} which is then folded around 1/2 the Nyquist frequency. This type of folding is due to improper sampling of the interferogram caused by a slight offset in the zero crossing detection of the HeNe reference signal. These types of artifacts are unique to FT-Raman and have no analogue in conventional Raman spectrometry.

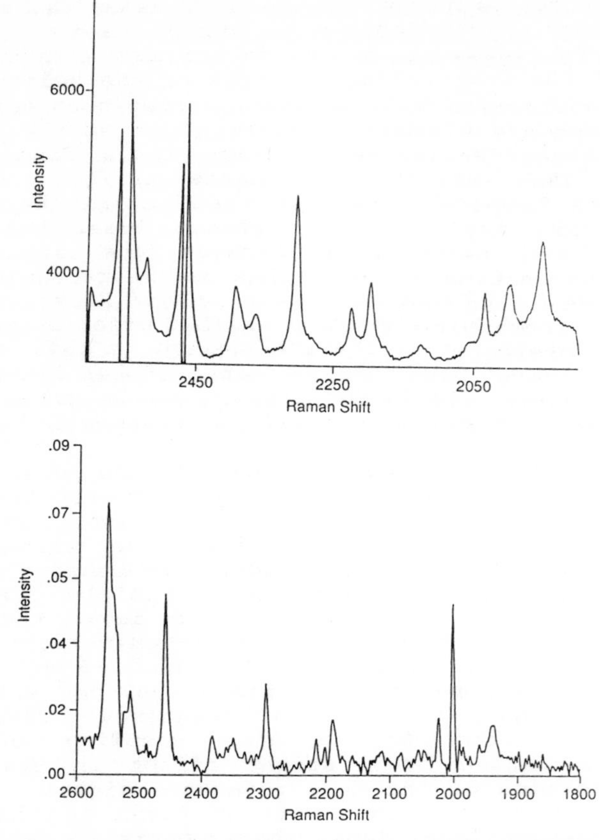

Figure 6. (a) Conventional Raman spectrum of benzene taken on a multichannel spectrometer (overtone region). (b) FT-Raman spectrum of benzene (overtone region).

The basic components of the FT-Raman experiment can be divided into four general areas, the source optics including laser, the collection optics and filters, the interferometer and the detector. Optimization of the performance in each area can be considered separately. The initial choice of a Nd/Yag laser still appears to be the optimum for FT-Raman spectroscopy. The current performance of charge coupled device detectors at wavelengths less than one micron dictates that Raman spectroscopy in this region will be done with multichannel spectrometers[8] . FT-Raman spectroscopy has been done using longer wavelength lasers, such as a 1.3 microns Nd/Yag and this can be useful in moving even further away from a fluorescent interference, but the loss in cross-section coupled with the increase in detector noise eliminates this approach from routine application. The contribution to spectral noise from the laser fluctuation is now being observed for strong Raman scatterers and several groups are incorporating laser stabilization schemes into the instrumentation.

Changes in collection optics are unlikely since most research groups are already collecting at close to f/1. Though even higher collection efficiencies are possible using deep dish mirrors or aspheric lenses, the distortions in image quality with these collectors will probably overcome any increase in collection efficiency. Improvements are being made in the Rayleigh line filtering. As mentioned previously, the use of chevron filters is increasing and the recent development of high extinction holographic filters offers the hope of replacing the dielectric filters currently in use. These dielectric filters have strong modulations in their transmission curve and produce an instrument response function shown in figure 7. These modulations

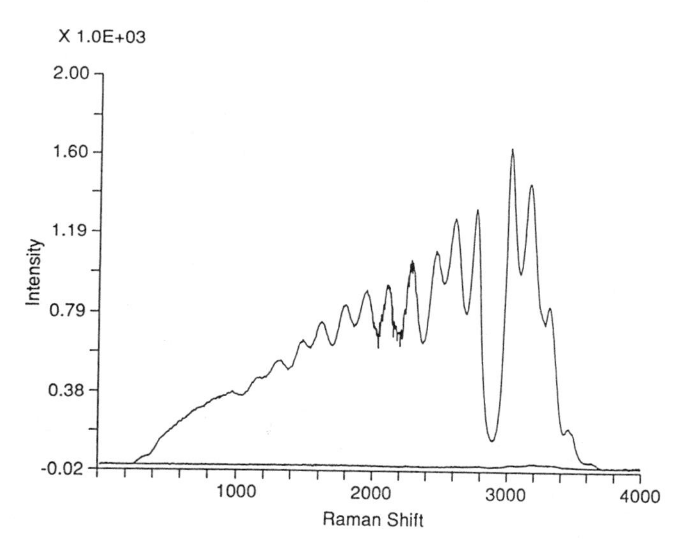

Figure 7. Instrument response function measured with tungsten source.

can severely distort the relative intensities of a Raman spectrum and ratioing to a white light spectrum is mandatory if one wishes to produce good relative intensities. The holographic filters are extremely flat and highly transmitting after the cut-on.

The third instrumental component is the interferometer, and in this area most users are forced to deal with what the manufacturer supplies. It is however, important to know where losses are occurring. A routine measurement of transmission and modulation efficiency should be made on any instrument being used for Raman spectroscopy.

The final area of instrumentation optimization concerns the detector. For FT-Raman spectroscopy, this is probably the most critical element. In a detector noise limited measurement, any decrease in detector noise will translate linearly to an increase in signal to noise of the spectrum. This assumes that other noise sources do not become dominant. The two detectors currently being used for FT-Raman are InGaAs and Ge. The NEP's of these units are 10^{-14} watts/$Hz^{1/2}$ or less. For moderate Raman scatterers, this is still in the detector noise limit. Figure 8 shows the spectrum of cyclohexane and the noise spectrum obtained with the laser off. Inspection of the region near 2000 cm^{-1} shows that the noise is independent of sample illumination. If either shot noise or source fluctuation noise were dominant, the noise level in the spectrum would be higher than the baseline noise without sample illumination.

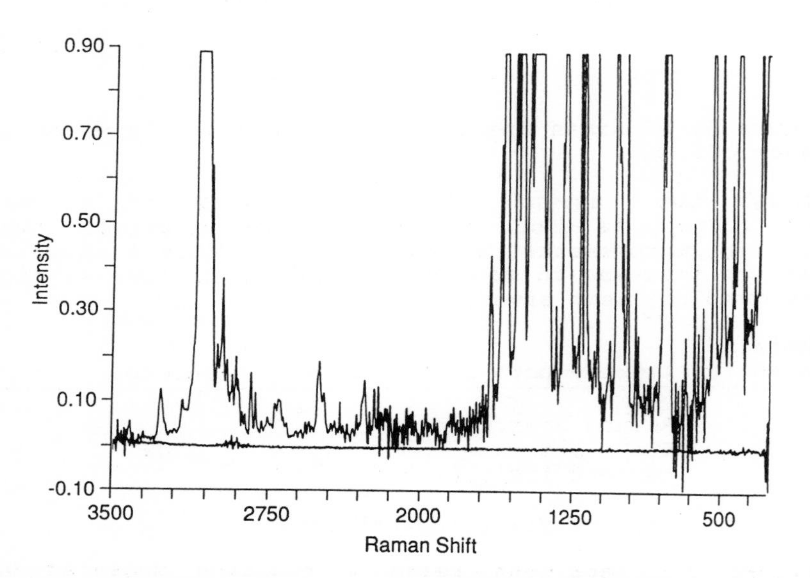

Figure 8. FT-Raman spectrum of cyclohexane, laser on and laser off.

For very strong Raman scatterers, this can be the case. Figure 9 shows a spectrum of anthracene and a baseline noise spectrum. Clearly, for this sample, the experiment is either shot noise or source fluctuation noise limited. This type of measurement should be made routinely when doing FT-Raman spectroscopy, to establish which noise is dominant.

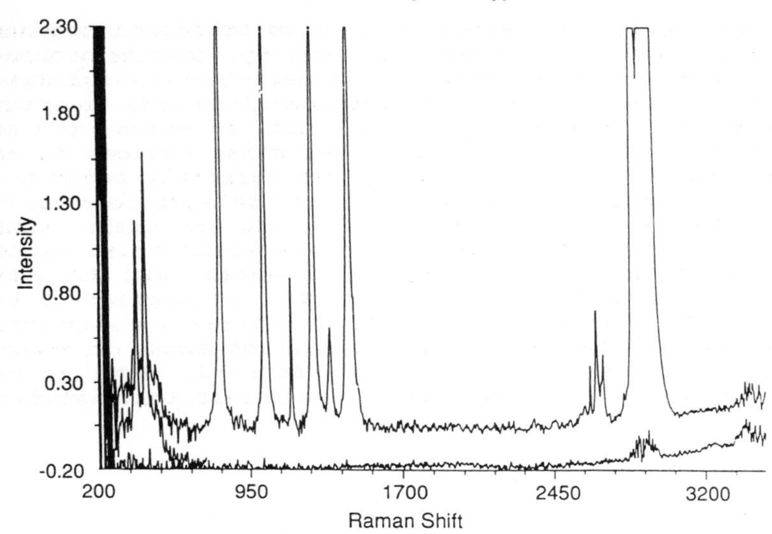

Figure 9. FT-Raman spectrum of anthracene, laser on and laser off.

Though the current performance level of FT-Raman spectroscopy is good, the question often arises, "What further improvements can we expect?" Table I shows the maximum increase in sensitivity that I believe we might see over the next several years.

Table 1.
Potential Improvements in FT-Raman Spectroscopy

1)	Reduce Source Noise, Stabilise Laser not current limitation	1.0x
2)	Improve filtering/collection	1.5x
3)	Improve transmission Improve modulation	2.5x
4)	Improve detectors to limit	8.0x
	Total potential improvement but introduction of shot noise or source fluctuation noise will limit this gain.	30.0x

The bulk of the improvement is in detector sensitivity, and the full gain will only be realised for weak scatterers or low laser power, since high photon flux will force the experiment into the shot noise limit.

The successful results in Raman spectroscopy using near infrared lasers and interferometers has partially driven new results in multichannel detection schemes. The use of charge-coupled device (CCD) detectors and diode lasers has been shown to be a highly sensitive approach to Raman spectroscopy which enjoys relative freedom from fluorescence. Currently manufacturers are working to increase the quantum efficiency of these devices in the near infrared. An alternate scheme involves multichannel detectors with InGaAs, Ge or PtSi elements. These devices are sensitive beyond 1 micron and can be used for Raman spectroscopy with Nd/YAG lasers. Several research groups are now evaluating these devices. These developments, coupled with the success enjoyed by FT-Raman promise an exciting future for Raman spectroscopy, both in the research environment as well as the industrial analytical lab.

REFERENCES

1. Chantry et al., *Nature*, 1964, **203**, 1052.
2. Hirschfeld T. and Chase B., *Applied Spec.*, **40**, 1986, 133.
3. Zimba, Hallmark, Swalen, and Rabolt, *Applied Spec.*, **41**, 1987, 721.
4. Hendra, P. and Mould, *Int. Lab.*, Sept 1988, 43.
5. Parker, Williams, and Hendra, P., *Applied Spec.*, **42**, 1988, 796.
6. Savoie, R. *Private communication*.
7. *Spectrochimica*, Part A, 1990, **46A(2)**.
8. McCreary and Wang, *Anal. Chem.*, 1989, **61(23)**, 2647.

INFRARED AND RAMAN SPECTROSCOPIES OF AN ACETYLENIC MALEIMIDE

Stewart F. Parker, Donald L. Gerrard, Heather J. Bowley, John N. Hay and Judith A. Lander

BP RESEARCH CENTRE, CHERTSEY ROAD, SUNBURY-ON-THAMES, MIDDLESEX, TW16 7LN

An attractive route for the synthesis of high temperature resins is from acetylene containing prepolymers. This type of resin cures through an addition reaction without the evolution of volatiles. Acetylene terminated polyimides show good mechanical properties and oxidative resistance [1]. In order to better understand the cure process in these materials we have studied the thermal cure of a low molecular weight model compound 1 (see Figure 1) by Raman and infrared spectroscopy.

The Raman and infrared spectra of 1 are shown in Figures 1 and 2 respectively. Of particular interest in the present context are;

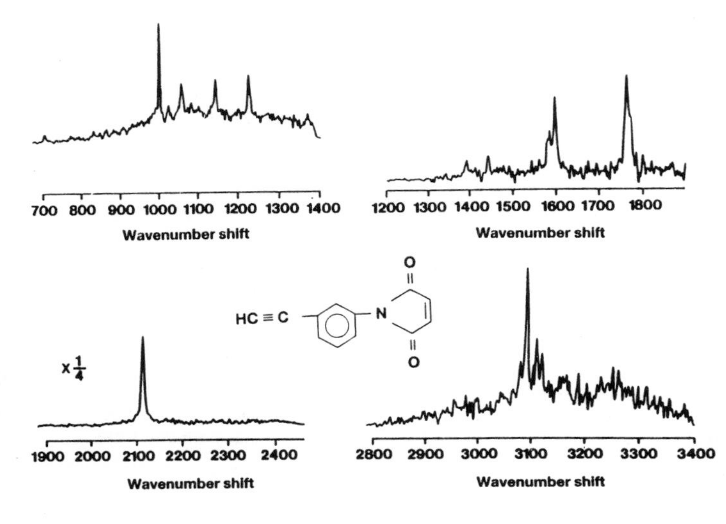

Figure 1 Raman spectrum of 1

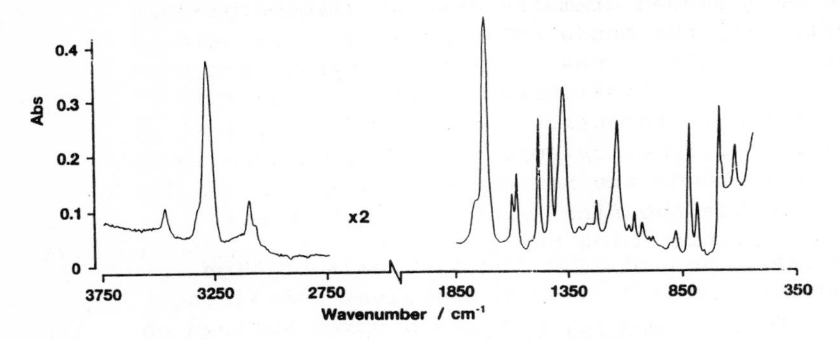

Figure 2 Infrared spectrum of 1

the acetylenic C-H stretch at 3266 cm^{-1}, the C≡C stretch at 2115 cm^{-1}, the maleimide C=C stretch at 1576 cm^{-1} and the maleimide C-H bending mode at 832 cm^{-1}. (The assignments are based on previous work on phenylacetylene [2] and N-phenylmaleimide [3]). The availability of both infrared and Raman

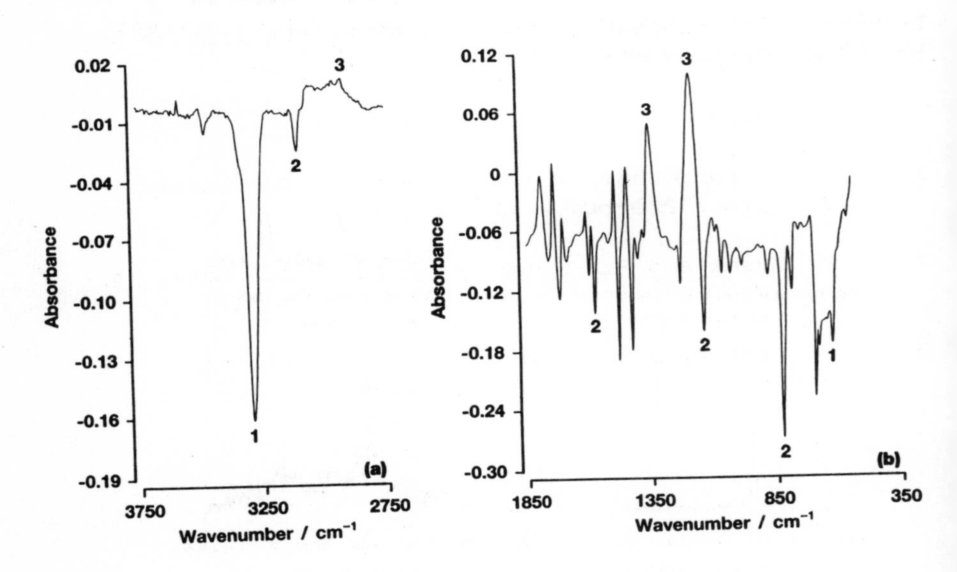

Figure 3 Difference spectrum of (sample at end of reaction) - (initial sample at 130°C)

data allows these key modes to be located with certainty.

Raman and infrared spectra recorded during the thermal cure of 1 showed dramatic changes (Figure 3). In particular, the bands noted previously as being characteristic of the terminal acetylene and the maleimide functionalities had all considerably diminished in intensity and weak bands typical of saturated aliphatics appeared. The changes in the maleimide bands are not due to rupture of the imide ring because the final product displays the characteristic imide bands at 1783, 1716, 1374 and 1176 cm^{-1}.

The isolated position of the C≡C stretching mode present in the Raman spectrum of 1 makes this an ideal measure of the residual acetylenic content at any stage during the cure process. Percentage conversion plots were obtained from temperature ramped and isothermal studies. These have revealed that reaction occurs mainly in the melt and is very temperature dependent.

ACKNOWLEDGEMENT

Permission to publish has been given by the British Petroleum Company plc.

REFERENCES

1. P.M. Hergenrother, S.J. Havens and J.W. Connell, ACS Polymer Preprints, 27, 408 (1986).

2. J.C. Evans and R.A Nyquist Spectrochim. Acta, 16, 918 (1960).

3. S.F. Parker, S.M. Mason and K.P.J. Williams, Spectrochim. Acta A, 46, 315 (1990).

EVOLVED GAS ANALYSIS BY FTIR OF THE CURE OF PMR-15

Stewart F. Parker and Edward Gimzewski

BRITISH PETROLEUM RESEARCH CENTRE, CHERTSEY ROAD, SUNBURY-ON-THAMES, MIDDLESEX, TW16 7LN

Advanced composites have become established as efficient high performance engineering materials in the past two decades. These materials consist of fibres or particles in a common matrix. An important class of these materials consist of carbon fibres in a polymer matrix, usually an epoxy. However, their use is limited to 130°C and there are many aerospace applications that require use temperatures well in excess of this. The NASA developed polyimide PMR-15 [1], is a frontrunner for use in composites for high temperature applications.

The polymerisation of PMR-15 is illustrated in Figure 1. It is complex and the properties of the cured product are very dependent on the cure conditions and the derived chemistry [2-4]. In the present paper we have followed the gases evolved during the cure of PMR-15 with a view to better understanding the cure chemistry.

The system used for the work reported here employed a Stanton Redcroft 671 DTA interfaced to a Nicolet 60SX FTIR spectrometer. The evolved gas from the DTA was passed via a heated transfer line (Severn Scientific Limited) to an in-house built heated gas cell. Both the transfer line and the cell were maintained at 200°C. The cell was positioned in the sample compartment of the Nicolet 60SX. Under the conditions employed for the DTA (10°C/min ramp and 50-100ml/min gas flow) the residence time of the gas in the cell was approximately one minute. Spectral data were accumulated for ≈15 seconds (50 scans at 4 cm^{-1} resolution) in each minute. This was sufficient to

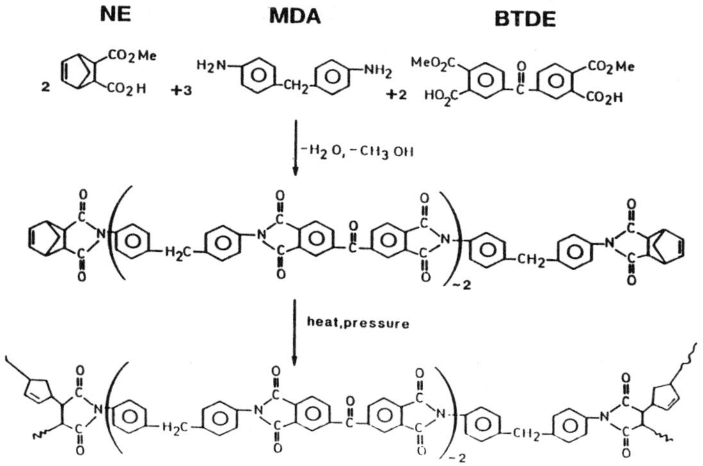

Figure 1 PMR-15 chemistry

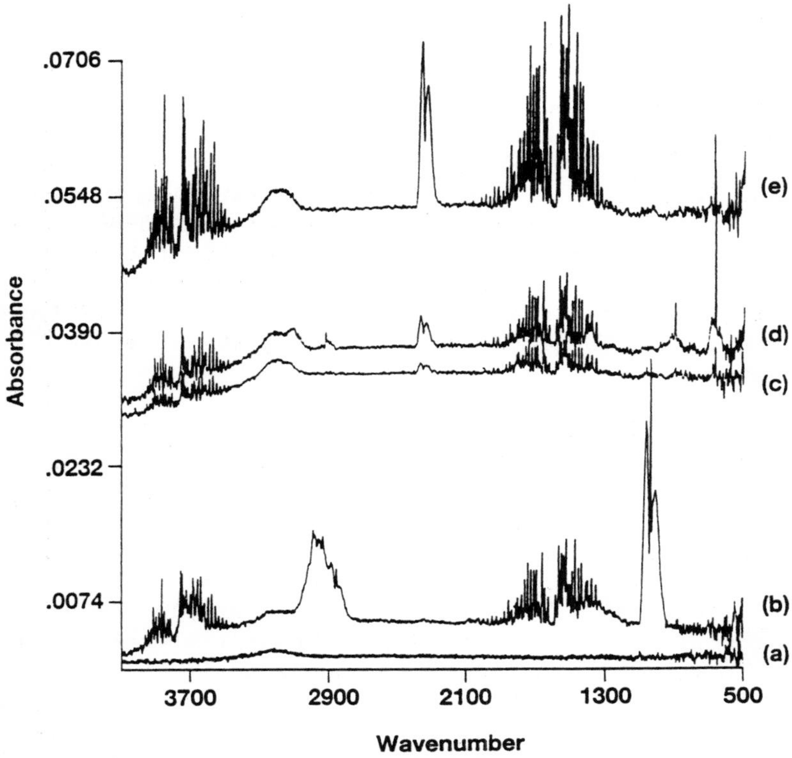

Figure 2 Evolved gases from the cure of PMR-15. Sample at (a) 27°C, (b) 147°C, (c) 260°C, (d) 333°C and (e) 450°C

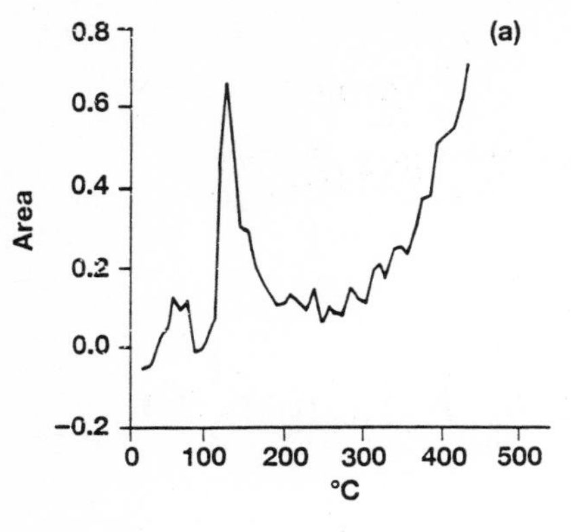

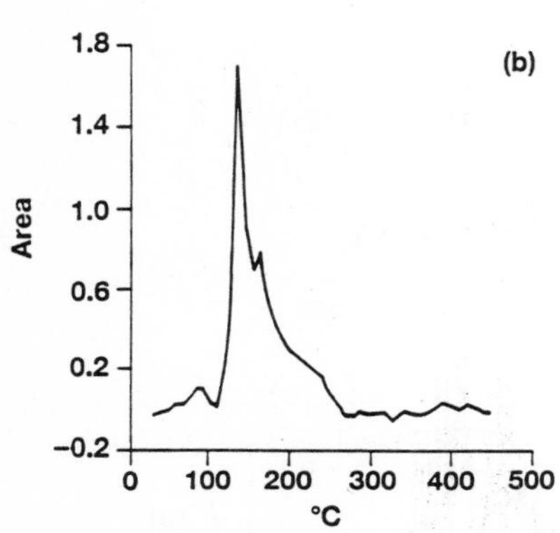

Figure 3 Thermograms of the cure of PMR-15 (a) for the region 4000-3400cm^{-1} (b) for the region 110-900cm^{-1}

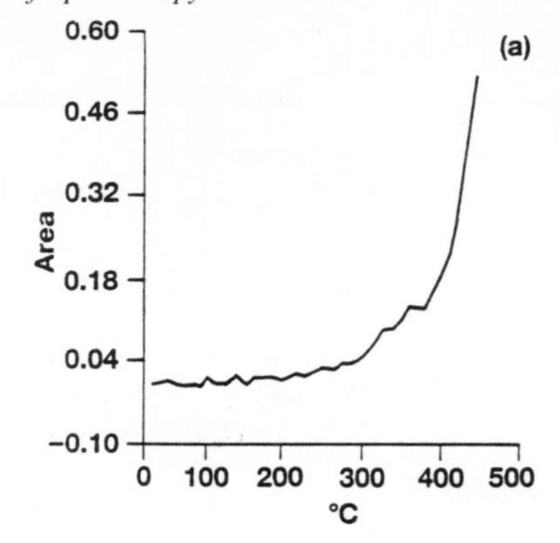

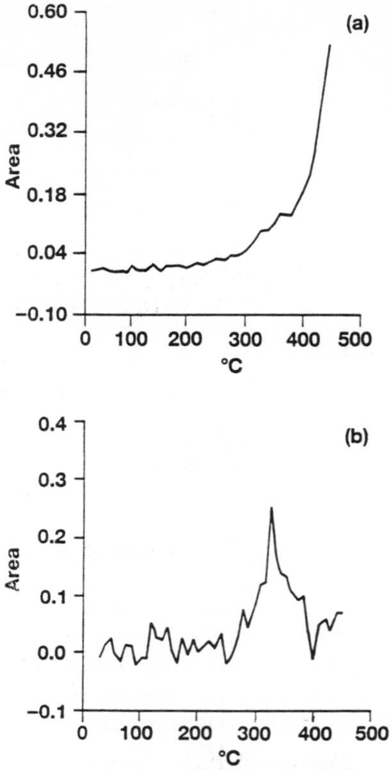

Figure 4 Thermograms of the cure of PMR-15 (a) for the region 2400-2250cm^{-1} (b) for the region 720-600cm^{-1}

generate excellent quality spectra. Immediately before and after each spectrum was recorded, the temperature of the DTA sample cup was measured and the value stored with the spectrum. Typically between 30 and 70 spectra were generated by each experiment. The sample consisted of 5.50 mg of a physical mixture of PMR-15 monomers (5-norbornene-2,3-dicarboxylic acid monomethyl ester (NE), 3,3′, 4,4′-benzophenone tetracarboxylic acid dimethyl ester (BTDE) and methylene dianiline (MDA) in a molar ratio of 2:2.087:3.087).

Figure 2 shows a stackplot of the spectra obtained at 27, 147, 260, 333 and 450°C. The integrated intensity versus temperature (thermograms) for the regions 4000-3400, 1100-900, 2400-2250 and 720-600 cm^{-1} are shown in

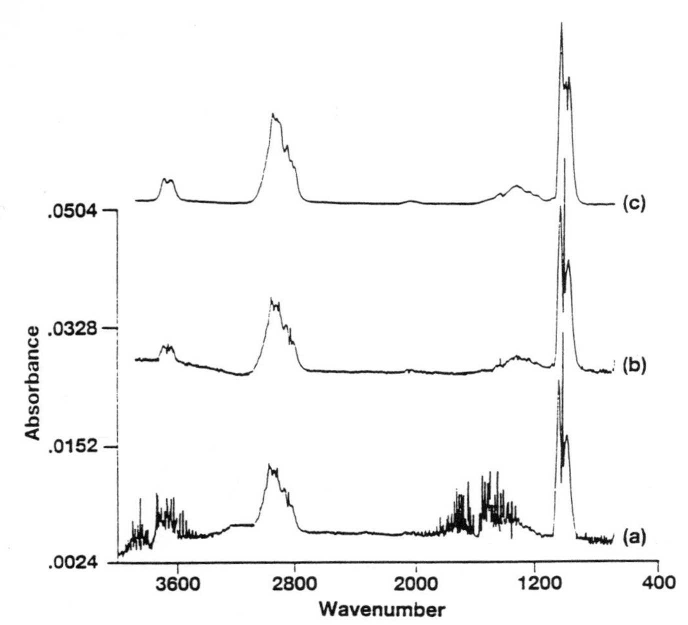

Figure 5 (a) Gas evolved at 147°C, (b) as (a) after subtraction of water vapour, (c) best match from on-line spectral search

Figures 3a, 3b, 4a and 4b respectively. These regions are where water, alcohols, CO_2 and unsaturated hydrocarbons have characteristic adsorptions.

The results can be understood in terms of the known [2-4] cure chemistry of PMR-15. The first stage in the cure is the condensation of the aromatic amine with the ester and acid groups, with the consequent evolution of methanol and water. These are clearly seen in the 147°C spectrum, Figure 2b. Inspection of Figure 3 shows that the condensation is taking place over a fairly wide temperature range and that it occurs in discrete bursts. These may be accounted for by assuming that the condensation occurs mainly in the melt; hence as the polymerisation proceeds the molecular weight increases and the growing polymer solidifies stopping the reaction. The condensation continues when the temperature has risen sufficiently to remelt the polymer.

The presence of methanol is readily confirmed by a computer search of the small spectral library available

UNIVERSITY OF GREENWICH

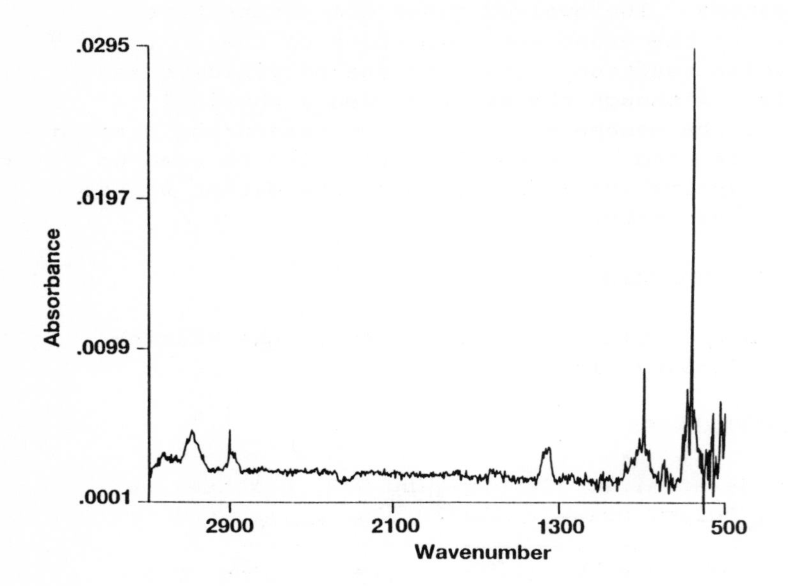

Figure 6 Spectrum of gas evolved at 333°C after subtraction of water vapour

on the Nicolet 60SX. Figure 5a shows the spectrum obtained at 138°C, 5b after computer subtraction of the water vapour and 5c is the best match from the spectral library. (The water vapour spectrum was obtained from a spectrum recorded with the sample at 250°C. The exact match of the envelopes of the water vapour spectra at the two temperatures, apart from a scaling factor, demonstrates that by the time the gas has reached the infrared cell, it is in thermal equilibrium with the heated transfer line. This may allow quantitative work).

The condensation reaction is essentially complete by 200°C as shown by the absence of methanol in the 250°C spectrum, Figure 2c. Between 300 and 400°C a third gas is evolved as seen in Figure 4b and the 333°C spectrum, Figure 2d. The spectrum is shown in Figure 6 after subtraction of water vapour. It is identified as cyclopentadiene from the retro Diels-Alder reaction that is involved in the final stage of the cure. Above 400°C, Figure 2e, the cured resin begins to decompose as evidenced by the rapid increase in water and CO_2 that it also apparent in Figures 3a and 4a.

The ability to monitor cure cycles has applications to other systems. The evolved gases are a sensitive indicator of the onset and completion of the condensation reactions common to the polyimide based materials. Although the example used a physical mixture of the monomers, there is no reason why prepreg could not be used. The method could also be used to provide quantitative information on the extent of the cure in staged materials.

ACKNOWLEDGMENT

Permission to publish has been given by the British Petroleum Company plc.

REFERENCES

1. T.T. Serafini, P. Delvigs and G.R. Lightsey, J. Appl. Polym. Sci. 16 (1972) 905.

2. D. Wilson, J.K. Wells, J.N. Hay, D. Lind, G.A. Owens and
F. Johnson, Proc. 18th SAMPE Tech. Conf. Oct. 1986, p.242.

3. J.N. Hay, J.D. Boyle, P.G. James, J.R. Walton and D. Wilson in "Polyimides: Materials, Chemistry and Characterisation", Eds. C Feger, M.M. Khojasteh and J.E. McGrath, Elsevier, Amsterdam, 1989, p.885.

IMMOBILIZATION OF HEPARIN ON POLYETHYLENE SURFACES: AN FTIR-ATR STUDY

Lars Bertilsson and Bo Liedberg

LABORATORY OF APPLIED PHYSICS, LINKÖPING UNIVERSITY, S-581 83 LINKÖPING, SWEDEN

ABSTRACT

We have used the FTIR-attenuated total reflection (ATR) technique to study the immobilization of heparin on polyethylene (PE) surfaces. The overall process involves activation of the PE surface followed by formation of a sandwich structure with alternating layers of polyethyleneimine and dextran sulphate, and finally covalent attachment of the heparin molecule. The infrared results reveal that the amino and sulphate groups are important binding sites between the polyethyleneimine and dextran sulphate layers. The amount of heparin present on the PE surface is also determined. The surface concentration Γ is found to be about 2.5 $\mu g/cm^2$, which is in good agreement with the results from a chemical analysis.

1 INTRODUCTION

Most synthetic surfaces show trombogenic properties in contact with blood. In medical devices this causes problems and proper surface modification is needed. It has been shown that a pentasaccharide sequence of heparin possesses tromboresistant properties and a method to immobilize partly degraded heparin has been developed (1).

In this study the surface sensitive FTIR-ATR method has been used to follow the different steps in the CBAS (Carmeda BioActive Surface) chemical modification process (2), were polyethylene is covered with heparin.

2 THE SURFACE

The PE surface is etched with H_2SO_4 and $KMnO_4$ to provide enough negative binding sites for the interaction with the first polyethyleneimine (PEI) layer.

The three PEI layers interact with the etched polyethylene (PE) surface and the dextran sulphate layers via van der Waals and electrostatic forces. The negatively charged dextran sulphate molecules interact with the positively charged amine sites in the PEI and a sandwich structure is formed.

Heparin is covalently bonded to the third PEI layer. Degradation of the heparin molecule, by nitrous acid, provides reactive terminal aldehyde groups, which are coupled to primary amines in the last PEI layer by reductive amination. This coupling procedure involves no other functional groups in heparin and results in an end-point attached molecule, which leaves the tromboresistant action intact.

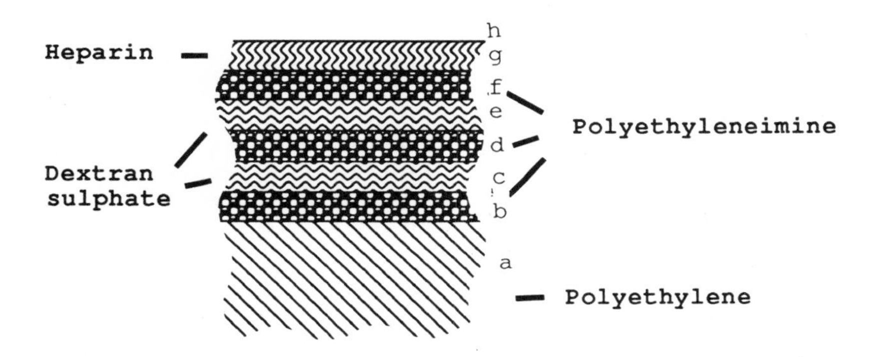

Figure 1 The modified surface.

3 INFRARED SPECTROSCOPY

To be able to follow the chemical modification process of the PE surface ATR measurements have been used. Samples were taken out after each step in the process and analysed in a Bruker IFS 113v spectrometer with a Wilks model 9 ATR unit. The spectra were obtained by averaging 200 inteferograms at a resolution of 4 cm^{-1}. Each difference spectrum was obtained by subtracting the spectrum of the previous reaction step from that of the latest step. In this subtraction procedure the symmetric CH_3 deformation of PE at 1370 cm^{-1} was chosen to cancel in the difference.

Reference spectra of the chemicals used, diluted in KBr, were taken in diffuse reflectance mode with a Spectra Tech Collector™ unit.

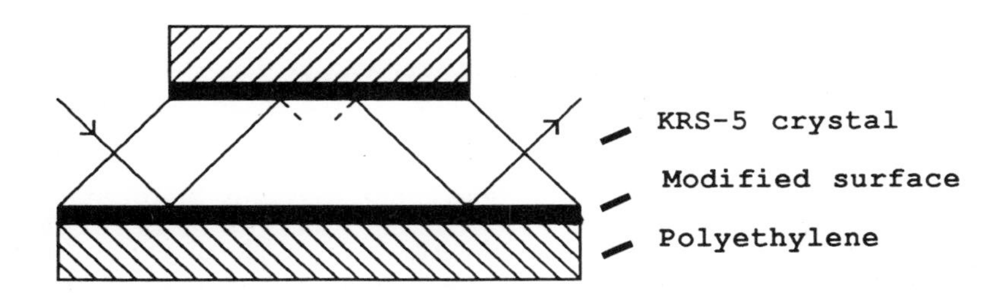

Figure 2 The ATR unit.

4 RESULTS

Qualitative evaluation

The etched PE surface, fig.1a, shows $O-SO_3^-$ absorption at 1255, 1215 and 1066 cm^{-1}, and COOH absorption at 1712, 1670 and 941 cm^{-1}.

The three PEI layers, fig.1b,d,f, show mainly N-H deformations near 1650 and 1550 cm^{-1}. Also seen is a negative peak at 1275 cm^{-1}, indicating interaction with the underlaying $O-SO_3^-$ groups.

The dextran sulphate layers, fig.1c,f, show characteristic $O-S-O_3^-$ absorption near 1240 cm^{-1} ($\nu_{as}S=O$) and 1060 cm^{-1} ($\nu_s S=O$), and C-O-S absorption near 980 and 820 cm^{-1} (ν_{as}, ν_s C-O-S), (3). In the spectra negative N-H def. peaks are also seen, indicating interaction with the underlaying PEI in analogy with above. Note also the intensity enhancement in the second layer.

The heparin layer, fig.1g, shows mainly the same spectral features as dextran sulphate except for the asymmetric and symmetric COO^- stretching modes seen at 1612 and 1419 cm^{-1}, respectively.

Figure.1h shows the surface after total degradation of the heparin layer by nitrous acid. The difference is between the spectrum of the degraded surface and that of the surface before the heparin was reacted, and the difference spectrum indicates that small fragments of heparin are left on the surface, and a loss of PEI (negative peaks).

Quantitative estimation

The quantitative estimation of heparin on the surface was made according to the thin film model developed by Harrick (4). The model is based on the fact that the thin film on the polymer surface is much thinner than the penetration depth, d_p, of the evanescent field in the polymer. This so-called "thin film " approximation implies that the electric field can be regarded as constant within the film, and that one can derive a simple expression for the absorption $A_{//}$, in the thin film, caused by incident radiation polarized parallel to the plane of incidence.

$$A_{//}=\varepsilon c d_{e//} N \qquad (1)$$

Above the effective thickness $d_{e//}$ is related to the geometrical thickness d as:

$$d_{e//} = n_{21} d E_{0//}^2 / \cos \Theta \qquad (2).$$

$A_{//}$ is the observed integrated absorbance value, $E_{0//}$ the normalized electric field at the ATR-crystal/polymer interface and ε the integrated extinction coefficient, as determined from a integrated absorbance *vs* concentration

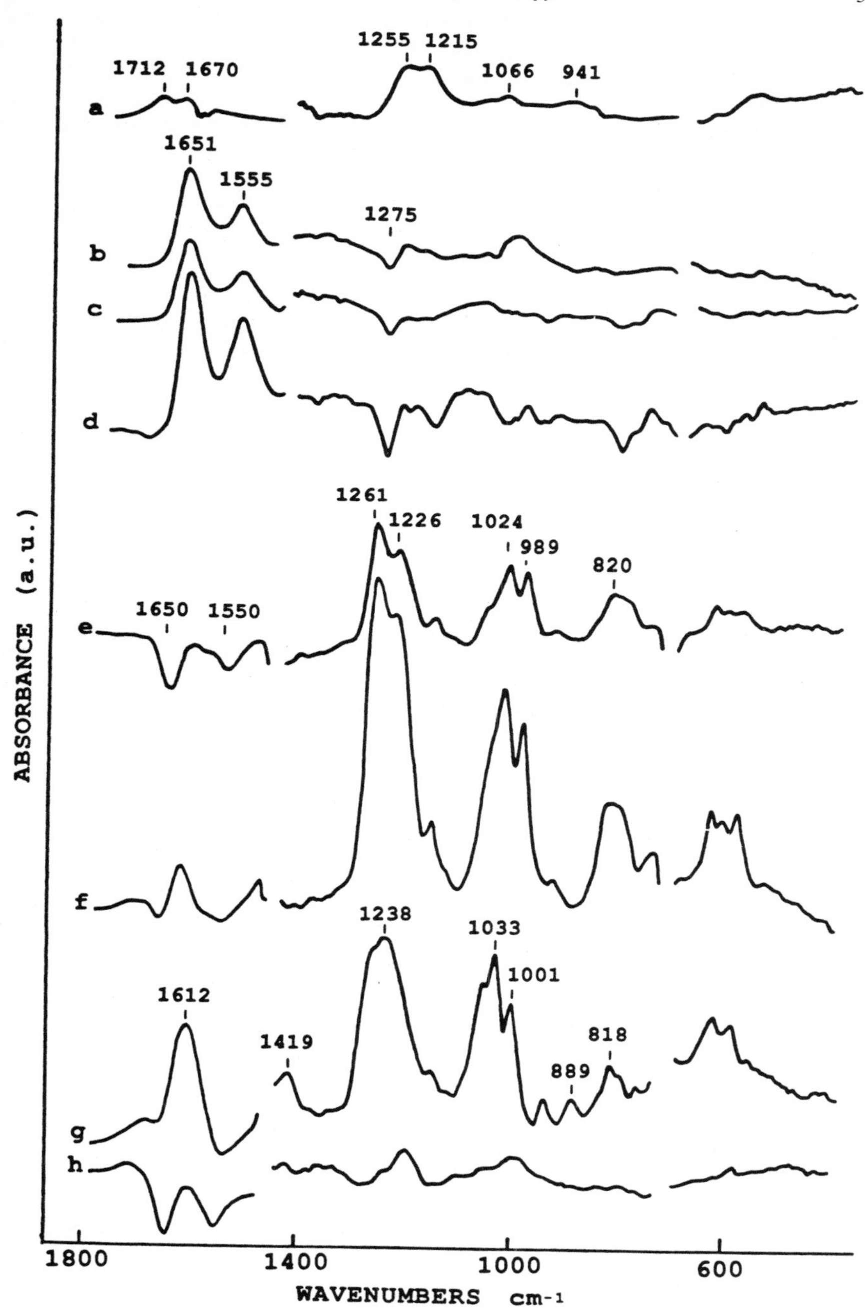

Figure 3 Infrared difference spectra of the etched surface (a), the polyethyleneimine layers (b-d), the dextran sulphate layers (e,f), the heparin layer (g) and the surface after total degradation of the heparin layer (h).

plot of heparin in KBr pellets. The refractive indices used for the ATR crystal (KRS-5) n_1, the heparin layer n_2 and the PE film n_3, were chosen to be 2.37, 1.5 and 1.5, respectively. Above n_{ij} equals n_i/n_j, N is the number of reflections used, in this case 25, Θ is the angle of incidence and c the sample concentration.

Using this model the surface concentration, Γ was calculated from different integrated peak areas in the difference spectra of the heparin layer. The results obtained, 2.4-2.7 μg/cm^2, were in good agreement with quantitative chemical analysis (5). The choice of peak area used in the calculation is however critical because the corresponding vibrational modes are more or less sensitive to the molecular environment.

5 SUMMARY

The ATR technique is found to be a powerful method to study complex sandwich structures on polyethylene. Each layer is analyzed with difference spectroscopy and interactions between the layers are detectable. Quantitative calculations are possible through reference measurements in transmisson mode. The choice of peak area for the calculation is critical and vibrations insensitive to bonding and interaction with charged environments must be chosen.

ACKNOWLEDGEMENT

This work was supported by Carmeda AB.

REFERENCES

1. J. Hoffman, O. Larm and E. Scholander, Carbohydrate Research, 1983, 117, 328.

2. O. Larm,R. Larsson, and P. Olsson, Biomaterials, Medicals, Devices, and Artificial Organs, 1983, 11, 161.

3. F. Cabassi, B. Casu, A. S. Perlin, Carbohydrate Research, 1978, 63, 1.

4. N.J. Harrick, 'Internal Reflection Spectroscopy', Interscience, New York, 1967.

5. J. Riesenfeld, and L. Roden, Anal. Biochem., In press.

A DEVICE FOR THE PRODUCTION OF CONSTANT THICKNESS FILMS

Gordon W. Tregidgo

SPECAC LIMITED, LAGOON ROAD, ST. MARY CRAY, ORPINGTON, KENT, BR5 3QX

Introduction.

Plastics play a very important part in all areas of modern living but due to the often uneven shapes of moulded or extruded products these present difficulties to the analyst. The traditional methods of sample preparation prior to analysis currently used include:-

a) Dissolution of the polymer in a suitable solvent followed by solution analysis.

b) Casting of a film onto an IR transmitting substrate from solution by evaporation of the solvent.

c) Dissolving or melting of the polymer and subsequent casting of the film using a metal loop.

d) Hot pressing of the polymer between heated platens.

These traditional methods are often long and tedious, involve the use of chemicals or result in non reproducible film thickness.

Operation.

The Constant Thickness Film Maker Kit CTFM (Fig 1) offers a quick and reproducible alternative method for sample preparation enabling both qualitative and quantitative studies to be made. The film maker provides a means of producing samples of thermo melting plastics 32 mm diameter with a thickness of 15 to 500 um, depending on the spacer selected, for infrared analysis.

A water circulated cooling unit is included which considerably reduces sample preparation time. Used in conjunction with a hydraulic press and heated platens with an automatic temperature controller (illustrated) films can be routinely produced within 8-12 minutes depending on the softening point of the polymer.

<u>Experiment 1.</u>

To demonstrate the reproducibility and precision of the CTFM, samples of polypropylene, a commonly used industrial polymer, were pressed with each of the spacer rings.

The standard range of spacer rings included are 15, 25, 50, 100, 250 and 500 um labelled A to F respectively. The discs were prepared at 180°C (polymer softening temperature) under 4-5 ton pressure. An aluminium foil disc was placed on the lower platen of the CTFM to ease releasing when cold (the thickness of the Al disc is accounted for). Each polypropylene disc produced was mounted in a Specacard sample holder and analysed using an FT-IR Spectrometer operated at 2 cm -1 and set to 30 scans.

Fig.1: CTFM Kit with Heated Platens

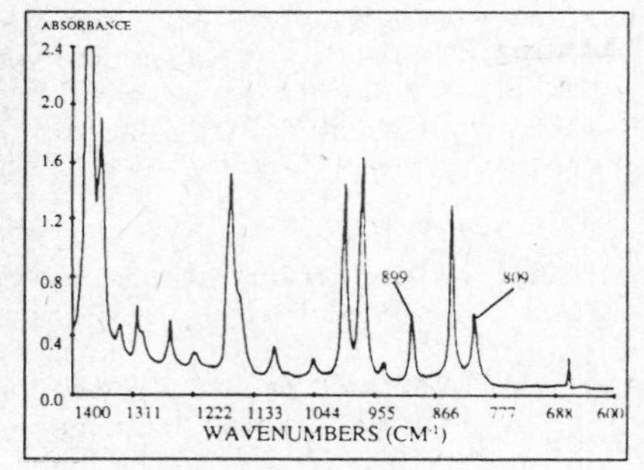

Fig 2. IR Spectrum of Polypropylene Film

<u>Results and Discussion.</u>

The absorption peaks at 899 and 809 cm -1 were selected for examination of film reproducibility (Figure 2). These bands were chosen because they gave an absorbance of less than 1.0 for all the different thicknesses (except for ring F) and hence should obey Beer's law for calibration purposes. The results of the reproducibility analysis by peak height measurements (after baseline correction) for all the rings showed a percentage relative standard deviation ranging from 1.7 to 3.6% at both analysis wavelengths. Table 1 lists the average infrared absorbance values for all the rings (A-F) at 899 cm -1 and 809 cm -1.

Table 1 Theoretical ring thickness and measured infrared absorbance values of polypropylene films.

Ring thickness (Microns)	IR absorbance value 899cm-1	809cm-1
A (15)	0.047	0.050
B (25)	0.069	0.074
C (50)	0.119	0.135
D (100)	0.251	0.281
E (250)	0.792	0.870
F (500)	1.550	1.646

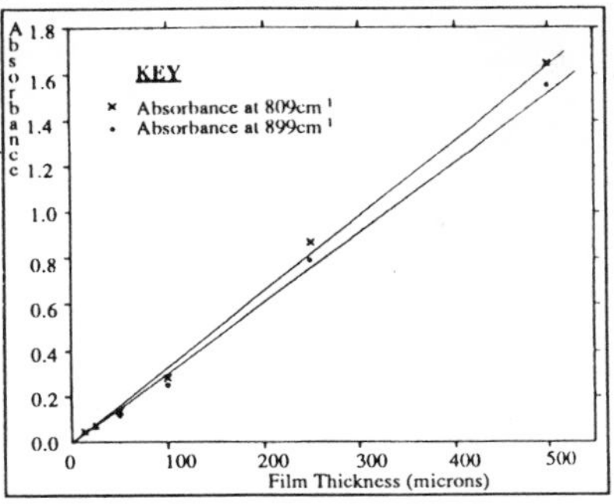

Figure 3. Calibration plot of film thickness vs IR abosrbance

Figure 3 shows a plot of absorbance values against theoretical thickness for rings A to F at 899 and 809 cm -1. A simple linear regression was carried out to examine the correlation of the data to a linear fit. For measurements made at both wavenumbers the correlation coefficient was better than 0.9985 for 30 data points each indicating a high degree of linearity. The error in the slope of the regression lines (- 2.1%) was within the percentage relative standard deviation values calculated earlier. It is, however, worth noting that in some circumstances the characteristics of the polymer may be such that shrinkage or expansion may occur after film preparation which may result in higher deviation from stated film thickness. This should cause no problem in the calibration process since such changes will be uniform across the various rings. In such situations a calibration plot will result in a greater shift from the origin. However, where absolute values are required for such polymers, it is recommended that the true thicknesses of the films produced are measured with a micrometer. Some polymers may yield under pressure. Generally, good results can be obtained by maintaining the applied pressure until the polymer ceases to yield prior to transfering the accessory to the cooling chamber.

Conclusion.

It has been ascertained from the small relative standard deviation results that the films prepared using the Constant Thickness Film Maker are reproducible. A good correlation was attained between disc thicknesses and the magnitude of infrared absorbance peaks. Precise analytical data can be obtained from a single disc, provided a calibration plot has been established.

Experiment 2.

To demonsrate further the quantitative performance of the CTFM, several samples of polyethyelene were acquired doped with different levels of an amide wax (used as a releasing agent).

To enhance the presence of the amide wax in the polyethylene, discs of 225um thick were prepared by using two aluminium foil discs (measured with a micrometer). Two peaks were monitored, 3360cm-1 and 1636cm-1 which are due to N-H stretch and C = 0 stretch respectively.

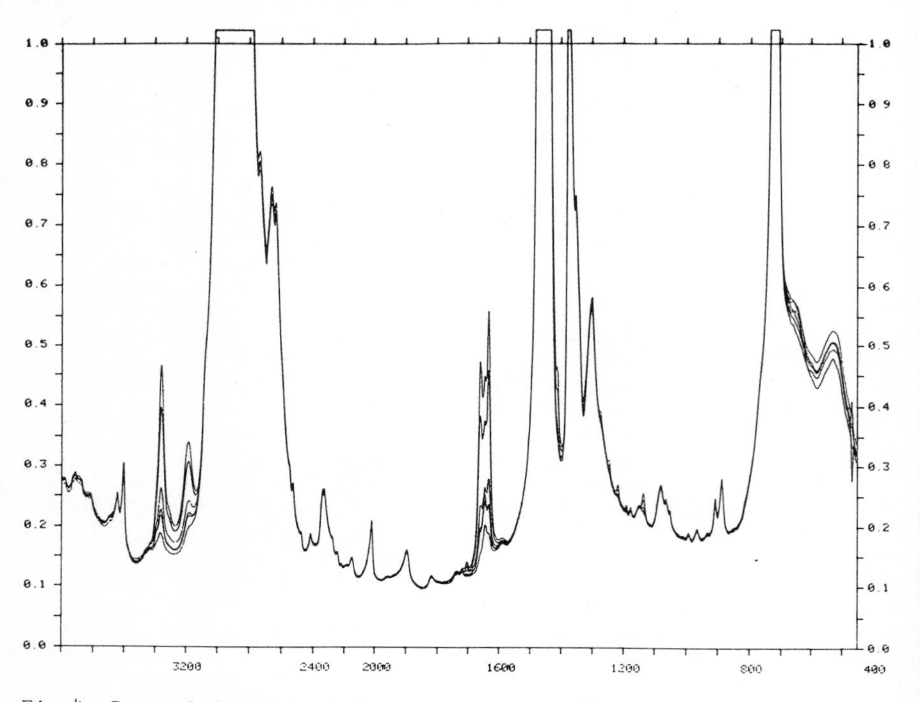

Fig 4. Several Overlaid Spectra of Amide Wax Doped Polyethylene

Fig. 4 - 8 shows spectra of polyethylene doped with varying levels of amide wax. Even though samples are subject to slight thickness change due to relaxation, the resulting discs were all 225um $\pm$ 1%. The reproducibility can also be shown by the overlaid spectra in Fig. 4 (includes 0.1%).

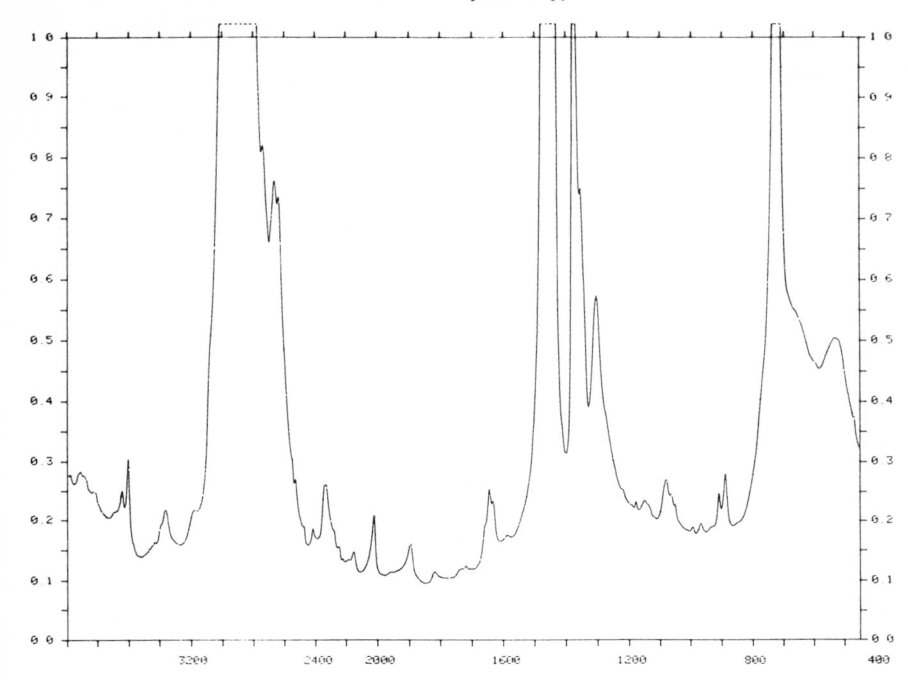

Fig. 5: 0.2% Amide Wax

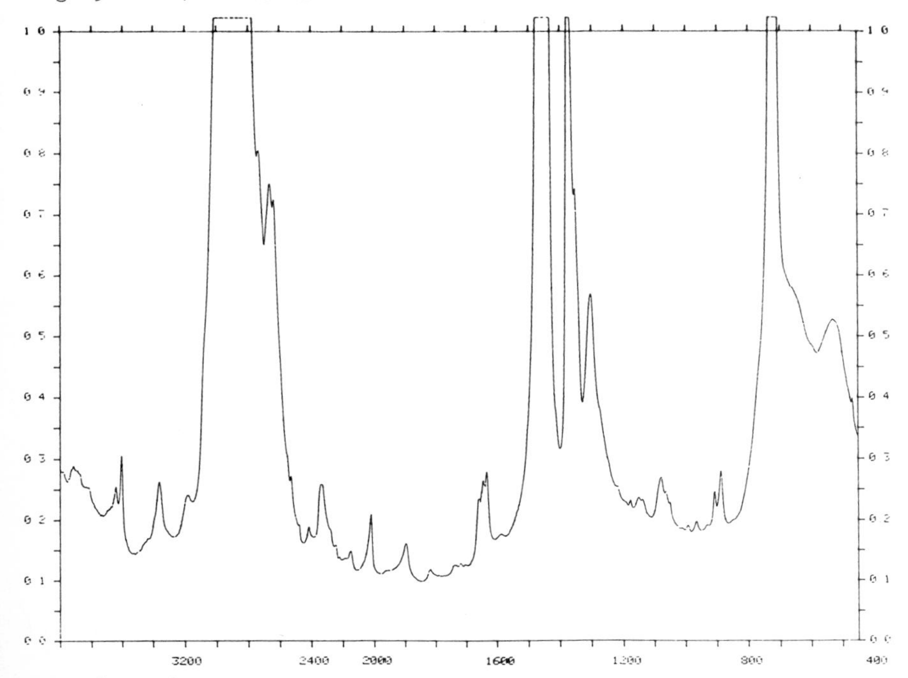

Fig. 6: 0.4% Amide Wax

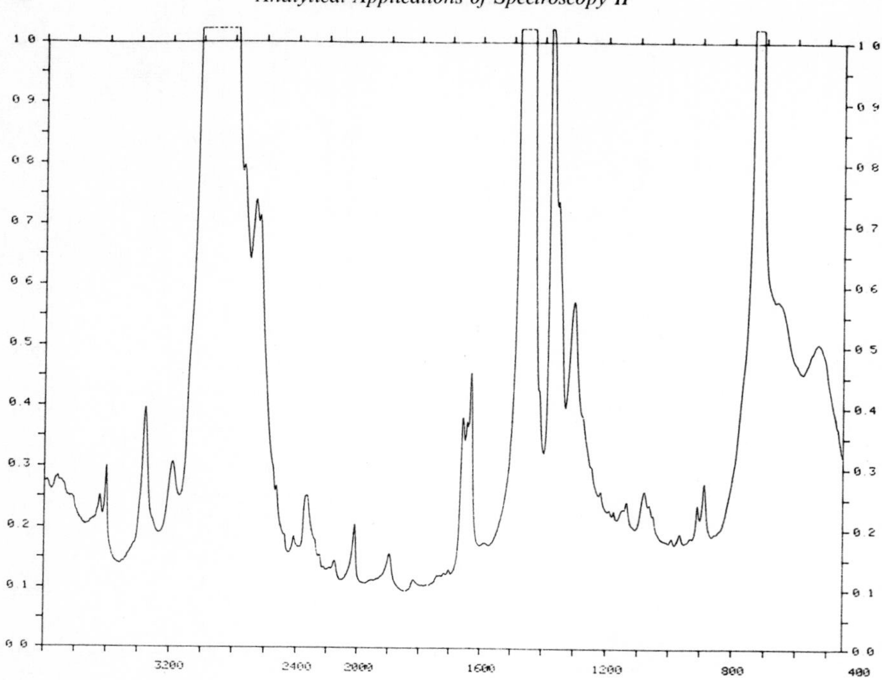

Fig. 7: 0.8% Amide Wax

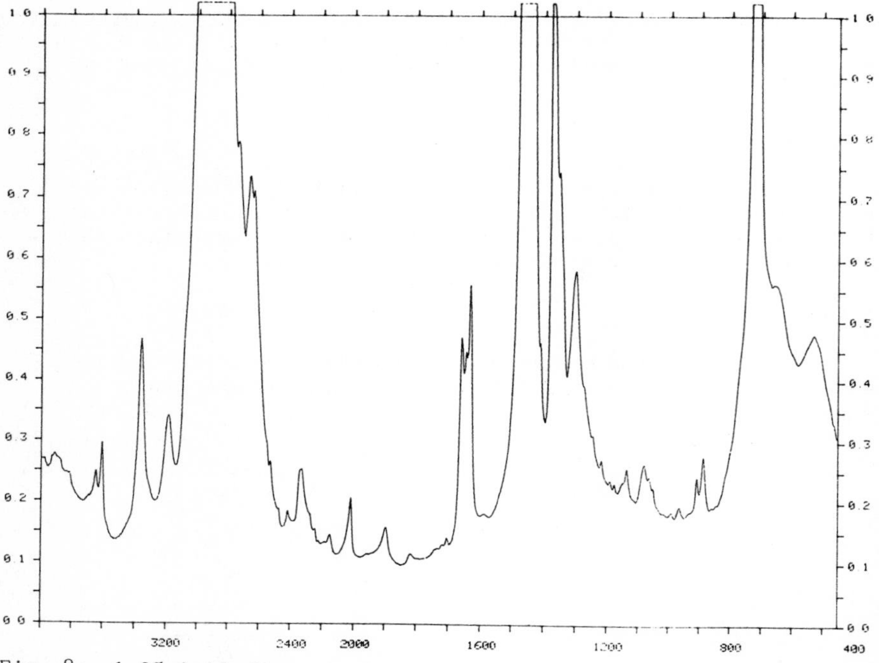

Fig. 8: 1.0% Amide Wax

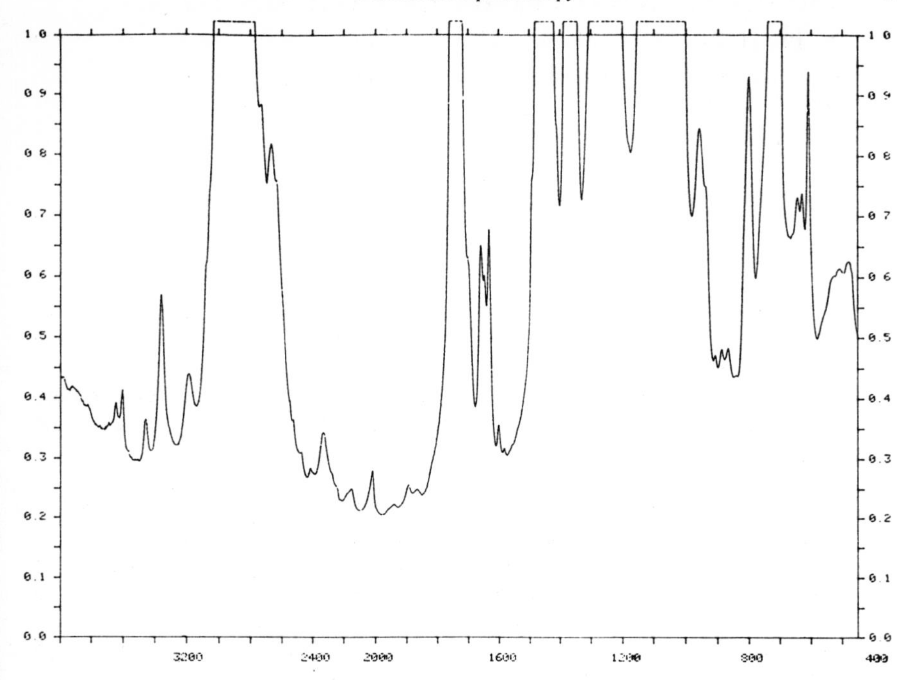

Fig. 9: Polyethylene with unknown additives

Fig. 9 shows a spectrum of an unknown material doped with the same amide wax plus further additives.
Note peaks at 3360cm-1 and 1636cm-1 plus a major peak at 1735cm-1.

Conclusions.

As can be seen from the spectra, excellent correlation was observed between the concentration of the amide wax and the peak heights at 3360cm-1 and 1636cm-1 illustrating the importance of such an accessory in polymer studies.

Acknowledgements.

The author would like to acknowledge the contribution made by Dr J E Newbury, Thames Polytechnic London and Dr A Afran, Specac Ltd.

ON-LINE MOISTURE ANALYSIS BY IR

S. H. Bruce and H. K. Dhaliwal

SERVOMEX UK LTD, CROWBOROUGH, EAST SUSSEX TN6 3DU

Many important industrial applications for the measurement of water in liquid process streams exist in industries varying from petrochemicals and refining, through to fine chemicals, pharmaceuticals, plastics, fragrances and synthetic fibres.

Several technologies exist for making moisture measurements at various concentration levels in liquids, for example, refractive index monitors (typically coarse ranges), N.I.R. reflectance from liquid surfaces, and of course, classical wet chemical methods such as Karl Fisher, although this latter method is not particularly practical for continuous on-line monitoring. The major and most obvious measurement technology to use is I.R. absorbance, due to the availability of the various extremely specific water absorbance bands, both strong (in the I.R.) and weak (in the N.I.R.), and the well established and reliable nature of conventional industrial on-line analysers. No other technology yet matches the simplicity, specificity and ruggedness of this type of equipment (Fig 1).

Historically, extensive low % level moisture measurements and high ppm level measurements have been made using N.I.R. absorption with simple transmission cell sampling. Pathlengths used are typically of the order of one or two mm, which ensures that simplicity and ruggedness of construction is easily achieved at a low cost. A typical industrial liquid transmission sampling cell as shown in Fig 2. These cells can be manufactured in corrosive-resistant metals (eg stainless steel, Monel, Hastelloy C-276 etc) and have a very high performance. Liquid pressures of up to 75 psig (5 barg) produce negligible alteration to the pathlength and the tested failure pressure is typically 1800 psig (120 barg).

A typical low % level moisture measurement is illustrated in Fig 3. A fragrance extraction process utilises methanol which is then recovered for recycling. The recovered solvent may contain significant amounts of water and therefore an on-line analyser ranged 0 - 10 wt % H_20 in methanol is required. The spectrum of 10% wt H_20 in MeOH shows a very strong absorbance at 1.93 um, as expected, but this is not ideal since it is on the upper limit of the working absorbance range of the analyser. The weaker band at 1.45 um, although superimposed on a shoulder of the MeOH bands, is ideal since the absorbance for 10% wt H_20 (full scale) is now in the centre of the analysers working range. The configuration of this unit therefore requires standard 2% bandwidth solid

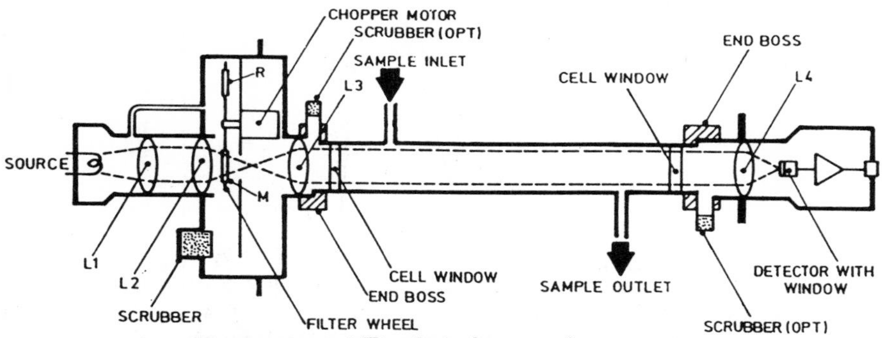

Fig. 1 Process I.R. Analyser Layout

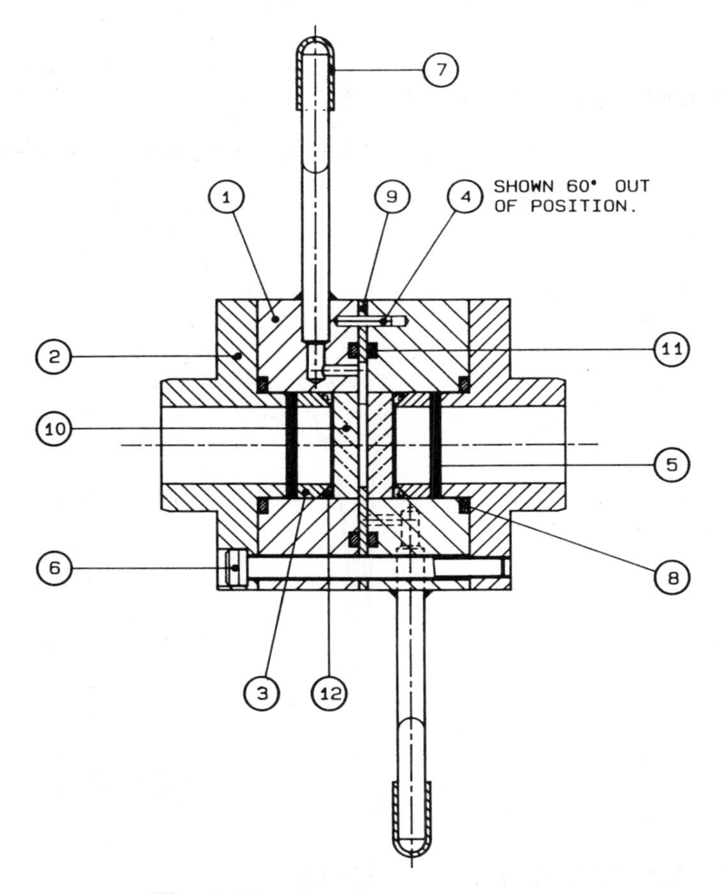

Fig.2 Transmission Sampling Cell

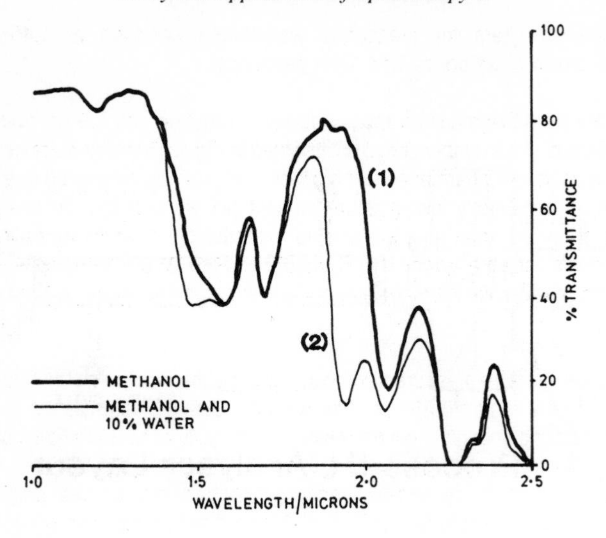

Fig.3 NIR Spectra of (1) Methanol
(2) Methanol and 10% Water

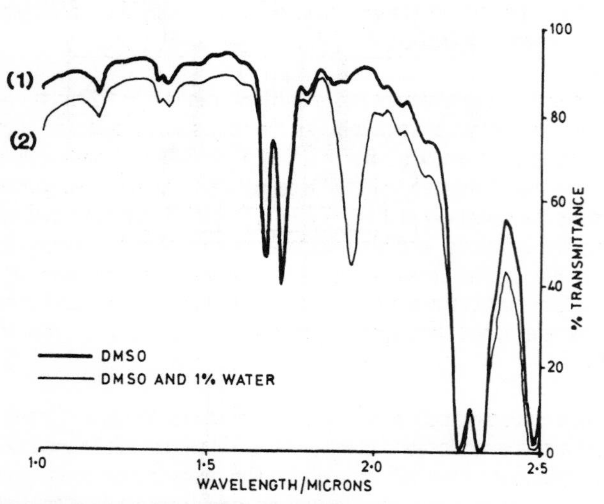

Fig.4 NIR Spectra of (1) DMSO
(2) DMSO and 1% Water

state interference filters for measuring wavelength selection at 1.45 um, plus conventional crown glass optics, and 1mm pathlength.

Where a more sensitive moisture measurement is required, the 1.9 um water band is naturally selected. An example of this is illustrated in Fig 4. Dimethyl sulphoxide solvent [$(CH_3)_2SO$] is used in a pharmaceutical process and must be recovered in a distillation column with an acceptably low moisture content (<1 wt % H_20). To ensure correct column operation, an analyser calibrated to measure 0 - 1 wt % water in DMSO is installed, and this uses the "strong" N.I.R. water band at 1.93 um, selected with standard 2% bandwidth interference filters to perform the measurement. Again, a 1mm pathlength is employed.

A more arduous moisture measurement requirement is the monitoring for ultra-low levels of water, i.e. in the region 0-200 ppm by weight. Clearly, water bands in the N.I.R. region are not strong enough to give this kind of sensitivity at sensible liquid pathlengths and so the fundamental absorbances in the mid-I.R. must be used. Several industrially important solvents are still reasonably transparent at transmission path lengths of 1 - 2 mm in the region of the fundamental O-H stretch, and hence this absorbtion around 2.7 um is the best for high sensitivity moisture measurement.

A typical example is the measurement of trace moisture in 1,2-dichloroethane (Ethylene dichloride or 'EDC') which is a vitally important measurement in the production of vinyl chloride - the feedstock for the manufacture of PVC. In this process ethylene is chlorinated over a catalyst to produce EDC. This is scrubbed with dilute NaOH to remove HCl and dried, before being fractionated to remove impurities and cracked in a furnace to yield HCl and vinyl chloride. The intermediate drying process is essential since traces of moisture in EDC cause the cracked products to be extremely corrosive towards stainless steels which are generally used to construct these plants. Consequently, to prevent catastrophic damage to the plant, the trace moisture levels in EDC after the driers and in storage must be permanently monitored, and measures taken to keep it well below 20ppm.

To achieve a useful measurement range of 0 - 50ppm H_20 in 1,2-dichloroethane using a transmission pathlength of 2mm requires the very specific selection of the water O - H stretch around 2.72 um (Fig 5) with an ultra-narrow band pass interference filter, typically of less than 0.5% band width. Once installed however, this narrow band width filter enhances the sensitivity of the analyser in that the water band closely matches its full transmission envelope, and also yields excellent selectivity, since other low level impurities in the solvent rarely exhibit any significant absorbances at this specific wavelength. Other special measures must also be observed such as thermostatting the incoming sample since temperature affects the accuracy of the measurement significantly.

This technique is readily applicable to most chlorinated solvents and many other liquids although the actual exact location of the water O - H stretch (and hence selection of the ultra-narrow bandpass filter) will vary, depending on hydrogen bonding etc. Notably, some organics such as propylene oxide which are exceptionally difficult to analyse by

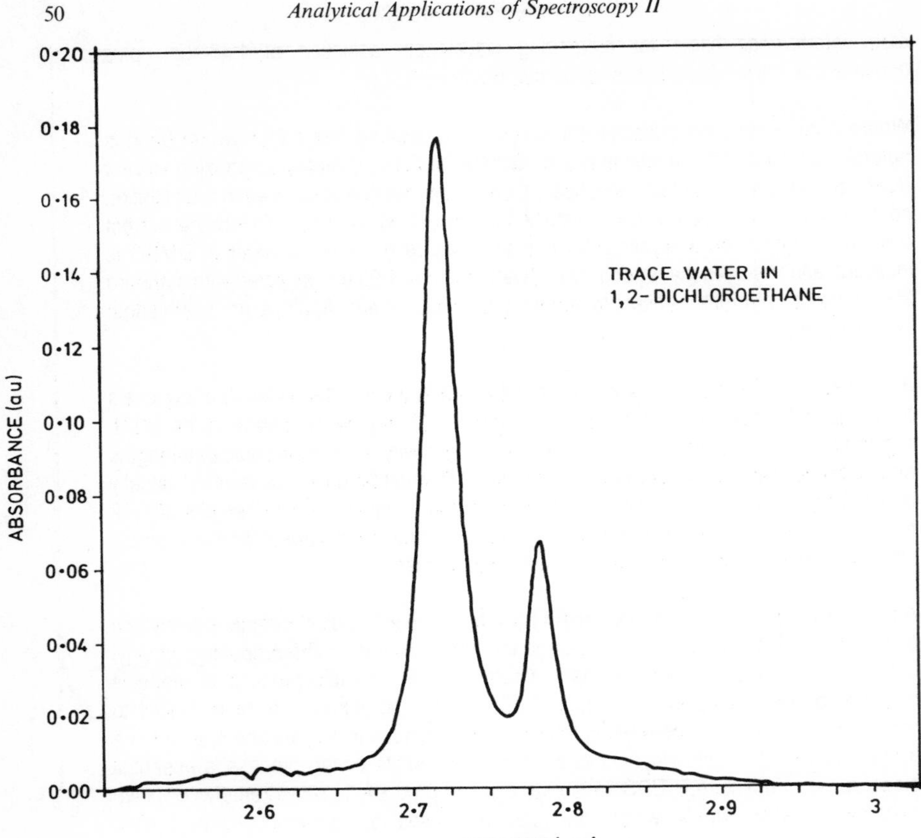

Fig.5 Trace Water in 1,2-Dichloroethane

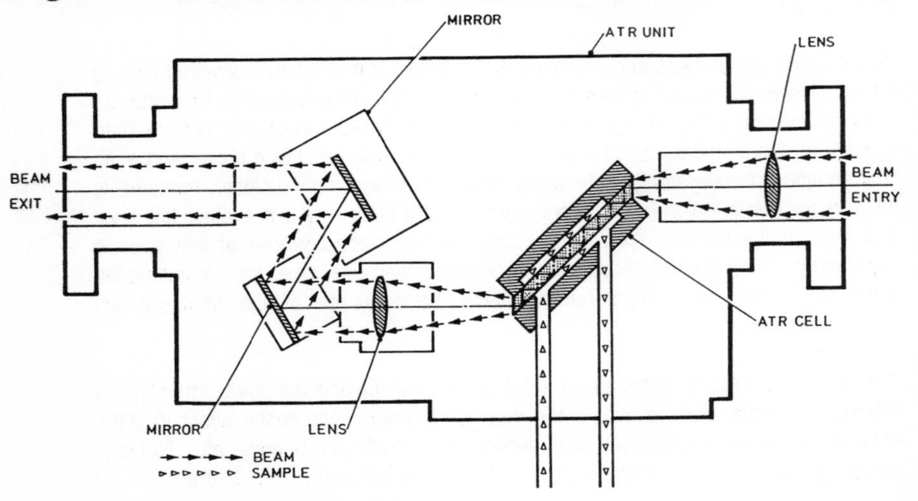

Fig.6 Process ATR Cell

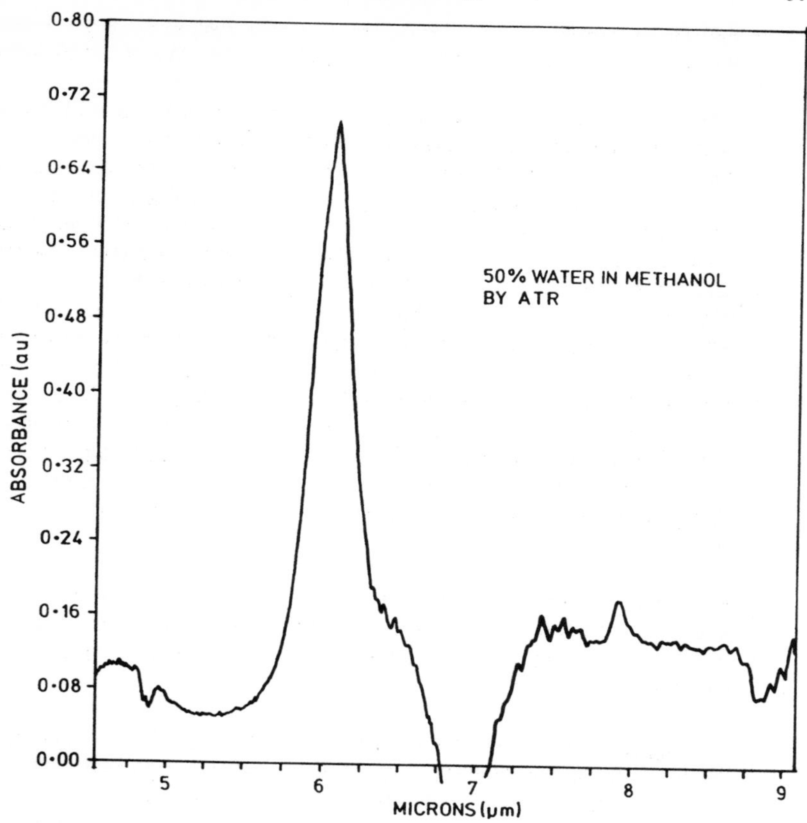

Fig.7 50% Water in Methanol by ATR

Karl Fisher titration do lend themselves to on-line analysis by this I.R. technique, provided of course an acceptable initial calibration of the analyser can be performed.

Where normal transmission sampling for moisture measurements in liquids becomes unsatisfactory is at the high % level end of the scale, where either the weak 1.4 um band becomes too intense or where pathlengths of 2 mm are considered too restrictive for the process stream to be analysed (it may be heavily contaminated or highly viscous). Under these circumstances, the Attenuated Total Reflectance (ATR) technique, used extensively in the laboratory, may be applied to the on-line analyser. This technique uses an infrared-transmitting crystal with a high refractive index through which the infrared beam is passes. The liquid sample surrounds the crystal and as the beam is reflected several times along the liquid/crystal interface, a very short effective pathlength in the liquid is generated, typically the order of only 25-50 um. Fig 6 illustrates the process ATR cell now being introduced. The design is an industrially rugged one, capable of withstanding sample pressures up to 10 barg (150 psig) and temperatures up to 150^{0}C, physically protected by a rugged housing. This may be fitted with a variety of ATR crystals, the material depending on sample stream compatibility and the configuration depending on the amount of effective pathlength required for the application.

A typical water measurement suited to an ATR sample cell is that in methanol at high levels. A range of 0 - 50 % water in methanol can be very practically accommodated (Fig 7), once more using simple interference filters at standard 2% band widths and an ATR crystal of zinc selenide.

Other process ATR applications include the measurement of % level organics in water (for example the measurement of acetic acid in water at high levels, monitoring for losses of organics through heat exchangers into cooling water, measurements of surfactants in water such as in consumer detergent formulations), and the measurement of alcohol and dissolved CO_2 in beverages etc. In addition, many inorganic anions have absorbances in the mid-I.R. which enable their aqueous solutions to be measured by this technique.

In conclusion, despite the recent introduction of many more sophisticated on-line measurement techniques, the simple infrared analyser remains a most practical, versatile, reliable and cost-effective means to perform the widest range of analyses. The recent extension of this method to accommodate ATR sampling and the potential for using NIR fibre optic coupling to sample streams will ensure that an increasing number of process measurements will become practical in the future.

ACCURACY OF MULTIPLE ANALYSIS BY DESIR-NIRS ON LIQUIDS

G. Alfaro, M. Meurens, and M. Vanbelle

BNUT-AGRO, UNIVERSITY OF LOUVAIN, 1348 LLN, BELGIUM

1 INTRODUCTION

Multivariate statistical techniques are gaining importance in near-infrared spectroscopy and one of earliest in introducing NIRS to practical use was the U.S. Department of Agriculture[1]. All these techniques applied to analytical chemistry are involved in a branch of the science called Chemometrics. Recent articles about NIRS show that multivariate calibration methods, particularly multiple linear regression, principal components regression and partial least squares, are continuously evolved in their application to qualitative, quantitative and structural analysis.

Near infrared spectroscopy (NIRS) has becomeapopular method of rapid and nondestructive analysis of major components on powdered samples. A few years ago, Osborne[2] was one who gives to NIRS a considerable support for changing the low academic status of NIR spectroscopy. Shortly after, Norris and Williams[3] joined many specialists and presented theory, instrumentation and research in the world of NIRS.

Actually quantitative analysis of complex mixtures by NIRS can be performed with multivariate calibration methods. In this way, one important disadvantage in the development of NIRS calibration is the requirement of a reference or standard method. In practice, each analytical method includes several sources of error and comparison between them is not easy, overall with complex natural samples, because the true composition cannot be measured. Studies with artificial samples gives the possibility of reducing errors cited above and, at the same time gives a better comprehension of NIRS technique.

Dry-Extract System for InfraRed (DESIR) is a recent sample preparation technique for liquids in NIR Spectroscopy. Thanks to DESIR and glass microfibre filters make it possible to take, in three min., near-infrared reflectance spectra of non-volatile residues from liquid samples[4]. Precision of this technique was reported previously[5] and instrumentation for drying is now available [6].

Sugar is the major constituent of most liquid foods and beverages. Sucrose , fructose and glucose are three natural sugars of orange juice. The purpose of this study was to evaluate the accuracy of determination of glucose, fructose, sucrose and degrees brix in aqueous solutions using two different calibration methods: multiple linear regression (MLR) and partial least squares (PLS). These results were compared with those determined by HPLC in the same 30 prediction samples.

2 MATERIAL AND METHODS

Samples and analyses

Sixty aqueous solutions of mixtures of three sugars, glucose, fructose and sucrose, were prepared to randomise the distribution of the quantities of each sugar in the range 0-50 g/L. These laboratory concentrations were considered as "true" values. These samples were also randomly divided to form 30-sample calibration set and 30-sample prediction set. The following sugars were employed for preparing the artificial liquid samples. Anhydrous D(+)-glucose, BAKER. Extra pure D(-)-fructose, MERCK. Sucrose for biochemistry and microbiology, MERCK.

The dissolved solids, expressed as degrees Brix, were determined by measurement of refractive index at 20°C, with a table refractometer. High performance liquid chromatography (HPLC) has been applied to the analysis of sucrose, fructose and glucose. The Column was polyspher^R AOKC of MERCK and the detector was SP8430(Spectra-Physics) refractive index. Operating conditions were: eluent 0.01 N sulfuric acid, column temperature: 30 °C, flow rate: 0.5 mL/min, detector temperature: 25 °C and internal standard: erithrytol.

Sample preparation and spectral measurements

Whatman GF/A glass microfibre filters are impregnated with 0.45 mL of each liquid sample. Then they were dried for three minutes and placed in cells under a quartz window and over virgin glass microfibre filters. Dry-extract was obtained with solvent elimination by hot air flow in DESIR dryer (experimental model).

NIR Reflectance data were obtained in a NIRSystems (PSCO)-6250 near-infrared grating spectrophotometer with a lead-sulfide detector. Spectra were measured from 1100 to 2500 nm, the nominal spectral resolution was 2 nm. All spectral data were saved on a IBM-PS2/50Z microcomputer for later data processing. All dry-extract spectra were normalised with a virgin Whatman filter and transformed using second derivative data processing (segment size 20 nm and gap size 0 nm).

Calibration and prediction

The calibration and prediction on model sugar samples have been carried out with the NSAS software[7]. PLS and stepwise MLR, on second derivative spectra, were used to provide calibrations for degrees brix, total and individual sugars in the calibration set samples. Samples of validation set were analyzed by HPLC and by the NIRS (PLS and MLR) and their results were compared.

Standard error of calibration (SEC) is a slight variation of standard deviation of residuals between predicted and actual values in the calibration model. Multiple correlation coefficient (R) is used to describe the fit of the data in the multivariate model.

Errors between laboratory values and NIRS values are expressed as standard error of prediction (SEP). The mean of deviations (bias) and the simple correlation coefficient (r) between calculated values by NIRS and laboratory values (or values of reference method analysis) are also reported. The formulas for calculation of all of these statistics may be found in the manual of NSAS software.

Accuracy of NIRS (PLS and MLR) , for each component, was expressed in terms of the standard error of prediction (SEP), which is the standard deviation of differences between NIRS and "true" (laboratory) values. Accuracy of HPLC, for each component, was expressed in terms of the standard deviation of differences between HPLC and "true" values.

3 RESULTS AND DISCUSSION

Table I displays the analytical wavelengths, and statistical results by MLR in calibration samples set, and statistical results of validation on prediction samples set. Calibration and validation results of PLS method are summarized in table II.

Looking at the standard error of prediction in tables I and II satisfactory results are obtained using PLS analyses and ordinary stepwise MLR. Multiple analysis of liquids by DESIR-NIRS can be accomplished successfully with the two methods.

The residuals for MLR calibrations are slightly higher than the residuals for corresponding PLS calibrations. This is an indication that more significant spectral effects are explained by the PLS calibration. Figure 1 gives for sucrose, glucose, fructose and degrees brix, the scatterplots of "true" (laboratory) values vs. calculated values by NIRS-PLS with good level of

TABLE I : RESULTS OF CALIBRATION BY MLR AND PREDICTION ON AQUEOUS SOLUTIONS OF MIXTURES OF SUCROSE, GLUCOSE AND FRUCTOSE.

			WAVELENGTH SELECTION	CALIBRATION	STATISTICS	PREDICTION		STATISTICS
	UNITS	RANGE	(nm)	R	SEC	BIAS	SEP	r
SUCROSE	g/L	5–50	1712–1430–1234–1954	0.998	1.010	–0.165	1.660	0.994
GLUCOSE	g/L	1–50	2450–1868–1826–1796	0.995	1.770	–0.515	1.610	0.995
FRUCTOSE	g/L	0–50	2154–1740–2298–1228	0.999	0.936	–0.342	1.510	0.995
TOTAL	g/L	28–150	1320–1156–1322–2236	1	0	–0.487	2.960	0.993
BRIX	°brix	2.7–14.3	1318–2422–1316–1320	1	0	–0.012	0.234	0.995

SEC= STANDARD ERROR OF CALIBRATION
SEP= STANDARD ERROR OF PREDICTION
R= MULTIPLE CORRELATION COEFFICIENT

CALIBRATION SAMPLES = 30
PREDICTION SAMPLES = 30
r= SIMPLE CORRELATION COEFFICIENT

TABLE II: RESULTS OF CALIBRATION BY PLS AND PREDICTION ON AQUEOUS SOLUTIONS OF MIXTURES OF SUCROSE, GLUCOSE AND FRUCTOSE.

			CALIBRATION	STATISTICS	PREDICTION		STATISTICS
	UNITS	RANGE	R	SEC	BIAS	SEP	r
SUCROSE	g/L	5–50	1	0.254	–0.656	1.450	0.995
GLUCOSE	g/L	1–50	1	0.302	–0.225	1.220	0.977
FRUCTOSE	g/L	0–50	1	0.266	0.313	1.000	0.998
TOTAL	g/L	28–150	1	0.438	–0.656	1.450	0.995
BRIX	°brix	2.7–14.8	1	0.033	–0.003	0.181	0.997

SEC= STANDARD ERROR OF CALIBRATION
SEP= STANDARD ERROR OF PREDICTION
R= MULTIPLE CORRELATION COEFFICIENT

CALIBRATION SAMPLES = 30
PREDICTION SAMPLES = 30
r= SIMPLE CORRELATION COEFFICIENT

correlation for the 30 prediction samples. The prediction errors by PLS method are 1.2 g/L for glucose, 1.0 g/L for fructose, 1.5 g/L for sucrose, 1.5 g/L for total sugars and 0.2 for degrees brix.

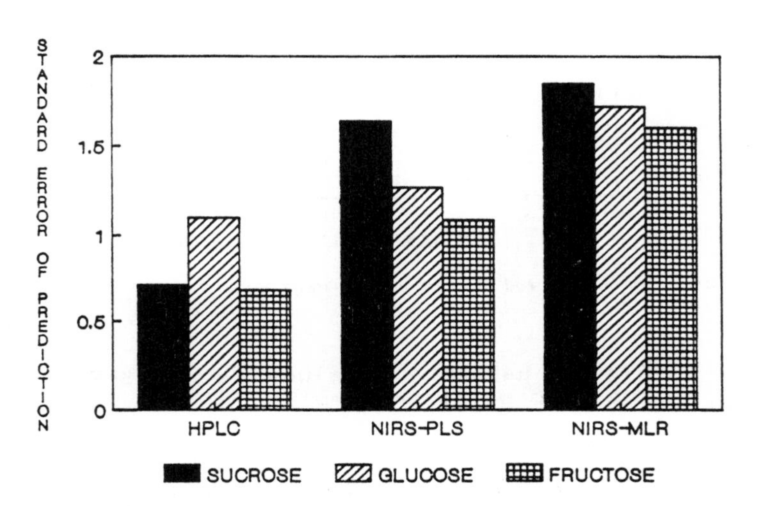

Fig. 1. Accuracy of NIRS and HPLC methods on aqueous solutions of sugars.

Accuracy of PLS, MLR and HPLC, for each component, was compared in terms of the standard error of prediction as is shown in Fig. 2. PLS method gives more slightly accurate calibrations than the MLR, because SEP values for glucose, fructose and sucrose are lower. HPLC showed slightly better accuracy than MLR and PLS.

Our sugar analysis by HPLC required 17 minutes. The speed of NIRS analysis, 30 sec, is the most advantageous factor. The DESIR technique is fast, simple and accurate. The multiple analysis by DESIR-NIRS can be performed on liquid sample in five minutes. This system of analysis for liquid samples is found to be well suited to quality control in a production cycle, where rapid analysis is a primary interest. In a paper of 3rd ICNIRS[8], the application of DESIR and NIRS on orange juice has been reported.

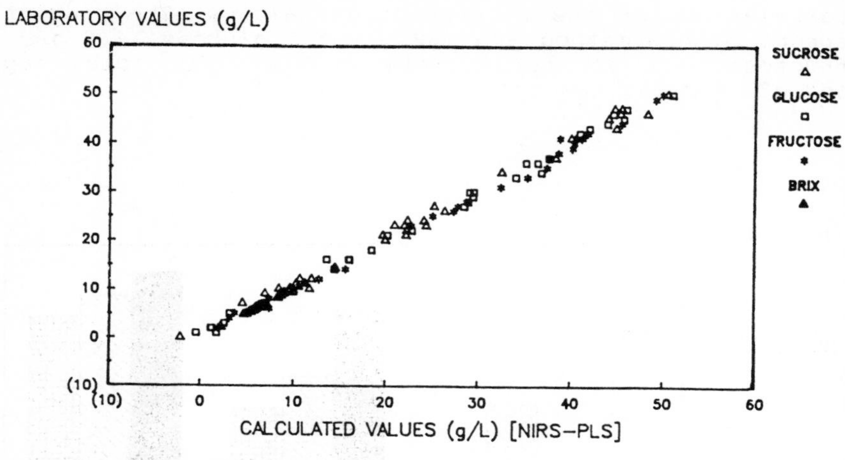

Units for dissolved solids are degrees Brix

Fig. 2 Laboratory values vs. NIRS-PLS values for the 30 prediction samples.

4 CONCLUSION

Rapid and multiple quantitative analysis of liquids by combination of DESIR and NIRS can be realized successfully with MLR and PLS calibration methods. PLS calibration method gives more slightly accurate calibrations than the MLR method.

5 REFERENCES

1. D.R. Massie and K.H. Norris, Trans. ASAE, 1965, 8, 598.
2. B.G. Osborne, and T. Fearn, Near-Infrared Spectroscopy in Food Analysis, (Longman, Wiley, New York, 1986).
3. K. Norris, and P. Williams, 'Near Infrared Technology in the agricultural and Food Industries', American Association of cereal Chemists, St Paul, MA, USA, 1987.
4. G. Alfaro and M. Meurens. 'Proceedings of 2nd international Near-Infrared Spectroscopy Conference', M. Iwamoto and S. Kawano. Eds., Korin Publishing, Japan, 1989, 204-212.
5. G. Alfaro, M. Meurens and G.S. Birth. Applied Spectroscopy, in press (for July 1990).
6. NIRSystems, Inc. A Perstorp Analytical Co., Silver Spring, MD. 20910, USA.
7. NSAS,(1990), Near Spectral Analysis Software of NIRSystems, Inc. A Perstorp Analytical Co., Silver Spring, MD. 20910, USA.
8. G. Alfaro, M. Meurens and P. Van Den Eynde. Application of DESIR and NIRS on orange juice, proceedings of 3rd ICNIRS, June 25-29, 1990, Brussels, in press.

Section 3

MICROSCOPY

FOURIER TRANSFORM INFRARED MICROSCOPY AND RAMAN MICROSCOPY, THE TECHNIQUES AND THEIR APPLICATIONS

B. W. Cook
ANALYTICAL AND PHYSICAL SCIENCES GROUP, ICI C&P LTD, RUNCORN, WA7 4QD

1 INTRODUCTION

The application of a microscope to the examination of surfaces and solids is a standard practice and has been since the 1600's. The relatively modern techniques of Fourier transform infrared (FTIR) and Raman spectroscopies are used to identify and fingerprint compounds. Each technique on its own provides valuable information about the nature and distribution of compounds on surfaces, within materials or in composites. The coupling of a microscope with a vibrational spectrometer has however become routinely available over the last decade. This development has resulted in powerful problem solving systems. There are advantages for both the microscopist and the spectroscopist from such combinations because it provides each with a new dimension to their work.

The spectroscopist gains from the sample alignment and manipulation properties inherent in the design of a microscope for the handling of small samples. The ability to focus the light onto particular areas and probe the sample to gather information about the small portion that has been displayed.

The microscopist gains from the molecular information available from the spectroscopy for the identification of the visual image. This information is able to be accessed directly from libraries stored in the computer system of FTIR spectrometers.

2 RAMAN MICROSCOPY

The results of Raman microscope experiments were first published at the IV International Conference on Raman Spectroscopy in 1974.(1,2) The paper from Delhaye described a micro Raman system based around a microscope and so created a totally new spectrometer the MOLE

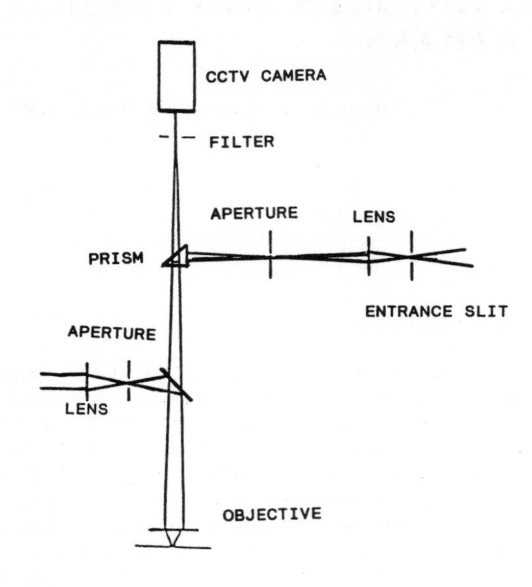

FIGURE 1
SCHEMATIC OF A FOCUSED RAMAN MICROPROBE SYSTEM.

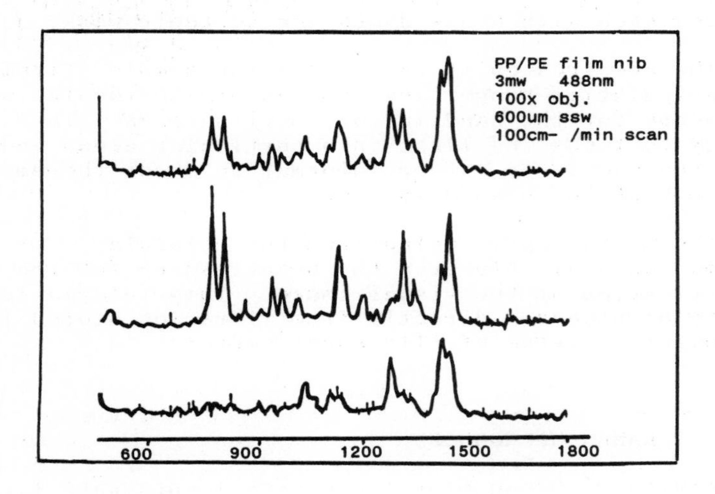

FIGURE 2
SPECTRA OF A SMALL INCLUSION IN A POLYPROPYLENE FILM. THE INCLUSION IS IDENTIFYABLE AS POLYETHYLENE.

(Molecular Optical Laser Examiner). A microscope attachment for a Raman spectrometer was reported by Cook and Louden (3) and now several manufacturers offer microscope attachments to conventional Raman systems. Raman microscopes maybe constructed from any research grade microscope with transmission and reflected light illumination facilities. Designs based on infinity corrected or focused systems have been built. The laser Raman - microscope combination is often called the Raman microprobe. See FIGURE 1. The source of light used in Raman spectrometers is the laser. This is an ideal source because it is a collimated beam of monochromatic light. The laser most commonly used is the Argon ion which emits at 514.5nm or 488nm. Other lines maybe obtained by coupling this to a dye laser. These are in the visible region of the spectrum and so maybe focused by glass lenses and reflected and directed by mirrors. The advantage of a visible source is that you can see exactly where it is probing the sample provided that you do this safely using a CCTV or an image screen. (Eyepieces particularly binocular ones should never be used to look at laser beams, no matter how weak you believe them to be!).

The microscope has been designed over the years to direct visible light onto a small area of a sample and collect the light to produce a visual image. The introduction of a new visible source into the microscope only adds a probe spot to the image. It is not used to illuminate all of the sample, although it maybe moved across the full area and a map or distribution of Raman intensities produced.

The laser enters the microscope through the epi-illuminator, and is then directed by a beamsplitter down to the objective lens. The objective lens of the microscope is used to focus the laser onto the sample and also to collect the Raman scattered light. This is the same lens that is used to produce the visual image so no refocusing of the sample is necessary to tune up the Raman signal. This is the major difference between the microprobe and the 90 collection Raman spectrometers.

The beamsplitter has a 10/90 or 20/80 reflection/ transmission ratio. The laser power has normally to be limited to between 0.5 and 20mw to prevent damage to the sample so a 10/90 beamsplitter attenuates the laser whilst still allowing most of the Raman scattered light to pass to the monochromator.

The laser is focused by microscope lenses to a diffraction limited spot which has a diameter of about 1um, and a focal depth of about 1um. The Raman scattered light is directed into the monochromator by a prism or mirror mounted on a slide mechanism. The coupling to the monochromator is through an achromatic

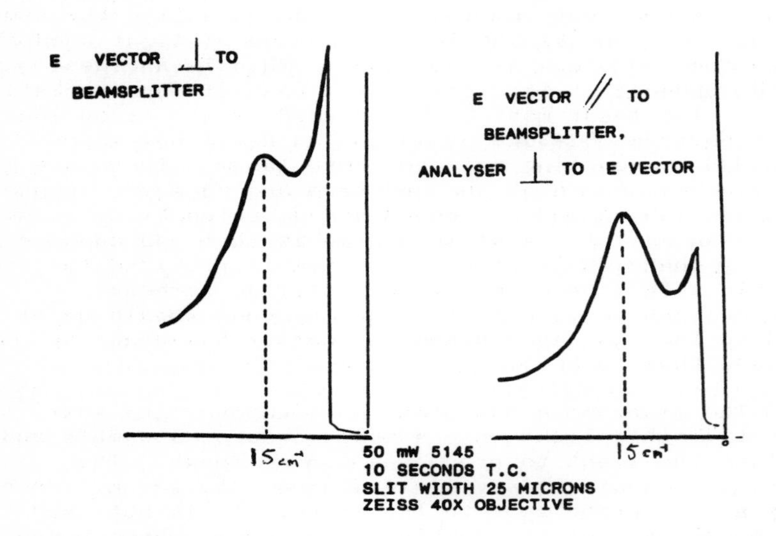

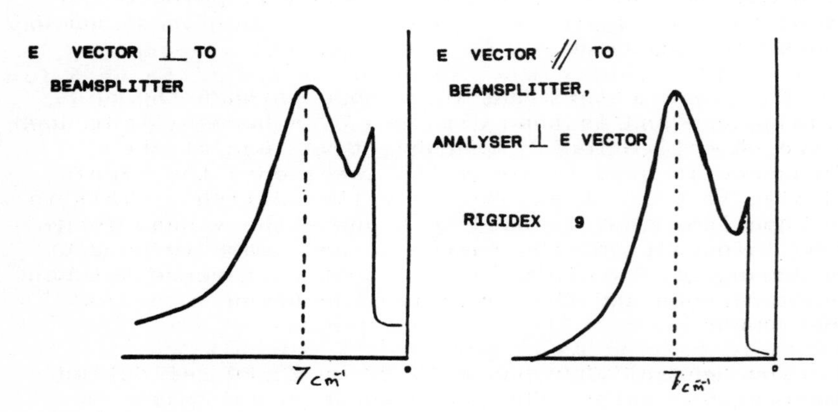

FIGURE 3 THE EXAMINATION OF THE BULK PROPERTIES OF POLYMERS, ACORDIAN MODE IN POLYETHYLENE.

doublet whose focal length is calculated to illuminate the whole of the grating and to be focused on the entrance slit of the monochromator. This coupling is important because it produces a confocal system which allows depth profiling of a sample to be performed. This confocal system therefore gives the Raman microprobe a three dimensional probing capability, but not of the same definition as the visible confocal microscopes. This difference is due to the Raman scattered light coming from all of the focused volume and being collected by the lens and transferred to the spectrometer.

Depth profiling through a multilayer polymer laminate in the form of a nib is a relatively easy task for the Raman microprobe work (see FIGURE 2). There is no sample preparation necessary whereas for infrared the sample would have to be microtomed across the sample and this is difficult.

The laser beam is polarised and therefore its passage through the beamsplitter is not uniform in all orientations, and ideally it should be aligned with its E vector perpendicular to the plane of the beamsplitter. In this orientation the maximum amount of Raman scattered light is transmitted through the beamsplitter and may therefore give an apparently better collection efficiency when measured at the Raman detector.

A microscope is not just an instrument used to examine small samples but to examine a small area of any size of sample that can be placed on the microscope table. This enables the Raman microscope to be used for solids, liquids and gases (4,5) and for bulk analysis problems such as the determination of the accordian mode of polymer lamellaes (see FIGURE 3).

Raman spectroscopy gives complementary spectra to those found in infrared spectroscopy. The bands that are strong in Raman are usually weak in the infrared. Inorganic samples often give good Raman spectra that have sharp lines. The infrared spectra of the same inorganic sample has broader bands and this is illustrated in FIGURES 4 AND 5 It is important that the microscope is able to be fully used in all modes of illumination so that the widest possible number of problems maybe tackled. A Raman microprobe is well suited to utilise these applications since visible light is used to produce an image and to probe the sample.

3 INFRARED MICROSCOPY

FTIR - microscopy is a relatively new hyphenated technique which combines the visualising capabilities of a light microscope with the analytical power of FTIR spectroscopy. The microscope is seen as an accessory to

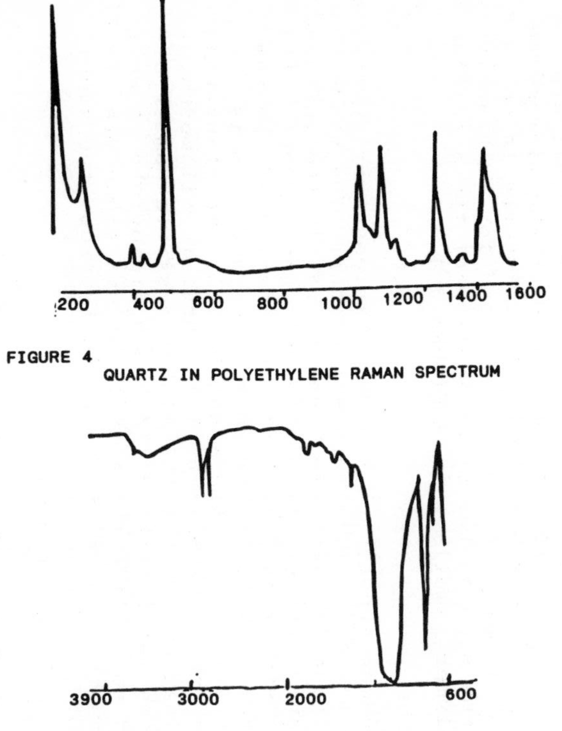

FIGURE 4
QUARTZ IN POLYETHYLENE RAMAN SPECTRUM

FIGURE 5
QUARTZ IN POLYETHYLENE INFRARED SPECTRUM

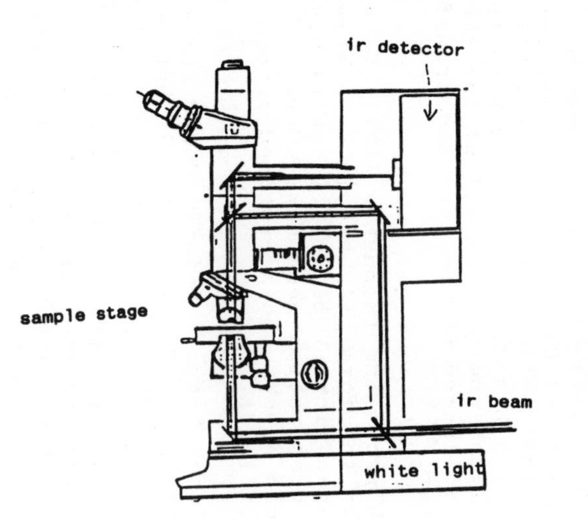

FIGURE 6
THE OPTICAL SCHEMATIC FOR AN INFRARED MICROSCOPE

a FTIR spectrometer, not as the replacement sample compartment, as in Raman spectroscopy. The optical layout of an infrared microscope is schematically very similar to the Raman microscope the main difference being that there are now two light sources which have to be co-linear and parfocal at the sample.

Unlike the Raman microscope the infrared microscope cannot be used with glass lenses. The light beam is directed from the FTIR spectrometer through the sample and to the detector by the use of front coated mirrors including the use of reflecting Cassegrainian objectives and condensers. These objectives and condensers are the same ones as are used for the white light, therefore once the sample has been viewed it is also in focus for the infrared light. The supports for the sample have to be of infrared transmitting materials such as KBr or BaF.

To probe the sample and examine a small portion of it (now in the order of 10-300 microns) a different means of limiting the analysis area is necessary. An aperture is placed in the light path such that it maybe viewed at the same time as the sample is viewed and both are in focus, this now serves as an optical mask. When two appertures are used to limit the light falling onto the sample and also to limit the light incident onto the infrared detector this is called Redundant Aperturing(TM) and this is the trade mark of Spectra-Tech, Inc. The apertures are four adjustable knife-edges each of which may be moved independently to mask down to the appropriate rectangular area.

The method of operation for the FTIR microscope is therefore to visualise the sample under white light, mask down if necessary to the area of interest and then to change to the infrared optics.See FIGURE 6

The energy through the system is governed by the aperture size, therefore it is imperative to measure both the sample and the background through the same aperture. The best way to do this experiment is to measure the sample, then move the sample out of the way and measure the background through the same aperture in the same optical configuration.

Since infrared spectroscopy is an absorption technique, sample thickness is an important parameter, and there is a finite usable range from which interpretable spectra maybe obtained. An area where microscopy techniques can make an important contribution to sample preparation is in microtomy.

This method of preparing samples is standard to a microscopist but is a skill that needs to be aquired by the spectroscopist. A variation for spectroscopy is that it is preferable if the sample is not embedded in a resin but is held between sheets of a polymer such as

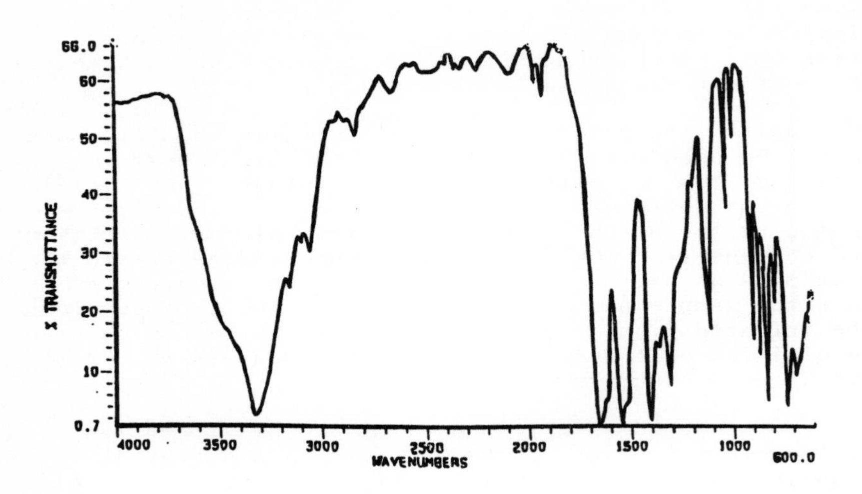

FIGURE 7
KEVLAR 12 MICRON FIBRE

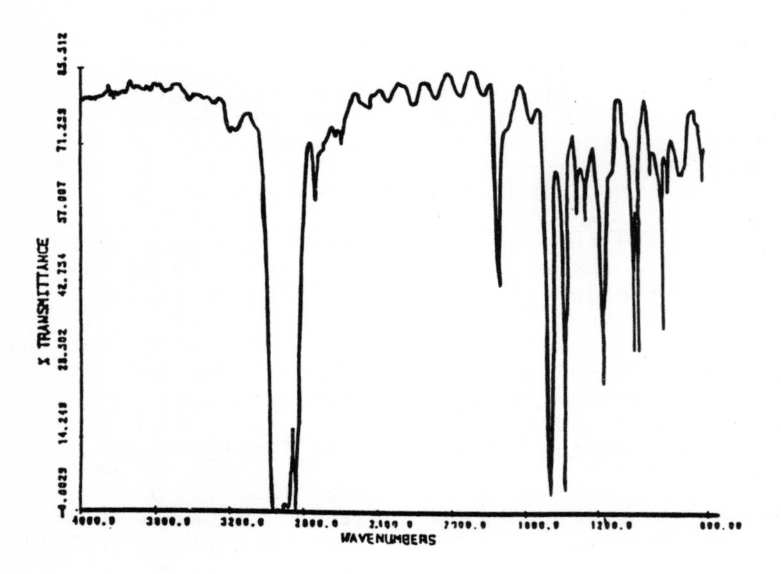

FIGURE 8
SPECTRUM OF UNSEALED FILM.

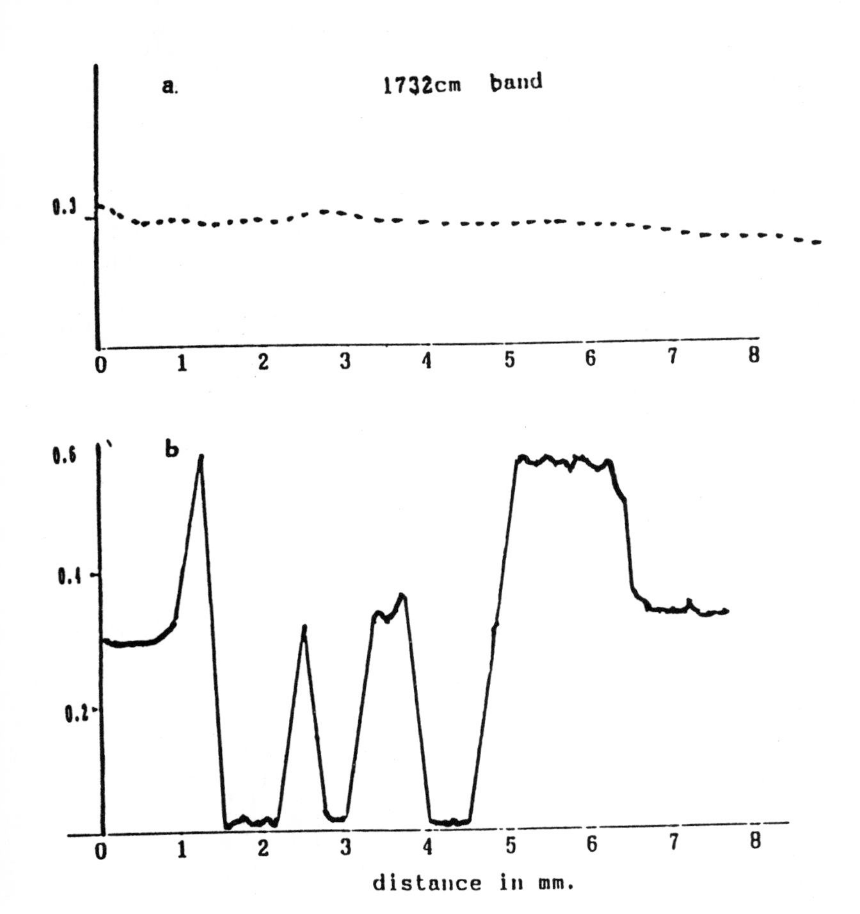

FIGURE 9
a, CARBONYL REPEATABILITY OF A HEAT SEAL ADHESIVE IN AN UNBROKEN BOND.
b, CARBONYL OF A HEAT SEAL ADHESIVE IN A BROKEN BOND

polypropylene or polystyrene.

Forensic science is an area of analytical work that is making use of the power of the hyphenated techniques. Single fibres are one of the easiest types of samples to examine on a microscope, however to get good spectra sample preparation is often necessary. The sample is naturally round and the best shape for infrared absorption measurement is rectangular so pressing or use of the roller knife is recommended, the spectra may then be compared with known library standards(6). An example of a single fibre is shown in FIGURE 7.

Both industrially and forensically paint chips and painted surfaces are a rich source of analytical problems. The sample preparation necessary to obtain good spectra from multi-layer paints is to microtome 2μm thick sections at an angle so that the layers are apparantly thicker and easier to mask down onto.(7,8). This method of sampling is very successful and identification of the paint is eased because there is a commercial library of 1800 paint chip spectra available.

The infrared microscope maybe used to investigate the microscopic distribution of functional groups in different fields of studies such as polymers/ material science(9) and biological sciences (10) . There are advantages to be gained by studying the distribution rather than single shot spectra from surfaces. The extra equipment necessary is a stepping stage , this will allow small repeatable movements to be made accurately.

The experimental details for a one or two dimentional functional group map are as follows:-
the wavenumber of the group under study and a reference point in the spectrum,
the step size in the x and y directions, and the number of steps,
the aperture size through which the examination is being made.

Once the spectra have been collected the intensity of the band at the selected wavenumber is measured relative to the reference and plotted versus the position of the xy stage. This will give the distribution relative to the step size, however it is preferable to match the aperture and the step sizes.See FIGURES 8 AND 9. These show the carbonyl in a polypropylene film and the distribution of the band plotted as a function of distance moved across a broken seal.

4 CONCLUSIONS

The effective spatial resolution of Raman microscopy, 1 micron spot size, is better than the FTIR microscopy minimum of 25-100 square microns. However there are papers which describe 2-3 micron resolution by the use of improved experimental and mathematical treatments (11,12) in FTIR microscopy. The development of both techniques is continuing. The sensitivity of the two spectroscopies are different for inorganic and organic compounds and complementary information is obtained. More commercial samples maybe tackled by FTIR microscopy than by Raman but the spectroscopy appropriate to the subject under investigation should be selected for the best results. Sample preparation is often necessary for FTIR. Raman microscopy can be used to probe a sample in the xy and z directions FTIR is only used in the x and y direction.Infrared and Raman microscopies are excellent problem solving techniques. The information obtained from such hyphenated techniques is better than that obtained by a combination of individual examinations.

5 REFERENCES

1 M. Delhaye and P. Dhamelincourt, IV International Conf. Raman Spectroscopy. Brunswick, ME, USA. 1974

2 G.J. Rosasco, E.S. Etz and W.A. Cassatt, IV International Conf. Raman Spectroscopy. Brunswick, ME,USA. 1974

3 B.W. Cook and J.D. Louden, J. Raman Spec. 1979, 8, 249.

4 C.G. Pandey Analyst, 1989, 114, 231.

5 J.D. Louden, ch 6 Practical Raman Spectroscopy, ed. D.J. Gardiner, P.R. Graves. Springer-Verlag 1989.

6 J.D. Louden, ch 22 Laboratory Methods in Vibrational Spectroscopy 3rd. edition ed. H.A. Willis, J.H.van der Maas and R.G.J. Miller, J. Wiley 1987.

7 J.C. Shear, D.C. Peters and T.A. Kubic Trends anal. Chem. 1985, 4, 246.

8 W. Stoecklein and M. Gloger Nicolet Spectral Lines 1988, 1, 2.

9 M.A. Harthcock, L.A. Lentz, B.L. Davis and K. Krishnan, Applied Spect. 1986, 40, 210.

10 M.P. Fuller and J. Rosenthal ch 12 Infrared Microscopectroscopy Theory and Applications ed. R.G. Messerschmidt and M.A. Harthcock M. Decker Inc. (Practical spectroscopy series vol 6).

11 G.A. Massey, J.A. Davis, S.A. Katnik and E. Omon Appl. Optics, 1985, 24, 1498.

12 D.H. Burns, J.B. Callis, G.D. Christian, and E.R.Davidson, Appl. Optics, 1985, 24, 154.

QUALITATIVE AND QUANTITATIVE EXAMINATION OF THE STRUCTURE OF CETOSTEARYL ALCOHOL AND CETRIMIDE EMULSION USING RAMAN MICROSCOPY AND FOURIER TRANSFORM INFRARED MICROSCOPY

J. D. Louden and R. C. Rowe*

ICI C&P LTD, THE HEATH, RUNCORN, CHESHIRE, WA7 4QD
*ICI PHARMACEUTICALS, ALDERLEY PARK, MACCLESFIELD, CHESHIRE, SK0 2TG

INTRODUCTION

The mixed emulsifier system of cetostearyl alcohol and cetrimide is frequently used in the formulation of antiseptic creams and a lot of work has been done in an attempt to determine the composition of the various phases within the cream using various microscopic techniques.[1-3] Raman microscopy and Fourier Transform Infrared (FT-IR) microscopy have the capability of examining microscopic areas of interest (down to 1μm for Raman and down to 10μm for FT-IR) and identifying the chemical species present.

The structure of these creams is very complex consisting of spherical droplets of liquid paraffin together with particles of cetostearyl alcohol dispersed in a gel network consisting of bilayers of cetostearyl alcohol swollen with water.[3-5] In this work we have used Raman and FT-IR microscopy to qualitatively analyse the structure of the emulsions and quantitatively analyse the oil droplets for cetostearyl alcohol content.

Experimental

Preparation of the Emulsion. All materials were of Pharmacopetal grade. An emulsion was prepared according to the formula in table 1. The liquid paraffin and cetostearyl alcohol were heated to 80°C and then dispersed in aqueous cetrimide solution at the same temperature. The mixture was then stirred gently with a paddle stirrer for 1 hour, allowed to cool to approximately 60°C, and then the dispersion homogenised using a Silverson multipurpose high speed mixer for a period of 15 or until the setting point of the cream was reached. The emulsion was allowed to stand for at least two weeks before being analysed.

Table 1 Emulsion formulation used in study.

	Concentration % w/w
Cetostearyl alcohol	10.0
Cetrimide	0.5
Liquid Paraffin	20.0
Purified water to	100.0

Raman Microscopy. The Raman microscope used in this study consisted of a modified Nikon Optiphot optical microscope coupled to a Jobin Yvon HG2S double monochromator. The detection system was a RCA C31034A gallium arsenide photomultiplier tube with an Ortec Brookdeal 5C1 photon counting system. Laser excitation was the 488nm blue line of a Coherent Innova 90-5 Argon ion gas laser. A 60x objective was used to obtain the spectra with approximately 50mW of laser power at the sample, 10cm-1 resolution, coadding 16scans.

Fourier Transform Infrared Microscopy. The FT-IR microscope used in this study was developed by AIRE Scientific Ltd. and was coupled to a Nicolet 510 FT-IR spectrometer. The microscope used a 0.5mm MCT (mercury cadmium telluride) liquid nitrogen cooled detector and a choice of 15x, 36x, and 52x cassegrain mirror objectives. The infrared energy was transmitted through the sample onto the MCT detector and the IR spectra were recorded at 8cm-1 resolution and 1000scans using the 52x cassegrain objective, and an aperture of 10 x10 microns.

The sample for analysis was in the form of a 20 micron thin film between two glass coverslips. Each sample was first examined on the microscopes using white light illumination on a close circuit television viewing system. The features of interest (gel network, polyhedral particles and oil droplets) were located centrally in the field of view and Raman and IR spectra obtained from the different regions. In the case of the FT-IR microscope, the specific area of interest (typically 10 x 10 microns) was then isolated by means of the remote aperture (4 knife edge blades) located at the primary image position. Since the features were typically of the order of 20 microns diameter this aperturing eliminated any interference from the surrounding medium.

RESULTS.

Figure 1 shows a photomicrograph of the emulsion showing the three features of interest. (A) oil droplet.
(B) polyhedral particle.
(C) gel network.

Figure 2 shows the Raman spectra of the reference materials in the C-H stretching region showing clear differences which can be used for identification purposes.

Figure 3 shows the Raman spectra obtained from the three features in the photomicrograph showing that the polyhedral particle is essentially cetostearyl alcohol , the gel network consisting of cetostearyl alcohol and water , and the spherical droplet is essentially liquid paraffin.

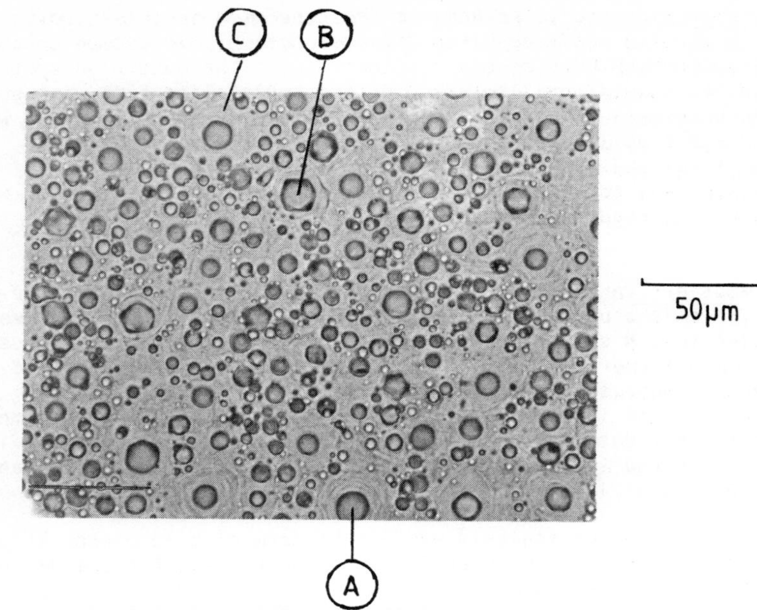

Fig.1 Photomicrograph of emulsion

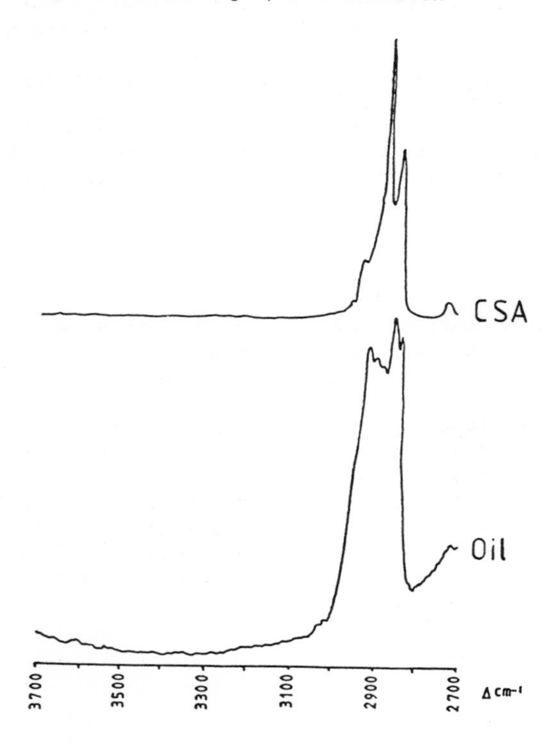

Fig.2 Reference Raman spectra

Figure 4 shows the IR spectra of the reference materials in the C-H stretching region showing clear differences which can be used for identification purposes.
Figure 5 shows the IR spectra obtained from the three features in the photomicrograph showing that the polyhedral particle is cetostearyl alcohol, the gel network consists of cetostearyl alcohol and the spherical droplet is essentially liquid paraffin.

Raman and IR measurements on the same features give the same results and therefore the features of interest can be positively identified.

However, while the IR spectrum of the oil droplets is very similar to the reference spectrum of pure liquid paraffin,closer examination of the IR spectrum reveals a shift in the peak position of the anti symmetrical methylene C-H at 2924cm-1.This is consistent with the presence of some dissolved cetostearyl alcohol.

In order to quantify the amount of cetostearyl alcohol present in the oil droplets, a series of standards were prepared of cetostearyl alcohol dissolved in liquid paraffin in the range 0 to 30%
FIgure 6 shows the linear relationship between % cetostearyl alcohol and shift in peak position of the antisymmetrical methylene C-H.

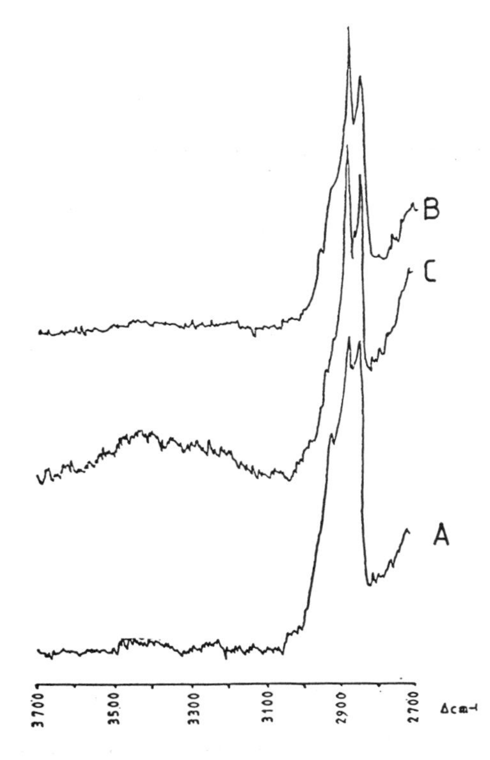

Fig.3 Raman spectra of the three features A,B, and C in figure 1.

Several oil droplets were examined in the emulsion by the FT-IR microscope and it can be seen that inhomogeneity at the "microlevel"was present.Table 2 list the peak positions and % cetostearyl alcohol for a number of oil droplets in the emulsion.A similar inhomogeneity in the gel network has been previously inferred from dielectric relaxation measurements.[6]

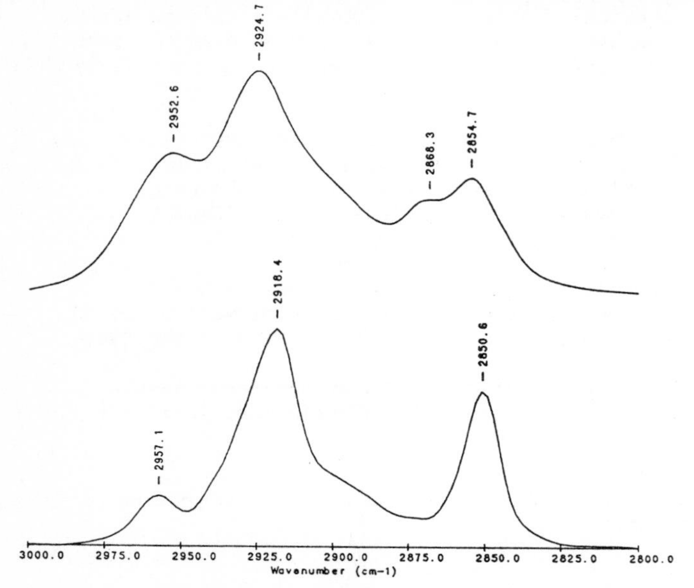

Fig. 4 IR spectra of the reference materials
top-liquid paraffin bottom-cetostearyl alcohol

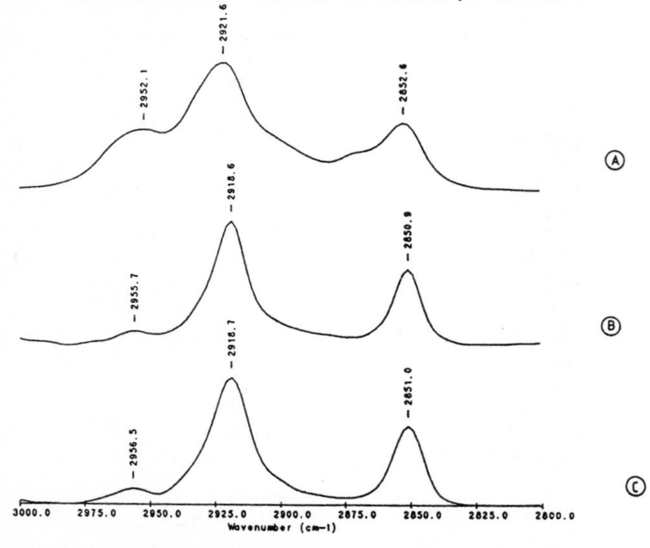

Fig.5 IR spectra of the three features in figure 1.
top A spherical droplet,middle B gel network,
bottom C polyhedral particle.

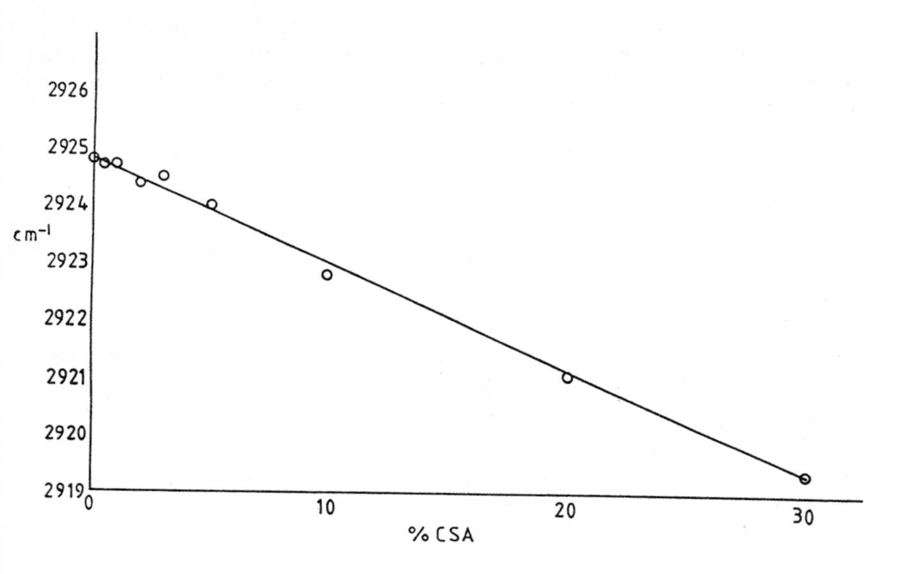

Fig. 6 % cetostearyl alcohol versus peak position showing linear relationship.

Table 2 Peak position and % cetostearyl alcohol of oil droplets in the emulsion.

Peak position cm-1.	Concentration % w/w
2922.6	12
2922.1	15
2921.7	17
2921.6	18
2921.6	18
2920.7	23
2920.7	23
2920.6	23
2920.6	23
2920.6	23
2920.5	24
2920.4	25
2920.3	25
2920.3	25
2920.2	26
2929.8	27

CONCLUSIONS.

The results clearly show the potential of the two relatively rapid, non destructive techniques of Raman and FT-IR microscopies in the qualitative and quantitative analysis of these complex emulsions and in the understanding of the mechanisms of formation.

REFERENCES

1. B.W. Barry and G.M. Saunders, J.Colloid Interface Sci., 1970 34 300.
2. G.M. Eccleston, Properties of fatty alcohol mixed emusifiers and emulsifying waxes. In Florence, A.T. (Ed.) Materials used in Pharmaceutical Formulation. Critical Reports on Applied Chemistry. 1984 6 124.
3. H.K.Patel R.C.Rowe J. McMahon and R.F.Stewart Int. J.Pharm 1985a 25 13.
4 R.C. Rowe and J.McMahon Colloids and Surfaces 1987 27 367.
5 M.D. Barry and R.C.Rowe Int. J. Pharm., 1989 53 139.
6 L.A.Dissado R.C.Rowe A. Haider and R.M.Hill J.Colloid Interface Sci. 1987 117 310.

THE USE OF FTIR-MICROSPECTROMETRY FOR THE MEASUREMENT OF CRYSTALLINITY IN POLYETHYLENE WELDS AND HELICAL CONTENT IN POLYPROPYLENE WELDS

S. M. Stevens

THE WELDING INSTITUTE, ABINGTON HALL, ABINGTON, CAMBRIDGE, CB1 6AL

1 INTRODUCTION

Thermoplastics such as polyethylene (PE) and polypropylene (PP) are widely used, for example in gas and water distribution and injection moulded components for industry, as their flexibility, toughness, corrosion resistance, and low weight enables them to compete with metals in a variety of applications. The mechanical properties of semicrystalline polymers are dependent on the thermal history and are expected to be influenced by welding.

Few data exist on the fundamental changes which occur when a thermoplastic undergoes a welding cycle, and FTIR-microspectrometry is well suited to the examination of the small areas encountered in welds.

This paper describes the quantitative determination of crystallinity (X) and helical content (H) in PE and PP welds, respectively. The accuracy of the method was assessed using parent materials and comparison with results from density, differential scanning calorimetry (DSC), and X-ray diffraction (XRD).

2 THEORY

FTIR

In PE, the crystallinity was calculated using the relative absorbances, A_c and A_a, of bands at $1894 cm^{-1}$ from crystalline regions, and $1303 cm^{-1}$ from amorphous regions, respectively[1] and the following equation[2].

$$\%X = \frac{100}{1 + \frac{A_a}{A_c}\frac{E_c}{E_a}} \tag{1}$$

where E_c and E_a are the specific extinction coefficients and were taken to be $6.2 cm^2 g^{-1}$ and $26.535 cm^2 g^{-1}$ respectively[1,3,4].

There are no purely crystalline and amorphous bands in PP, and the helical content was calculated from the ratio of the bands at $998cm^{-1}$, associated with structural regularity, and $974cm^{-1}$ from molecules in disordered arrays[5,6,7], according to the equation[7,8].

$$\%H = 100 \left(\frac{A_{998}}{A_{974}}\right) - 31.4 \quad (2)$$

The $974cm^{-1}$ band is relatively insensitive to conformational order and is widely used as an internal reference band. There is a direct relationship between %H and %X since only isotactic chains can assume a helical conformation, and thus there cannot be more crystalline material than there is isotactic material.

Density

The crystallinity was calculated as follows [9]:

$$\%X = \frac{\rho_c}{\rho} \left(\frac{\rho - \rho_a}{\rho_c - \rho_a}\right) 100 \quad (3)$$

where ρ, ρ_c, and ρ_a are the densities of the sample, 100% crystalline material, and 100% amorphous material, respectively. Values of $1.0045gcm^{-3}$ and $0.855gcm^{-3}$, respectively, were used for ρ_c and ρ_a for PE[10], and $0.936gcm^{-3}$ and $0.853gcm^{-3}$ for PP[11].

Differential Scanning Calorimetry

The ratio of the heats of fusion (ΔH) required to melt a known weight of the sample and of a 100% crystalline material was used as follows:

$$\%X = \left(\frac{\Delta H_{sample}}{\Delta H_{100\%\ crystalline}}\right) 100 \quad (4)$$

Values of $290Jg^{-1}$ (PE)[12] and $153Jg^{-1}$ (PP)[11] were used for 100% crystalline material.

X-ray Diffraction

The crystallinity was determined from the ratio of the integrated crystalline scattering and the total scattering[9].

3 EXPERIMENTAL

Details of the density, DSC, and XRD equipment are described elsewhere[13].

Materials

Parent materials for method assessment included low, medium and high density, and ultrahigh molecular weight, PEs, and PP homopolymers

Table 1 Reproducibility and precision results

PE %X

	Sample compartment			Microscope		
	Reproducibility		Precision	Reproducibility		Precision
	Mean	SD	SD	Mean	SD	SD
L1	48.7	0.3-0.7	0.2	48.4	0.3-0.7	0.1
M1	62.6	0.2-0.3	0.3	62.7	0.2-0.3	0.1
H6	79.4	0.2-0.3	0.1	80.3	0.2-0.3	0.1

PP %H

	Sample compartment			Microscope		
	Reproducibility		Precision	Reproducibility		Precision
	Mean	SD	SD	Mean	SD	SD
PP4	65.1	0.3-0.7	0.4	64.5	0.4-0.8	0.3
PP7	61.1	0.3-0.7	0.5	59.3	0.4-0.8	0.3
PP11	55.6	0.3-0.7	0.2	54.5	0.4-0.8	0.2

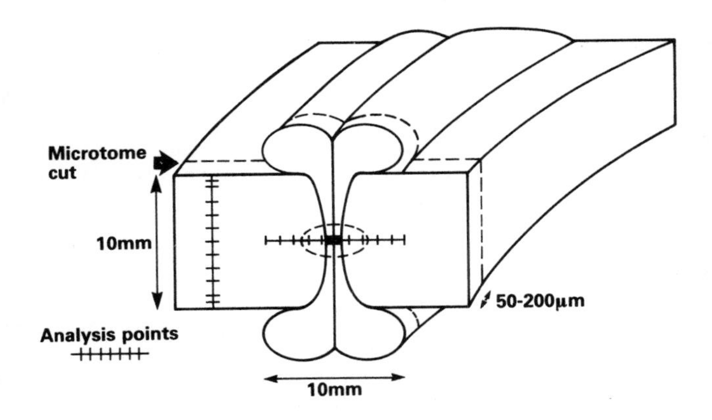

Fig.1 Schematic diagram of hot plate weld.

and copolymers (with ethylene).

Hot plate welds were made using medium density PE and PP copolymer pipes 180mm diameter and 10mm wall thickness.

FTIR-Spectrometer and IR Microscope

Infrared spectra were recorded using a Mattson Polaris/ICON spectrometer with a DTGS detector, and a Spectra Tech IR-PLAN microscope with a 0.25mm MCT detector. Spectra obtained in the spectrometer sample compartment were recorded at 16 scans and $4cm^{-1}$ resolution, and microscope spectra at 360 scans and $8cm^{-1}$ resolution, apart from the PP welds and precision tests where 36 scans were used.

4 ANALYTICAL TESTS

Method Assessment Using Parent Materials

FTIR measurements were carried out on 200 micron (PE) or 50 micron (PP) thick microtomed sections. Analysis areas were 5mm diameter (sample compartment) or 150 x 150 microns (microscope). The density, DSC, and XRD tests are described elsewhere[13].

Reproducibility data were obtained on low, medium, and high density PE (L1, M1, H6) and PP homopolymer (PP7) and copolymer (PP4, PP11). Six microtomed samples from each material were analysed on six separate occasions.

Instrumental precision was evaluated using six consecutive analyses on one sample from each material.

FTIR Measurements on Weldments

An analysis area of 188 x 50 microns was used, and sample thicknesses were as above. Parent pipe sections were analysed at various locations through the thickness, and weld sections were analysed along an axis perpendicular to the weld centreline, Figure 1.

5 RESULTS AND DISCUSSION

Parent Material Tests

For PE, the reproducibility and precision, Table 1, of samples analysed in the microscope are normally equal to or better than for samples analysed in the sample compartment, but for PP the reproducibility decreases slightly for microscope measurements while the precision is better than or equal to sample compartment tests.

The FTIR, density, DSC, and XRD results are compared in Figure 2 which shows that the methods correlate well for PE, even though the slope of the DSC line is altered by a low value obtained on L1. This behaviour of low density PE has been reported by other workers[14]. For PP, the overall agreement between the methods is quite acceptable taking into account the fact that FTIR is measuring %H and the others %X.

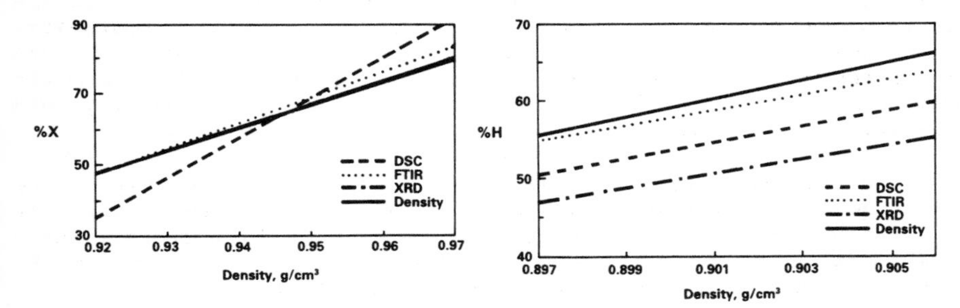

Fig.2 Correlation between techniques.

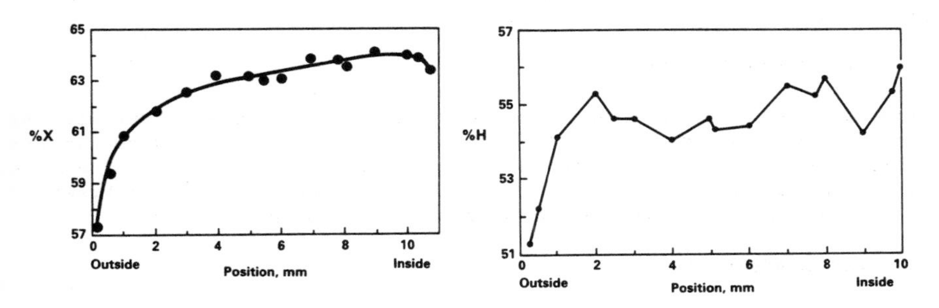

Fig.3 Through-thickness crystallinity/helical content in parent pipe.

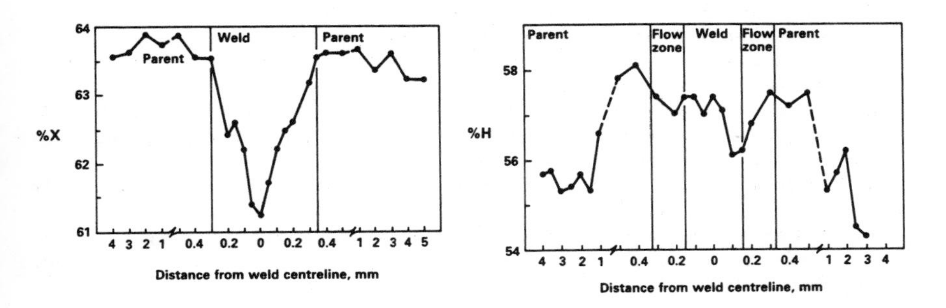

Fig.4 Crystallinity/helical content variation in welds.

Weldment Measurements

The through thickness results on parent materials, Figure 3, show that, overall, the %X and %H increase from the outside to the inside of the pipe, consistent with a slower cooling rate at the inside. The smooth curve for PE indicates a high degree of local homogeneity, while the PP curve exhibits considerable scatter, which is not unexpected as the reproducibility is less good for PP. Also, the presence of ethylene might contribute to the heterogeneity and increase the scatter.

The crystallinity in the PE weld, Figure 4, is less than in the parent pipe and decreases to a minimum at the centreline, while the opposite is true for the helical content in PP. In addition, there appears to be an annealing effect in the parent pipe adjacent to the weld flow zones, resulting in a higher helical content.

6 CONCLUSIONS

FTIR-microspectrometry is a suitably accurate and reproducible technique for the determination of crystallinity in PE welds, and helical content in PP welds.

7 ACKNOWLEDGEMENTS

This work was funded by Research Members of The Welding Institute and the Non-Metallic Materials Committee of the Technology Requirements Board, Department of Trade and Industry. Thanks are also due to Mr D Robson and Dr D C Puxley of British Gas for the DSC analyses on PE, and the XRD analyses, respectively, and to Dr H A Willis.

8 REFERENCES

1. T Okada and L Mandelkern, Journal of Polymer Science: A2, 1967, 5, 239.
2. G Kampf, 'Characterisation of Plastics by Physical Methods', Karl-Hanser, Munich, 1986.
3. B E Read and R S Stein, Macromolecules, 1968, 1, 116.
4. H Hendus and G Schnell, Kunstoffe, 1961, 51, 69.
5. P J Hendra et al, Polymer, 1984, 25, 785.
6. P Blais et al, Journal of Polymer Science: A1, 1972, 10, 1077.
7. R G Quynn et al, Journal of Applied Polymer Science, 1959, 2, 166.
8. R Zbinden, 'Infrared Spectroscopy of High Polymers, Academic Press, New York and London, 1964.
9. J F Rabek, 'Experimental Methods in Polymer Chemistry', J Wiley and Sons Ltd, London, 1980.
10. P J Hendra, Southampton University, Private Communication.
11. F M Mirabella Jr, Journal of Polymer Science Part B, Polymer Physics, 1987, 25, 591.
12. R L Blaine, DuPont Application Brief No TA 36.
13. S M Stevens, 'Advances in Joining Newer Structural Materials', Proceedings of International Conference, 23-25 July 1990, International Institute of Welding, Abington, 1990.
14. J Vile et al, Polymer, 1984, 25, 1174.

APPLICATIONS OF THE IR MICROSCOPE TO PROBLEM SOLVING

F. O. Cox

ANALYTICAL DEVELOPMENT LABORATORIES, BURROUGHS WELLCOME CO., P.O. BOX 1887, GREENVILLE, NORTH CAROLINA, 27835-1887, USA

1. INTRODUCTION

The molecular spectroscopy laboratory at any pharmaceutical company faces a wide variety of problems. The unequivocal structure determination of new chemical entities is necessary when registering new products for the marketplace. Generally this work is performed using a combination of IR and NMR spectroscopies, and mass spectrometry. Unlike NMR and MS, however, IR has traditionally required 1-2 mg of sample to obtain a spectrum, and frequently, the sample is prepared in a matrix where its recovery is difficult, if not impossible.

The IR microscope, which has been described in detail[1], has changed the way in which solid samples are approached. Infrared spectra may now be obtained on nanogram to microgram quantities of samples[2,3], and because the sample preparation methods are generally non-destructive, the samples may be recovered for further analysis by other techniques. This capability allows for the elucidation of the structures of isolated degradation products and microscopic contaminants, where frequently only microgram quantities of material exist. The capability of the IR microscope has rejuvenated the enthusiasm of organic synthetic chemists who previously were unwilling to sacrifice milligram quantities of sample for IR analysis.

2. EXPERIMENTAL

The infrared spectra were obtained using a Nicolet 20 DXB Fourier Transform Infrared Spectrometer. The microscope employed was a Spectra-Tech IR Plan equipped with a dedicated MCT detector and 15X REFLACHROMAT objective lens. The samples were manipulated and prepared for analysis under a standard stereoscope, which provides a much wider field of view for sample preparation. All samples were placed onto a 13 X 2 mm KBr window for analysis. The aperture size used for "masking" the sample depended on the sample being analyzed. Following the acquisition of an appropriate number of scans for each sample, an equal number of scans of the KBr window were obtained as the background using the same aperture size. The ratio of the sample spectrum to the background spectrum was taken to obtain a spectrum in % Transmittance.

3. RESULTS AND DISCUSSION

Polymorphism

In a general sense, polymorphs can be considered to be substances that have two or more forms which differ in physical properties, but have the same chemical properties in solution. The specific knowledge of which polymorph exists is essential to the pharmaceutical industry. In oral dosage forms, the differences in solubility between two polymorphs can have an effect on the bioavailability of the drug. In the development of a parenteral formulation, more of the polymorph having the higher solubility is desirable as a higher concentration of the drug can be prepared in the same volume. The IR microscope is a very useful tool for examining different polymorphs for several reasons. If the microscope is equipped with polarizing optics, both an optical and chemical picture of two different polymorphs can be obtained. The sample integrity, in many cases, is preserved and thus, the sample may be recovered for further analysis, such as X-Ray diffraction or hot-stage microscopy. Finally, the physical effects (such as grinding and compressing) of KBr pellet preparation are eliminated.

773U82 Hydrochloride. 773U82 Hydrochloride is a drug substance which is known to exhibit two different polymorphs[4]. Figure 1 shows a comparison of Forms 1 and 2 which represent forms of higher and lower solubility, respectively. Upon storage under stressful conditions, a vial of 773U82 Hydrochloride for Injection was observed to contain particles which averaged 30 microns in size. The contents of a vial were filtered, and one of the isolated particles was placed on a KBr window, flattened slightly, and the IR spectrum of a 30 x 30 micron area obtained. Figure 2 shows a comparison of the IR spectrum of the isolated particles and Form 2 of 773U82 Hydrochloride. From these data, it was determined that the particles were the lower solubility polymorph of the 773U82 Hydrochloride drug substance, and not a degradation product arising from storage under the stressful conditions.

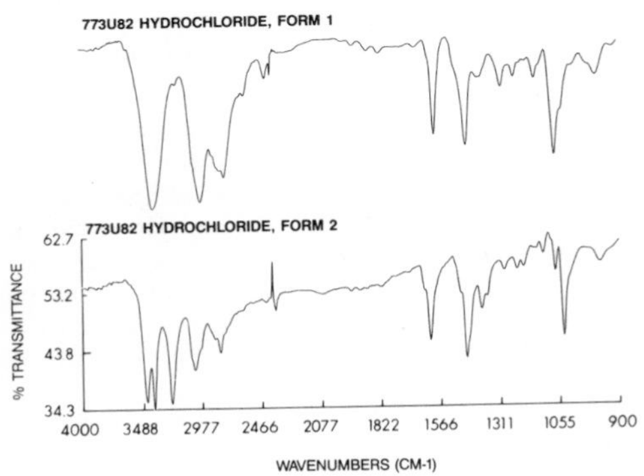

Figure 1 Spectral comparison of Form 1 (top) and Form 2 (bottom) of 773U82 Hydrochloride.

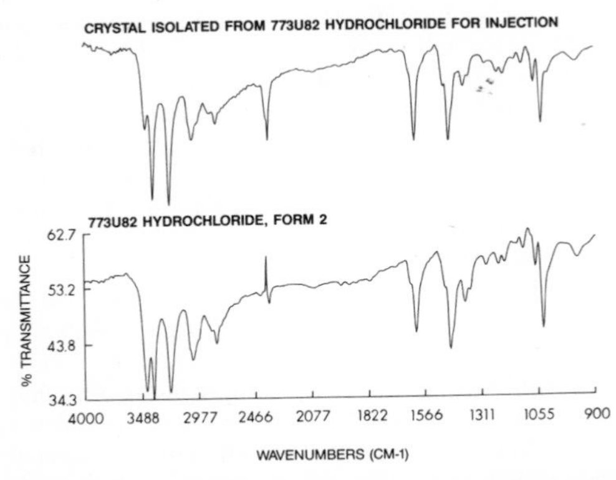

Figure 2 Spectral comparison of particulate isolated from 773U82 Hydrochloride for Injection (top) and Form 2 (bottom)

Contaminant Identification in a Pharmaceutical Binding Agent

Povidone is a binding agent used in tablet compression in the pharmaceutical industry. During the initial Quality Assurance inspection, small grey "flakes" were observed in one drum of Povidone. Examination beneath a stereoscope revealed that the Povidone was agglomerating around these grey flakes, and manual isolation of the grey matter itself was impossible due to the fine nature of Povidone powder. Several of the agglomerations were isolated, dissolved in water, and filtered to isolate the less soluble matter, which appeared on the filter paper as discreet droplets of gelatinous matter. A smear of this isolated grey matter was transferred to a KBr window with the aid of a sharpened tungsten needle, and the infrared spectrum of a 20 x 20 micron area of the smear was obtained. Figure 3 shows a comparison of the spectrum of the grey matter obtained using the microscope with that obtained by manually isolating several of the grey agglomerates and preparing a KBr pellet. The spectrum obtained from the pellet is virtually indistinguishable from that of Povidone. However, through the use of the IR microscope, the contaminant was identified as a food grade grease used by the Povidone manufacturer for lubrication, which was contaminating their manufacturing process.

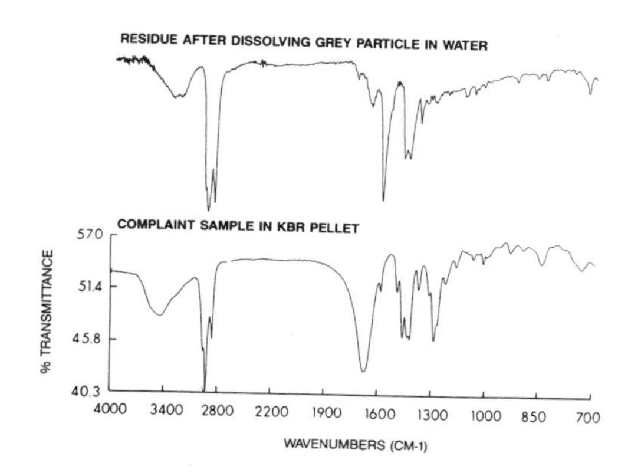

Figure 3 Spectral comparison of grey contaminant using the IR microscope (top) and KBr pellet preparation (bottom)

Contaminant Identification from a Dissolution Vessel

Dissolution testing is performed routinely by the pharmaceutical industry to measure rates of disintegration and release of drug from tablets and capsules. Following dissolution testing, a scientist noticed a very small ribbon-like contaminant floating in a dissolution vessel. This contaminant was isolated from the dissolution vessel, gently flattened on a KBr window, and a spectrum of a 20 x 50 micron area was obtained. This spectrum, along with the "best match" spectrum obtained from a spectral search against an Aldrich spectral library, is shown in Figure 4. Although an exact match was not obtained, the contaminant was suspected to be some kind of nylon, probably from the dissolution instrument itself, which has nylon gears and nylon dissolution paddles.

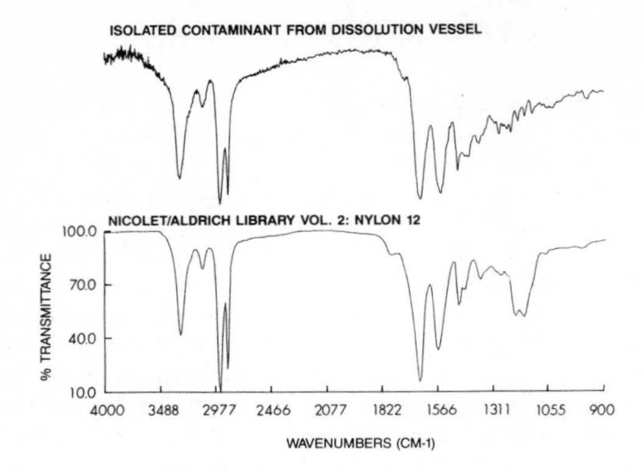

Figure 4 Spectral comparison of dissolution vessel contaminant (top) and Nylon from Adlrich spectral library (bottom)

HPLC Troubleshooting

HPLC autosampler valves in Quality Assurance were becoming plugged after a few months use. Examination of one of the components in the valve revealed both clear and brown fibers to be present. The fibers were isolated with a sharpened tungsten needle, sandwiched between two KBr windows, and their IR spectra obtained. In both cases, a 15 x 50 micron area of the fibers was scanned. One potential source of the contamination was the HPLC autosampler vials themselves, and as a result, the IR spectra of the packing materials used for shipping the vials were obtained. The spectra of the clear fiber and shipping cover placed over the vials during shipment from the manufacturer are shown in Figure 5, while the spectrum of the brown fiber compared with that of cardboard is shown in Figure 6. From these data, it was determined that the source of the plugging of the HPLC valves was the packing materials used for packing the HPLC autosampler vials, and as a result, QA now routinely cleans the vials prior to use by gently passing a stream of purified air over the vials. No additional failures of HPLC autosampler valves have been experienced since this procedure has been implemented.

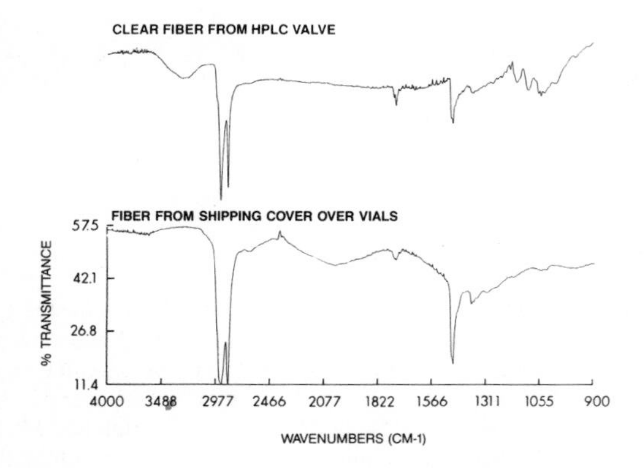

Figure 5 Spectral comparison of clear fiber from HPLC autosampler valve (top) and clear packing material (bottom)

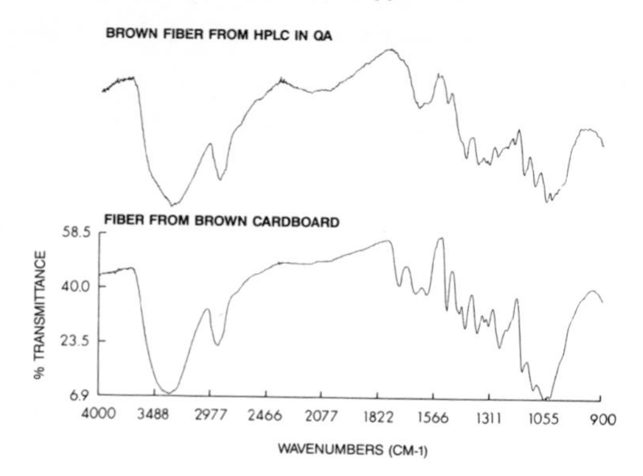

Figure 6 Spectral comparison of brown fiber from HPLC autosampler valve (top) and cardboard (bottom)

4. CONCLUSIONS

The IR microscope is a very useful tool to the spectroscopy laboratory for many reasons. It is fairly easy to use, and provides for both an optical and chemical picture to be obtained for a given problem. Quality IR spectra from minute quantities of sample can be obtained, and the sample can generally be easily recovered for further analysis by other analytical techniques. Finally, the sample integrity is preserved and the effects of KBr pellet preparation are eliminated, which is very important when studying polymorphic substances.

5. REFERENCES

1. P.B. Roush, "The Design, Sample Handling, and applications of Infrared Microscopes", ASTM Publications, Philadelphia, PA, 1987.
2. R. Cournoyer, J.C. Shearer, and D.H. Anderson, Analytical Chemistry, 1977, 49, 2275.
3. J.C. Shearer, D.C. Peters, J. Hoepfner, and T. Newton, Analytical Chemistry, 1983, 55, 874A.
4. R. Poust, D. Dutton, J. Chaber, F. Cox, and A. Denny, "Polymorphic Characterization of A773U Hydrochloride", Poster presentation at the American Association of Pharmaceutical Scientists, Orlando, FLA, 1988.

INFRARED MAPPING OF DETERRENTS (MODERANTS) IN NITROCELLULOSE BASED PROPELLANT GRAINS BY FOURIER-TRANSFORM INFRARED MICROSCOPY

J. D. Louden

ANALYTICAL AND PHYSICAL SCIENCES, RESEARCH AND TECHNOLOGY DEPARTMENT, ICI C&P LTD, THE HEATH, RUNCORN, CHESHIRE, WA7 4QD

J. Kelly

D & A RESEARCH, ICI NOBEL'S EXPLOSIVE COMPANY LTD, ARDEER WORKS, STEVENSTON, AYRSHIRE KA20 3LN

INTRODUCTION.

Moderants(deterrents) are materials that are diffused into single-based nitrocellulose (NC) and double-based nitrocellulose/nitroglycerine (NC/NG) small-arms propellant grains to modify the initial burning rate at the surface of the propellant and hence the rate of gas evolution avoiding overpressure early in the ballistic cycle. Ballistic performance is thought to be related to the concentration and depth of penetration of the deterrent into the propellant grains. Therefore reliable analytical methods for the determination of the concentration and depth penetration for the prediction of ballistic performance are required.

Established methods for measuring the penetration depth of moderants into nitrocellulose propellant grains include various staining and optical techniques.[1-4] However these techniques were not capable directly of measuring the concentration profile of moderants into the grains. Until recently, to measure the concentration profile of the moderant required the use of specially prepared deterred propellants containing 14C-radioisotope-labeled deterrents and autoradiographic,[5,6] and scintillation counting procedures[7]. A laser Raman microspectroscopic method[8] has been reported for determining the penetration depth, qualitative and quantitative diffusion profiles for methyl centralite in a nitrocellulose and nitrocellulose/nitroglycerine matrices, and an Infrared[9] microspectroscopic method has been reported for determining the penetration depth and qualitative diffusion profiles of dinitrotoluene, dibutyl phthalate and methyl centralite.

In this communication we report on an Infrared method for the mapping of methyl centralite diffused into nitrocellulose. The method can also be applied to other deterrent systems.

EXPERIMENTAL.

Deterring Process. The required quantities of propellant base grain, methyl centralite, and water are added to a Sweetie pan. Steam is injected directly into the rotating pan, allowing the temperature to rise to 90-95°C and held at this temperature for 30 minutes. After coating is completed the contents of the pan are placed in a hessian bag and subjected to a hot water steep (90-95°C) for 22hours. The powders are then air dried at 43°C.

Propellants. The samples studied were of nitrocellulose deterred with 3% w/w methyl centralite (N,N'-dimethyl-N,N' -diphenyl urea) (MC).

Microtomy. 10µm thick sections of the grains were prepared using a REICHERT 1050 microtome and glass knives. The propellant grains were clamped directly into the microtome chuck without embedding in any resin. Grains were microtomed to approximately two thirds of their original length and then the 10µm thick section taken for analysis by FT-IR microscopy.

Infrared Microscopy. An AIRE SCIENTIFIC universal infrared microscope coupled to a Nicolet 510 Fourier-Transform spectrometer was used in this study. A medium band MCT (HgCdTe) detector in the microscope gives high sensitivity in the 4000-600cm-1 range. Propellant grain cross sections were placed on 15mm diameter KBr windows and infrared spectra were taken at 10µm intervals from the outer edge inwards (external profile) and 10µm intervals from the perforation edge inwards (internal profile) by positioning the grain under the 10x10um aperture using an AUTOMATIC MEASURING SYSTEMS (AMS) QUICK STEP stepper motor driven automatic sample stage programmed to move horizontally in the X direction in 10µm steps.The aperture was calibrated with a standard micro-calibration slide. A 36x cassegrain mirror objective was used to obtain the infrared spectra. Infrared spectra were taken across the grain until no deterrent peaks were detected.For the mapping of the deterrent the stage was programmed to move in a raster fashion in 10 micron steps in the X and Y directions, and an area 340 by 70 microns was examined.The 1596 cm-1 band of the MC was used to determine the contour and axonometric plots of the distribution of the MC deterrent through the grain.The IR spectrometer was operated at a resolution of 8cm-1, 1000 scans of each 100 square micron area were acquired.

RESULTS

Method Requirements. The principal requirement in the development of an IR microscopy mapping method for monitoring deterrent profiles into smokeless powders is that the deterrent absorb at a frequency where there is relatively little absorption by the matrix. Figure 1 shows IR reference spectra (4000-650cm-1) of nitrocellulose base grain,methyl centralite and MC deterred nitrocellulose grain. It can clearly be seen that the

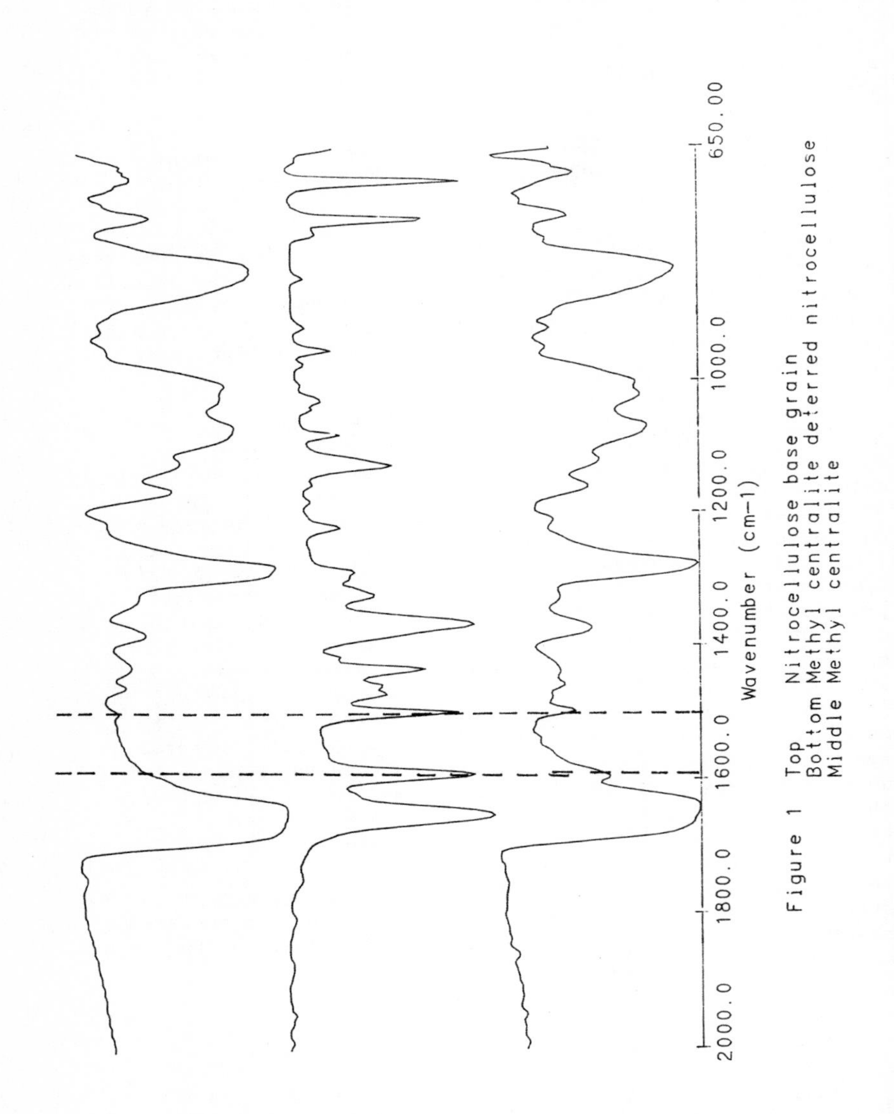

Figure 1 Top Nitrocellulose base grain
Bottom Methyl centralite deterred nitrocellulose
Middle Methyl centralite

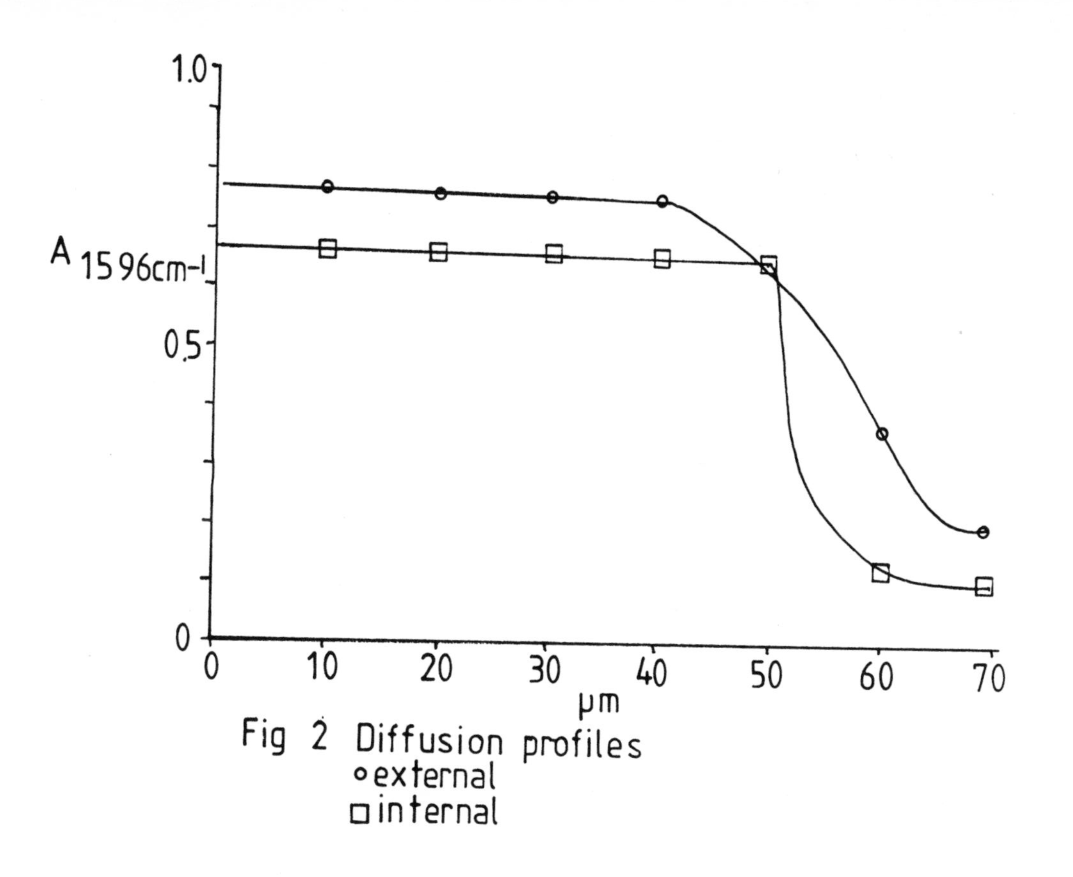

Fig 2 Diffusion profiles
∘ external
□ internal

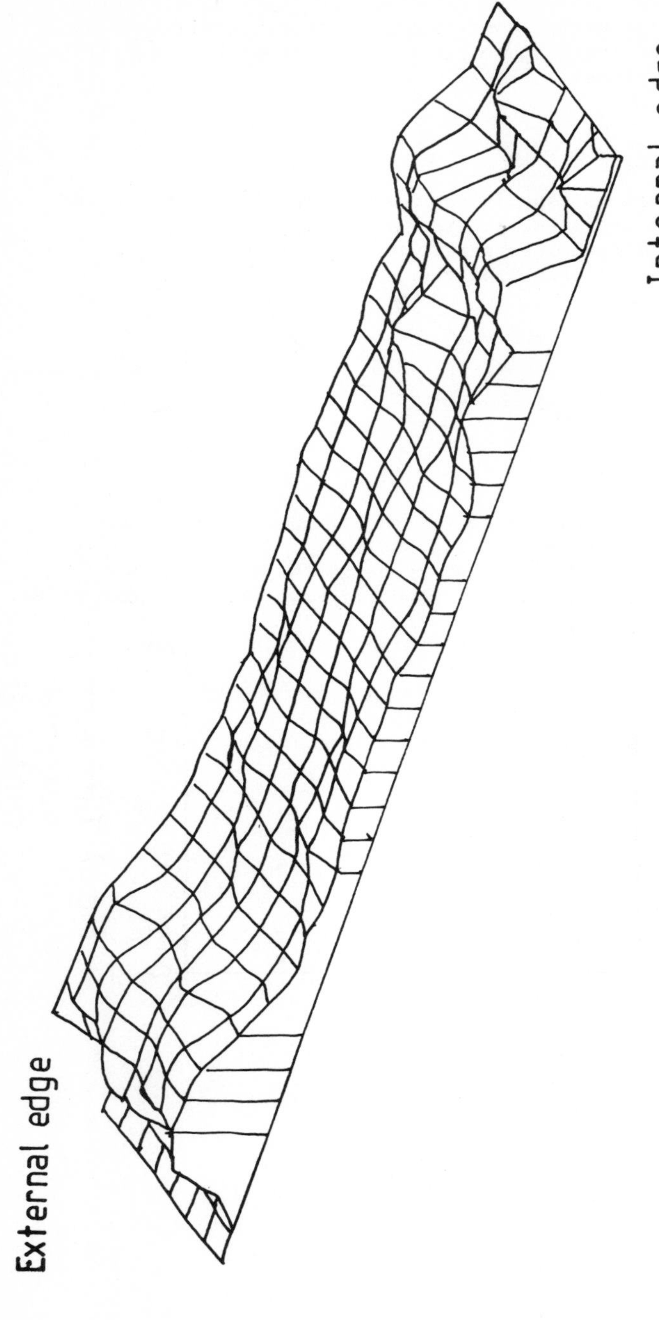

Fig. 3 Axonometric plot of Methyl centralite distribution

aromatic C-C bands at 1596cm-1 and 1496cm-1 of MC can be used as indicator bands to monitor and map the deterrent through the grain. Figure 2 shows the external and internal diffusion profiles of the methyl centralite in the nitrocellulose and figure 3 shows the axonometric plot of the methyl centralite in the nitrocellulose.

CONCLUSIONS

Infrared microscopy is capable of providing data on the concentration profiles, penetration depths and mapping of deterrents diffused into single based nitrocellulose or double based nitrocellulose/nitroglycerine matrix.

Acknowledgement.

This work was carried out with the support of the Procurement Executive, Ministry of Defence.

REFERENCES

1.J.B.Quinlan, Frankford Arsenal Report R-1302 Dec. 1955
2.J.B.Quinlan, Microscope, 14, 385 1965
3.M.F.Levy, Frankford Arsenal Report R1286 Sep. 1955
4.A.H.Milfoerd, Winchester-Weston Division Technical Report No. WWR-68-2, Jan.-Dec.1967
5.B.W.Brodman, M.P.Devine, R.W.Finch and M.S.MacClaren, J.Appl.Polym.Sci.,18,3739, 1974
6.Olin Corporation Final Progress Report for Period Aug.31-Nov.30 1970 contract DAAA-25-70-C-0140
7.H.Meier, D.Boesche, G.Zeither, E.Zimmerhackl and W.Hecker, Bundesministerium de Verteidigung Report BMVg-FBWT-79-14,Bonn, Germany, 1979
8.J.D.Louden, J.Kelly, ans J.Phillipson, J.Raman Spectrosc.,18 137 1987
9.E.Varriano-Marston, J.Appl.Sci.Chem., A8 837 1974

Section 4

MASS SPECTROMETRY

TANDEM MASS SPECTROMETRY OF HIGH MOLECULAR WEIGHT COMPOUNDS

Peter J. Derrick

WARWICK INSTITUTE OF MASS SPECTROMETRY AND DEPARTMENT OF CHEMISTRY, UNIVERSITY OF WARWICK, COVENTRY CV4 7AL

Tandem mass spectrometry or mass spectrometry/mass spectrometry (MS/MS) is a technique involving two mass spectrometers, one of which is used as a separating device and the other as an analysing device.[1,2] The sample of interest is converted to gaseous ions in the source of the first mass spectrometer, which is set to transmit ions of a particular mass-to-charge ratio (m/z). These ions are transmitted through the first mass spectrometer, separating them from all other ions of different m/z values. Typically, the selected ions are excited through collisions with gas or by other means (see below), and fragmentation is induced. The fragmentation is typically rapid, taking place before the ions have entered the second mass spectrometer. The second mass spectrometer serves to provide a mass spectrum of the fragment ions produced from the selected ions. There are variations on this pattern, such as effecting both stages with a single mass spectrometer (e.g. an ion trap) as discussed below.

There are several reasons why MS/MS has developed, but there are two which stand out above the others. One is the ever-growing importance of the needle-in-the-haystack type of problem, in which one component of a complex mixture is to be identified and possibly quantified more or less precisely. MS/MS is employed for this sort of task by selecting the molecule-ion or some characteristic fragment ion with the first mass spectrometer (or stage of mass analysis) and obtaining that ion's mass spectrum with the second. The first mass spectrometer is in effect being employed as a means of refining or purifying samples in this type of application. The information as to the component of interest is likely to be present in the mass spectrum obtainable from the first mass spectrometer, but this information is masked because of peaks from all the other components.

The other reason for the growth of MS/MS is associated with the rapidly growing importance of ionization techniques for non-volatile, thermally sensitive samples such as peptides, synthetic polymers and new materials. The ionization techniques include keV-ion or keV-atom bombardment,[3,4] plasma desorption,[5] matrix-assisted laser desorption[6] and electrospray.[7,8] To greater or lesser extents, all of these ionization techniques tend to yield mass spectra which are difficult to interpret and contain too little information. Fragmentation patterns in these mass spectra are not infrequently poorly reproducible, and typically are strongly dependent on precise experimental conditions including procedures for preparing samples. MS/MS circumvents these difficulties by providing reproducible and more easily interpretable mass spectra largely free of artifacts and without peaks due to impurities.

TYPES OF TANDEM MASS SPECTROMETERS

Most types of mass spectrometers have been paired with most other types of mass spectrometers for the purposes of tandem mass spectrometry. The various possible combinations have different advantages and disadvantages for high-mass studies, which are summarised in Table 1. The quadrupole ion traps,[9] is a device showing extremely exciting promise for MS/MS.[10] Conceptually, MS/MS on a quadrupole ion trap is similar to MS/MS with a Fourier transform ion cyclotron resonance (FT-ICR) spectrometer. One interesting difference is that the ion trap allows both negative and positive ions to be trapped simultaneously, whereas with an FT-ICR only ions of one or other polarity are trapped at the same time. The key common feature of both the ion trap and the FT-ICR is that the whole experiment takes place in the same region, i.e. there is only one spectrometer which effects both ion separation and fragment ion analysis.

Tandem mass spectrometers consisting of 2 magnetic sector mass spectrometers have been developed principally for use with sources providing continuous ion currents, such as fast atom bombardment (FAB), field desorption (FD) and electron impact (EI). The large commercially available four-sector tandem mass spectrometers (4ST/MS's) such as the HX-110/HX-110 (Jeol), ZAB T (VG) and the Concept/Concept (Kratos) have been developed for work in biological areas and more recently in the field of new materials. These top-of-the-range SECTOR + SECTOR instruments (Table 1) are used for the most part for the determination of molecular structure, rather than for mixture analysis. The second stage of mass spectrometry provides additional information as to molecular structure. For example, the FAB mass spectra of many peptides contain fragment peaks providing partial but not complete sequences, whereas the full sequences can be determined from their tandem mass spectra.[11] The large sector instruments give the required mass accuracy (better than ±0.3 m/z) in the measurements of fragment ions in the tandem mass spectra. The mass resolutions achieved in practice in tandem mass spectra from large 4ST/MS's rarely

TABLE 1

Characteristics of Possible Pairs of Mass Analysers

Analyser Pair	Continuous or Pulsed Source of Ions	Resolution in Selection of Precursor Ion	Resolution in Analysis of Fragment Ions	Mass (m/z) Range of Tandem Mass Spectrometer
SECTOR + SECTOR	continuous (pulsed source feasible with array)	high	high in principle	higher than is currently exploitable
SECTOR + QUAD ("hybrid")	continuous	high	low	not high
QUAD + QUAD ("triple quad")	continuous	low	low	not high
FT-ICR) RF ION TRAP)	pulsed	high in principle	high	high in principle
TOF + TOF	pulsed	selection not necessary	moderately high	unlimited in principle

exceed about 2000 when working with real samples, as distinct from model compounds. The seemingly low resolution is a consequence of the need to maintain sensitivity and of limited amounts of sample. With sector instruments, resolution is traded for sensitivity, and vice versa. The large 4ST/MS's equipped with array detectors have usable detection limits in the picomole range. Using FAB, the detection limits tend to be set by interference effects from matrix-derived ions, as much as by the absolute numbers of sample ions. The upper limits to the usable mass range of 4ST/MS's are not set by the characteristics of the analysers. The limits are set by the characteristics of the currently available ionization techniques and by the characteristics of the molecular excitation process between the two stages. FAB and field desorption both exhibit falls in ionization efficiencies as molecules of increasing molecular masses are considered, so that with most samples ion currents are generally low above m/z 3000 with either of these ionization techniques. Without an array detector, precursor ion currents above m/z 3000 are typically too low to permit acquisition of good-quality tandem mass spectra, even if the precursor ions could be effectively fragmented. The second limitation is that collisional activation becomes an ineffectual means of breaking up ions with masses above about 2500 u, if they are singly charged.[12,13]

Array detectors for the second stage of 4ST/MS's provide much improved sensitivities over single-point detectors (improvements by factors of 10^2, or 10^3 in optimum cases). These detectors collect some fraction of the mass spectrum simultaneously. Originally, the figure was 4%, and this has been increased in the last year or two to 40%. Array detectors are likely to find particularly effective usage with ionization techniques suffering fluctuations in ion currents with time (such as FD and FAB) and with pulsed ionization techniques (such as laser desorption).

A double-focussing mass spectrometer followed by a quadrupole mass filter is designated SECTOR + QUAD in Table 1, and generally referred to as a "hybrid".[14] Hybrids are very popular instruments, and most of the manufacturers are able to retrofit quadrupoles to their sector instruments. A significant feature of a SECTOR + QUAD is the necessity to retard the velocities of ions prior to analysis in the second stage, as quadrupoles are effective mass filters only at low energies (< 100 eV). SECTOR + QUAD combinations are typically less sensitive than SECTOR + SECTOR instruments, due to quadrupoles tending to compare unfavourably with sectors in this regard. Quadrupoles do not match double-focussing instruments in terms of mass resolution, and the upper limits of the mass ranges of quadrupoles are lower than those of large magnetic sectors. It seems to be quite widely agreed that hybrids are best suited to the mass range below m/z 1000, and that they tend not to perform well above about m/z 1200.

Triple quadrupoles[15] are ubiquitous and the most common of all the different types of tandem mass spectrometers, being relatively simple to operate and relatively inexpensive. In principle, a QUAD + QUAD could consist of just 2 quadrupole mass filters, but in practice a third quadrupole operated as an ion containment device (so called "RF only" mode) is almost always used as the excitation region between the mass-analysing quadrupoles. Sometimes, this middle quadrupole is replaced by some other ion lens, such as a hexapole or an octapole. The QUAD + QUAD combination is a well-balanced instrument, in that the mass resolutions of the 2 mass spectrometers match each other and also match the requirement of giving unit mass-resolution over the accessible mass range. QUAD + QUAD is primarily an instrument for the mass range below m/z 1200. Triple quadrupoles have been used very effectively indeed for electrospray ionization.[8,9] The QUAD + QUAD combination does not lend itself to parallel detection (e.g. array detectors).

Fourier transform ion cyclotron resonance (FT-ICR) using a superconducting magnet might seem to be the near-ideal tandem mass spectrometer.[16] Certainly mass resolution and accuracy of mass measurement in the tandem mass spectrum are very good, surpassing even the large 4ST/MS's. Selection of the precursor ion can be difficult to achieve with unit resolution, when the experiment is performed by ejecting all other ions from the FT-ICR cell. Using an external ion source and some separate mass analyser to select the trapped ion circumvents any difficulty with ejecting ions at high resolution, but obviously significantly increases the experimental complexity and loses the special feature of carrying out ion selection, fragmentation and detection in the same volume. The mass range of an FT-ICR is in principle very high, but in practice it tends to be difficult with organic ions to achieve good mass resolution at high masses (m/z > 1000).

The combination of 2 time-of-flight analysers might also seem to be ideal for tandem mass spectrometry, but to date the approach has not been much explored. In principle, TOF + TOF obviates the need to select a precursor ion. All ions can be taken into the excitation region without selection or rejection, and all fragments from all precursors can be detected in the tandem mass spectrum. The unravelling of the spectrum in terms of linking each fragment ion with its precursor is wholly possible in theory, but has been demonstrated only for a very simple model system.[17] The mass range of a TOF + TOF is not limited by the analysers. Mass resolution in the tandem mass spectrum could be reasonably high (10^3 to 10^4) and has the attractive characteristic of sometimes increasing as mass increases. TOF + TOF is a parallel-detection instrument, and can be regarded as the ultimate (i.e. 100%) array detector. Wastage of ion current, which is fundamental to the operation of 4ST/MS's (without arrays), hybrids and triple quadrupoles (at a resolution R, at most 1 ion in R ions is detectable) because of the need to scan the mass spectrum, is avoided, bringing significant advantages in potential sensitivity.

METHODS OF EXCITATION

The worth of tandem mass spectrometry in any given application hinges upon the effectiveness as regards inducing information-rich fragmentations, of the method of exciting the precursor ion. There are 4 different types of method of excitation which are normally considered (Table 2): collisional activation, surface-induced dissociation, electron impact and photodissociation. By far and away the most widely employed of these is collisional activation,[18] in which the ion is excited by collision with a gaseous atom or molecule. The gas is normally contained within a cell, through which the ions are passed. Mechanisms of excitation are still the subject of current research. There are some drawbacks to collisional activation, when the analyser is some sort of ion trap. With ion traps, the ions and any gas introduced for collisional activation occupy the same region of space throughout the experiment, and in practice possibilities arise for deterioration of mass-analysing performance and for artifacts in the mass spectrum. Collisional activation can be carried out at ion energies in the keV-range ((SECTOR + SECTOR) and (TOF + TOF)) or at much lower energies (<100 eV) (hybrids, triple quads and ion traps). There is evidence that with 4ST/MS's the effectiveness of collisional activation does not vary dramatically over a range of ion energies extending from 5 keV to 10 keV and quite probably further. The present concern is that at these ion energies the effectiveness of collisional activation may be very much lower for larger ions (m/z >2500) than for lighter ions (m/z <1500). It may be that this apparent limitation on the mass of ions which can be fragmented by collisional activation can be lifted by using higher ion energies (>10 keV). Collisional activation at the very low ion energies (<100 eV) is if anything even less efficient with larger ions than collisional activation at 5 - 10 keV.

Surface-induced dissociation[19] is a technique which has been explored for about 15 years without as yet convincing evidence of superiority over collisional activation being produced. This is not to say that the method will not become important, but there is at present no indication of how the method could be made reproducible given the difficulty of creating and maintaining clean surfaces. Surface-induced dissociation is particularly attractive for ion trap studies, because the need to raise the pressure in the trap is avoided. There is evidence

TABLE 2

Method of Excitation of Ions

Analyser Pair	Gas (collisional activation)	Surface	Electron	Laser (photodissociation)
SECTOR + SECTOR	currently method of choice	attractive; as yet unproved	attractive; unproved	inefficient with continuous source of ions
SECTOR + QUAD ("hybrid")	currently method of choice	probably not necessary	unproved	inefficient with continuous source of ions
QUAD + QUAD ("triple quad")	currently method of choice	probably not necessary	unproved	inefficient with continuous source of ions
FT-ICR) RF ION TRAP)	alright	extremely attractive; not yet adequately proven	extremely attractive; not yet adequately proven	currently method of choice
TOF + TOF	satisfactory	attractive; unproved	unproved	in principle, method of choice

that surface-induced dissociation in an FT-ICR is possible for very large ions (m/z >3000).[20]

Electron impact of ions has been little employed in tandem mass spectrometry,[21] but in principle the method ought to be effective. The difficulty seems to lie in obtaining sufficiently high electron currents. Electron impact appears to be well suited to ion traps,[22] and is an attractive possibility for 4ST/MS's because of the negligible scattering.[23] Having said this, there are no signs that electron impact is going to displace collisional activation as the method of excitation of choice for high-mass studies, not even in the case of ion traps.

Photodissociation is a relatively well developed method of exciting ions. Using lasers, cross-sections for fragmentation of ions can easily be high enough to be practical. The major drawbacks are the need for the ion to absorb at the wavelength employed and the complexity and expense of lasers. With sectors and quadrupoles,[24,25,26] there is the additional and most serious problem of "duty cycle". That is to say, with a pulsed laser and a continuous ion beam only a fraction of the ion beam is irradiated (e.g. 20 ns pulses at 100 Hz result in at most about 1 part in $5x10^5$ of the ion beam being exposed to the laser beam). Photodissociation is much better matched to ion traps,[27,28,29] because the continual presence of the same body of trapped ions means that the duty cycle of the laser is not a catastrophic problem.

METHODS OF IONIZATION

The compatibilities of the various analyser pairs with different ionization techniques are summarised in Table 3. Electron impact (EI)[30] matches perfectly with sectors and quadrupoles, and has no significant drawbacks with ion traps. There is a consideration with TOF + TOF, which is that ideally ions are formed in pulses and sample is not wasted due to large intervals between ionization pulses. Pulsing is necessary in order to provide a reference for the measurement of times of flight. To pulse the ionizing electron beam is a means of creating ions in pulses, but if the sample flows continuously through the ion source there is wastage of sample. To ionize continuously and subsequently "gate" the ion beam is an alternative approach, but again there is wastage (of ions in this case) and hence loss of potential sensitivity. "Bunching" ions following continuous ionization would be the preferred strategy. That is to say, ions would be formed continuously from a sample streamed through the ion source, and then packets of ions would be compressed spatially and hence also "in time". This procedure is feasible, but it is not clear to what degree "bunching" could be taken in practice. Could a 100 μs packet of ions

TABLE 3

Compatibilities of Ionization Techniques and Analyser Pairs

Analyser Pair	EI	CI	FAB	ES	FD	LASER
SECTOR + SECTOR	natural combination	good	good	potentially excellent	natural combination	potentially excellent with array detector
SECTOR + QUAD ("hybrid")	natural combination	good	good	potentially excellent	good	difficult
QUAD + QUAD ("triple quad")	natural combination	natural combination	good	excellent	satisfactory	difficult
FT-ICR) FT ION TRAP)	good	natural combination under certain conditions	external source needed	external source needed	external source needed	good
TOF + TOF	satisfactory	satisfactory	good if atom/ion beam is pulsed	unclear	potentially excellent	natural combination

be compressed to 1 μs, 100 ns or 10 ns? The answer to this question, which depends on the qualities of the ion beam before and after "bunching", will probably determine the worth of TOF's with EI <u>and</u> with chromatographic couplings.

Chemical ionization (CI)[30] differs from EI in that it is potentially extremely well suited to ion traps, as regards the ionization step. If the reagent gas can be used as the collision gas for collisional activation, the MS/MS experiment is straightforward. If the collision gas differs from the reagent gas, the latter must be pumped away before the former is introduced.

Fast atom bombardment (FAB)[31,32] in Table 3 refers to any experiment in which the sample is bombarded with keV atoms <u>or</u> ions. It is supposed that the sample is contained within a matrix, such as glycerol, so that ions can be produced more or less continuously. The presence of a matrix means that there are problems with background gas when using ion traps (at least with an FT/ICR). External sources can be employed with FT/ICR's, so that ions are formed by, for example, FAB outside of the analysis cell and are brought into to cell to be trapped.[33,34] TOF + TOF is well matched to FAB, if the ionizing atom (or ion) beam can be pulsed, ionization efficiency retained and ionizing pulse used as a time-reference.

Electrospray (ES)[7,8] has been shown to be well-matched to triple quadrupoles with their low voltages in the source. Electrospray has been shown to be compatible with sector instruments,[35,36] but more research is necessary to demonstrate that the combination is as reliable as the electrospray/quadrupole combination. Ion traps call for external ES sources, and both the desirability and feasibility of coupling ES and TOF have yet to be demonstrated convincingly. Field desorption (FD) can be carried out very effectively on sector instruments,[37] because both FD and sectors call for high potentials in the source. Hybrids provide the same good match in the source, but there is limitation in that the mass range of the quadrupole is too low to exploit the full capabilities of FD. Triple quadrupoles are not natural partners for FD, because the ions must be retarded in the source to achieve the low velocities necessary for mass analysis. FD seems to demand external sources for ion traps,[38] because the high potentials would distort the fields were ionization to be effected inside a trap. FD is an excellent source for TOF analysers, provided the ionizing field can be pulsed sufficiently quickly without reducing the efficiency of ionization. Ionization methods based on lasers are not normally well-suited to either sectors or quadrupoles, because of the duty-cycle problem. Parallel detection techniques such as arrays on sectors overcome the duty-cycle problem. The pulsed nature of most laser-based methods of ionization is on the other hand perfect for TOF, if the laser pulse can be used as the time-reference for mass analysis. Similarly, the pulsed ionization of lasers is not a problem with ion traps, and can be an advantage.

CONCLUSION

Over the next few years tandem mass spectrometry is likely to become commonplace, as regards molecular structure determination. It seems likely that all new larger mass spectrometers designed for molecular analysis will be tandem mass spectrometers. Larger magnetic sector instruments will almost as a matter of course and at the very least have quadrupoles or some other simple analysing device positioned after the two sectors. The sensitivities of these mass spectrometers will be in the sub-picomole range for ordinary mass spectrometry and sub-nanomole for tandem mass spectrometry.

As of 1990, analytical applications of tandem mass spectrometry are performed on SECTOR + SECTOR (4ST/MS), SECTOR + QUAD (hybrids) or QUAD + QUAD (triple quads). The excitation of ions is effected by collisional activation - excitation by collision with gas. Worldwide, there is a handful of 4ST/MS's with array detectors. Looking ahead to 1995, parallel detection will increasingly come to be regarded as an essential characteristic of the second of the two mass-analysis stages of a tandem mass spectrometer. Parallel detection may in the future be achieved through array detectors, but it would not be surprising to find they had be superseded or replaced by simpler devices achieving the same ends at less expense. Time-of-flight (TOF) techniques will almost certainly be more widely used in 1995 than they are now, and SECTOR + TOF combinations can be envisaged which could replace 4ST/MS's with array detectors. In 1995, TOF + TOF combinations will quite possibly be realistic alternatives to 4ST/MS's, hybrids and triple quadrupoles. RF ion traps will quite likely be realistic alternatives to today's tandem mass spectrometers, especially to triple quadrupoles. Excitation of ions ought to have advanced by 1995, but none of the other methods currently under consideration (Table 2) are obvious improvements over collision activation. Depending upon the method of ionization and the analysers, laser photodissociation could have become important for analytical purposes. Otherwise, likely improvements fall within the category of improved ways of effecting collisional activation. What seems to be certain is that, for determination of molecular structure, particularly of biomolecules, and for mixture analysis, tandem mass spectrometry as a subject and as a practised technique will be yet more important in five years time.

ACKNOWLEDGEMENTS

I am pleased to acknowledge the efficient assistance of Mrs. Margaret Hill in preparing this camera-ready copy, and the patience of Dr. Tony Davies and Dr. Colin Creaser in awaiting its delivery.

REFERENCES

1. F. W. McLafferty, Ed., "Tandem Mass Spectrometry", John Wiley and Sons, New York, 1983.

2. K. L. Busch, G. L. Glish and S. A. McLuckey, "Mass Spectrometry/Mass Spectrometry: Techniques and Applications of Tandem Mass Spectrometry", VCH Publishers, Inc., New York, 1988.

3. A. Benninghoven and W. K. Sichtermann, Anal. Chem., 1978, 50, 1180.

4. M. Barber, R. S. Bordoli, R. D. Sedgewick and A. N. Tyler, Nature, 1981, 293, 270.

5. D. F. Torgerson, R. P. Skowronski and R. D. Macfarlane, Biochem. Biophys. Res. Commun., 1974, 60, 616.

6. M. Karas, D. Bachmann, U. Bahr and F. Hillenkamp, Int. J. Mass Spectrom. Ion Proc., 1987, 87, 53.

7. J. B. Fenn, M. Mann, C. K. Feng, S. F. Wong and C. M. Whitehouse, Mass. Spectrom. Rev., 1990, 9, 37.

8. M. Mann, Org. Mass Spectrom., 1990, 25, 575.

9. D. Price, Ed., "Dynamic Mass Spectrometry", Int. J. Mass Spectrom. Ion Proc., 1990, 99.

10. J. N. Louris, R. G. Cooks, J. E. P. Syka, P. E. Kelley, G. C. Stafford Jr. and J. F. J. Todd, Anal. Chem., 1987, 59, 1677.

11. A. E. Ashcroft and P. J. Derrick in "Mass Spectrometry of Peptides", D. M. Desiderio, Ed., CRC Press Boca Raton, Florida 1990.

12. G. M. Neumann, M. M. Sheil and P. J. Derrick, Z. Naturforsch., 1984, 39a, 584.

13. D. L. Bricker and D. H. Russell, J. Am. Chem. Soc., 1986, 108, 6174.

14. G. L. Glish, S. A. McLuckey, T. Y. Ridley and R. G. Cooks, Int. J. Mass Spectrom. Ion Phys., 1982, 41, 157.

15. R. A. Yost and C. G. Enke, J. Am. Chem Soc., 1978, 100, 2274.

16. B. S. Freiser and R. B. Cody, Int. J. Mass Spectrom. Ion Phys., 1982, 41, 199.

17. D. R. Jardine, D. S. Alderdice and P. J. Derrick, Org. Mass Spectrom., to be published.

18. F. W. McLafferty, P. F. Bente, R. Kornfeld, S.-C. Tsai and I. Howe, J. Am. Chem. Soc., 1973, 95, 2120.

19. R. G. Cooks, D. T. Terwilliger, T. Ast, J. H. Beynon and T. Keough, J. Am. Chem. Soc., 1975, 97, 1583.

20. E. R. Williams, K. D. Henry, F. W. McLafferty, J. Shabanowitz and D. F. Hunt, J. Am. Soc. Mass Spectrom., 1990, 1, 405.

21. W. Aberth and A. L. Burlingame in "Biological Mass Spectrometry", A. L. Burlingame and J. A. McClosley, Eds., Elsevier, Amsterdam, 1990.

22. R. B. Cody and B. S. Freiser, Anal. Chem., 1979, 51, 547.

23. S. Tajima, S. Tobita, K. Ogina and Y. Niwa, Org. Mass Spectrom., 1986, 21, 236.

24. D. C. McGilvery and J. D. Morrison, Int. J. Mass Spectrom. Ion Phys., 1978, 28, 81.

25. E. S. Mukhtar, I. W. Griffiths, F. M. Harris and J. H. Beynon, Org. Mass Spectrom., 1980, 15, 51.

26. S. A. Martin, J. A. Hill, C. Kittrell and K. Bieman, J. Am. Soc. Mass Spectrom., 1990, 1, 107.

27. R. C. Dunbar in "Gas Phase Ion Chemistry", M. T. Bowers, Ed., Academic Press, New York, 1979, Vol. 2.

28. J. N. Louris, J. S. Broadbelt and R. G. Cooks, Int. J. Mass Spectrom Ion Proc., 1987, 75, 345.

29. E. R. Williams, J. J. P. Furlong and F. W. McLafferty, J. Am. Soc. Mass Spectrom., 1990, 1, 288.

30. F. H. Field, Adv. Mass Spectrom., 1986, 10, 271.

31. M. Barber, R. J. Bordoli, G. J. Elliott, R. D. Sedgwick and A. N. Tyler, Anal. Chem., 1982, 54, 645A.

32. A. Benninghoven and W. Sichtermann, Org. Mass Spectrom., 1977, 12, 595.

33. R. T. McIver, R. L. Hunter, M. S. Story, J. Syka and M. Labunsky, Proc. 31st ASMS Conf. Mass Spectrom. Allied Topics, 1983, 44.

34. C. F. Ijames and C. L. Wilkins, J. Am. Soc. Mass Spectrom., 1990, 1, 208.

35. C. K. Meng, C. N. McEwen and B. S. Larsen, Rapid Commun. Mass Spectrom., 1990, 4, 147.

36. R. T. Gallagher, J. R. Chapman and M. Mann, Rapid Commun. Mass Spectrom., 1990, 4, 369.

37. A. G. Craig and P. J. Derrick, Aust. J. Chem., 1986, 39, 1421.

38. S. C. Davis, G. M. Neumann, P. J. Derrick and M. Allemann, Org. Mass Spectrom., to be published.

MULTIPHOTON IONIZATION MASS SPECTROMETRY: TECHNIQUE AND PROSPECTS

C. Köster, M. Dey, J. Lindner, J. Grotemeyer

INSTITUT FÜR PHYSIKALISCHE UND THEORETISCHE CHEMIE, TECHNISCHE UNIVERSITÄT MÜNCHEN, LICHTENBERGSTR. 4, D-8046 GARCHING

1 INTRODUCTION

The arrival of lasers with different wavelengths in the IR, VIS and UV regions has yielded a large number of applications in mass spectrometry. Beside their use in ion activation (photodissociation)[1] lasers have also been introduced throughout the years for ion production. The first applications deal with simultaneous desorption and ionisation[2] and this technique has been widely applied in recent years to the desorption of large bioorganic ions, such as intact proteins, enzymes and nucleotides, from a matrix[3].

An alternative method for the formation of ions by the impact of laser light, multiphoton ionization, has been used in the last decade to study small as well as medium sized molecules in the gas phase[4]. Through its features multiphoton ionization (MUPI) added further dimensions to mass spectrometry. These are the possibilities of *soft ionization* with the formation of mainly molecular ions, *tuneable fragmentation* with an easy adjustment of the degree of fragmentation and the *wavelength* dependence of the mass spectrum.

Starting as far back as 1982, a different approach has been taken to investigate larger, involatile molecules in the gas phase[5] by multiphoton ionization. By separating the desorption from the ionization in space and time it is possible to ionize molecules without any formation of adduct ions. It should be noted that in this set-up the ionization of the laser evaporated neutral molecules can occur by various different ionization methods such as Electron Ionization (EI) or even Chemical Ionization (CI)[6], or, as in our experimental set-up, by multiphoton absorption of the neutral molecules.

A large number of mass spectrometric investigations are concerned with the coupling of chromatographic techniques. The most widely used technique is the coupling of gas chromatography to mass spectrometer systems. Here we show the first results of such coupling, which demonstrate the possibilities of multiphoton ionization. Besides the use of gas chromatography, supercritical fluids and high-pressure liquids used for chromatographic separation of mixtures in combination with mass spectrometric detection of samples has found growing interest[7]. It should be noted that use of supercritical fluid technology cannot only be made in chromatography, but as a general direct introduction method for different samples. We[8] and others[9] have succeeded in coupling a supercritical fluid inlet to a MUPI-MS. Some results will be discussed. Interfacing this inlet with multiphoton ionization yields mass spectra which display no interferences from the solvent used and show only sample ions.

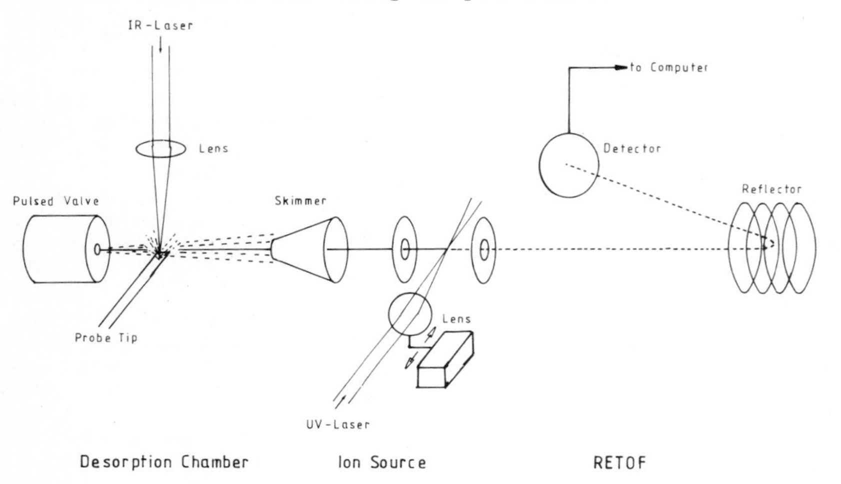

Figure 1 Scheme of the mass spectrometric system.

2 EXPERIMENTAL SET-UP

The mass spectrometer system used for the study consists of three different parts: the laser desorption unit, the laser ionization source and the Reflectron-Time-of-Flight (RETOF) mass spectrometer. For the investigations presented here either a home-build system or a Bruker TOF-1 has been used. The main difference between these mass spectrometer systems is the size of the desorption chamber with an adjustable probe tip as well as a different ion source with different electrostatic fields resulting in different ion lifetimes. In our experimental set-up neutral molecules are transformed into the gas phase separated from the ionization step. The complete system is shown schematically in Fig. 1. It is described in greater detail elsewhere[10].

3 MULTIPHOTON IONIZATION

Details of the multiphoton ionization and/or fragmentation procedure have been published in great detail in the past[11]. To ionize a molecule in the gas phase by a resonant absorption multiphoton procedure with light, two basic points have to be fulfilled: (a) the molecule must absorb the photons from the laser beam and (b) the sum of the photon energies must be higher than the ionization energy of the molecule. If these basic

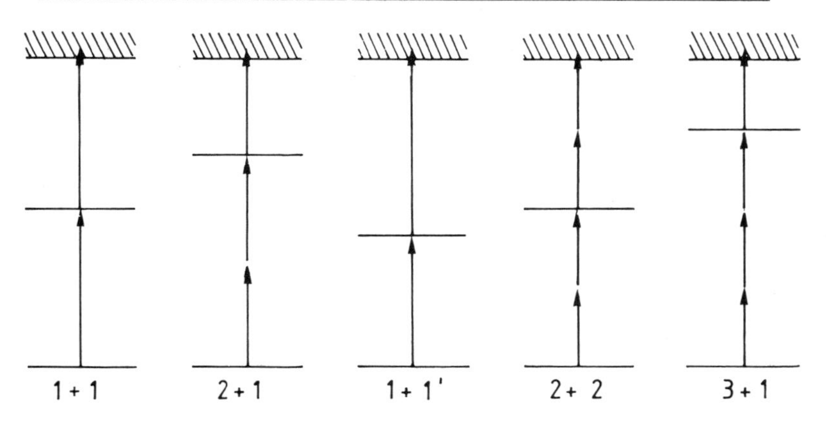

Figure 2 Absorption schemes used in multiphoton ionization.

requirements are met, ionization of the molecule occurs, thus forming a molecular ion. The number of molecular ions formed by the absorption process is linked to the absorption cross section leading to the excited intermediate state of the molecule as well as for the absorption of further photons in this excited state. Generally, the extinction coefficient of a simple UV-absorption spectrum can give a good indication whether an overwhelming ion signal or only a low ion current will be found. Activation of the neutral molecule can occur by different schemes. As shown in Fig. 2 several different photon absorption processes can be used for ionization. The 1+1 multiphoton ionization process (Fig. 2a) is widely used. Here the first photon promotes the molecule into its excited state, while the second photon leads to ionization. Higher absorption schemes such as 2+1, 3+1, or non-resonant absorption are known.

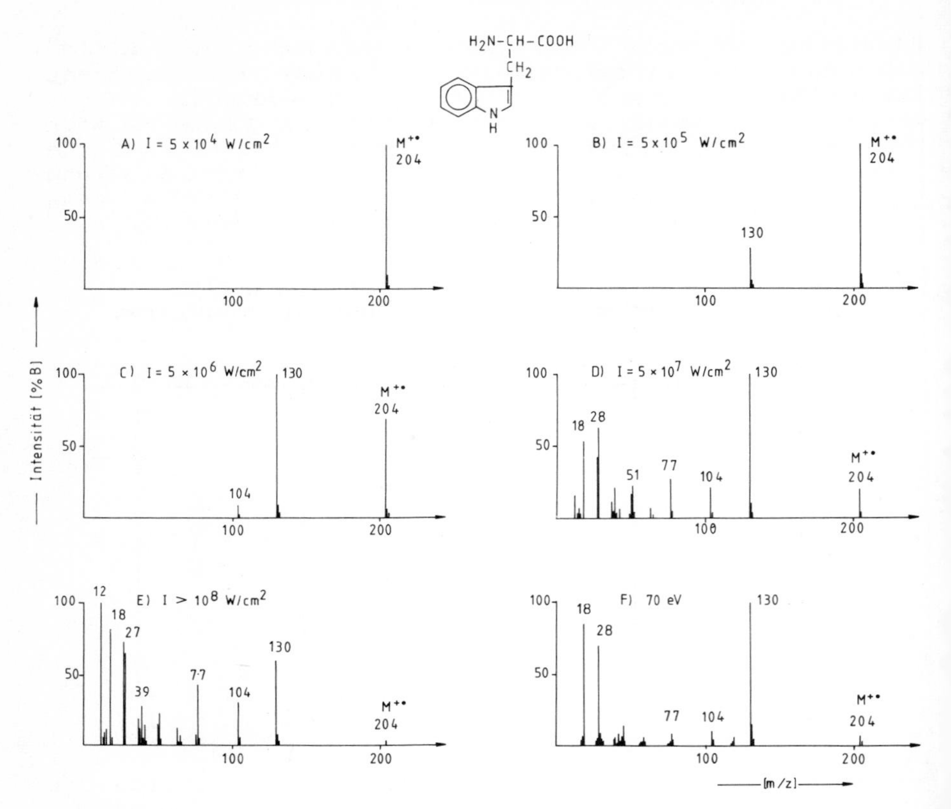

Figure 3 Comparison of soft (a), partially hard (b,c,d) and hard (e) multiphoton ionization mass spectra of *(dl)-tryptophan*. Spectrum (f) Electron impact ionization.

Once the molecular ion is formed, the formation of fragment ions from this ion can be induced by the intensity of the laser beam. If the intensity is kept low, no absorption of photons by the molecular ion after its formation can occur. By increasing the intensity of the UV-laser beam, further absorption of photons can take place. This results in an accumulation of energy in the molecular ion. If the energy is sufficient, fragmentation can follow, thus producing smaller ions. Now these ions can themselves absorb photons during the laser pulse and undergo further fragmentation. As a result, a strong intensity dependence of the mass spectrum is observed stretching from exclusive production of molecular ions to a total destruction of the molecular ion into its atomic parts. A typical example of this behaviour is shown in Fig. 3 for the molecule *(dl)-tryptophan*.

4 PRODUCING NEUTRALS BY VARIOUS METHODS

o fulfil the needs of the mass spectrometer system in onjunction with the multiphoton ionization we have nvestigated different inlet methods, such as the laser esorption of neutrals, liquid introduction etc. Some of he different principles are shown in Fig. 4.

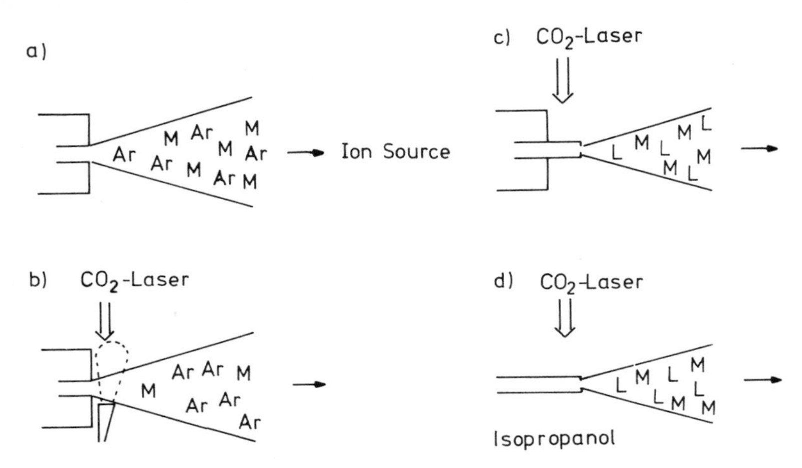

igure 4 Principles of the production neutral molecules in multiphoton ionization.

Samples with sufficient volatility can be introduced y seeding them in the gas stream used for the supersonic et before expansion. (Fig. 4a). For the desorption of olecules with low vapour pressures three different ethods can be used. Fig. 4b shows the standard laser esorption procedure, while Fig. 4c and Fig. 4d emonstrate possibilities of the interfacing of liquid nd supercritical fluids.

aser desorption of neutrals

o introduce involatile, thermally labile samples into he gas phase, different lasers can be used. In general, he best wavelength for generating neutral molecules in acuum is in the infra-red. Typically 10.6 μm from a O_2-laser is used.

The details of the desorption process are still ather poorly understood. But a couple of experimental bservations can be used in the understanding the rocess. Molecules, which are semi-transparent at the aser wavelength, are desorbed best[12], while absorbing amples lead only to fragments in the resulting mass pectrum. Here the molecules are destroyed in the esorption process. To overcome these problems we have investigated the influence of matrices on the desorption process for intact molecules[13]. From the work of Hillenkamp et al.[14] it is known that matrices can have a drastic influence on both the desorption of ions and neutrals by the impact of an UV-laser beam.

As with UV-light, matrices have a strong influence on the desorption of neutrals by IR-light. Figure 5 shows three multiphoton ionization mass spectra of a strongly infra-red absorbing sample, the fully protected mono-nucleotide, *N-benzoyle-5'-O-dimethoxy-trityl-adenonsine-3-'-(2-chloro-phenyl-cyanoethyl)-phosphate.* The influence of the matrices is clearly seen. If the sample is desorbed without any matrix material, a small molecular ion is detectable, while the observable fragment ions, m/z 303 especially, mainly stem from the pyrolytic decomposition of the molecule. Mixing a more strongly absorbing matrix with the sample yields a change in the mass spectrum. At the same laser intensity, molecular ions are formed in higher yield, but still most of the signals, especially the new contributions, are due to a non-ionic fragmentation during the desorption process. Adding a semi-transparent matrix, such as maltose hydrate, then leads to a multi-photon ionization mass spectrum, with the molecular ion as base peak. These results underline the importance of matrices in the desorption of neutral molecules from a surface.

The desorption laser (either a CO_2-laser (10.6 μm) or a Nd:YAG-laser (1.06 μm, 532nm 355nm, 266nm)) is focused to a 1 - 2 mm spot on a metallic sample holder on which the sample is distributed either neat or in a matrix. The samples are dissolved in a suitable solvent, normally methanol or dichloromethane, and deposited onto the probe tip in solution. After evaporation of the solvent, this preparation is repeated, until the sample size is about 10 - 500 μg. If matrices are used, the sample is mixed with the matrix material (1:100 parts) and distributed onto the probe tip after solvation in a suitable solvent or pressed as pellets. A part of this pellet is then deposited onto the probe tip. The power density on the sample holder in these experiments is about 10^4 - 10^5 W/cm^2. Due to this low power density only few ions, if any, are formed in the desorption process.

The probe tip is positioned some millimeters below the nozzle of a pulsed valve. After the IR-laser pulse, the valve is opened and the evaporated particles are entrained and cooled by the expanding jet of noble gas.

Introduction of supercritical fluids

One of the characteristic properties of a supercritical fluid is the ability to dissolve relatively large and non-volatile molecules as liquids can, dependent on the

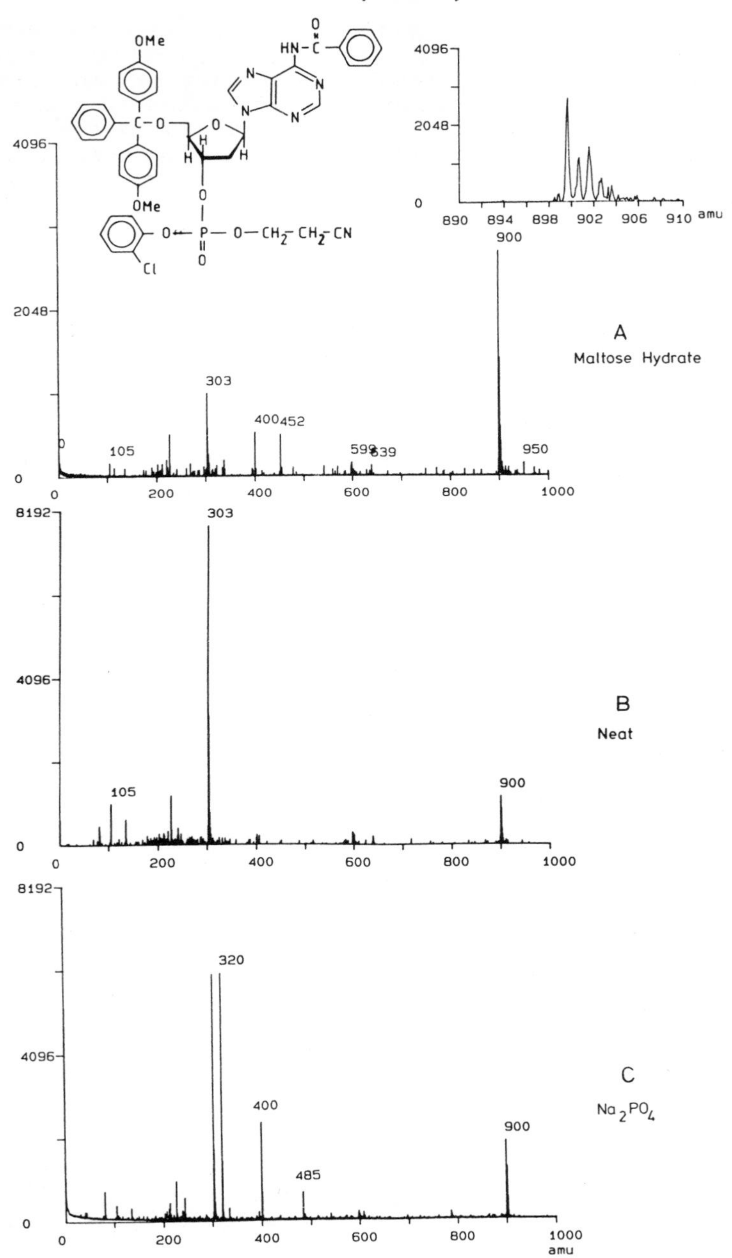

Figure 5 MUPI mass spectra of the protected N-Bz-DMT-adenosine-phospate. a) with maltose hydrate as matrix, c) Na_2PO_3 as matrix added, b) neat.

polarity of the solute and the density (pressure) of the fluid. Other properties like viscosity and diffusion are similar to gases or something between gases and liquids. These properties make supercritical fluids an interesting phase for sample introduction into a vacuum system of an MS. For instance carbon dioxide is a commonly used supercritical fluid with the critical parameters, T_c = 31°C (critical temperature) and p_c = 73 atm (critical pressure). At a pressure of 400 atm and a temperature of 60°C the CO_2 density is nearly 1 g/cm^3. This density correlates with a very good solvent-power especially for nonpolar substances. The problem of the supercritical fluid introduction - mass spectrometer (SFI-MS) interface is to introduce the supercritical solution with a pressure usually over 100 atm into the mass spectrometer with vacuum conditions of 10^{-3} mbar and lower.

To overcome this problem Guthrie and Schwartz[15] developed an integral pressure restrictor. With an integral restrictor it is possible to create a pressure drop over a very short distance allowing the introduction of fluid and solute into the vacuum. The restrictor is made out of a silica capillary with 100 µm i.d. and 400 µm o.d. One end is fused by an oxygen gas-burner and ground, so that a nozzle of about 1 or 2 µm i.d. is opened up. To maintain supercritical conditions up to the very end of the restrictor, 3 or 4 turns of heating wire are wound around the end and held at a temperature of 300°C to compensate for the cooling effect of the adiabatic expansion.

With this continuous working interface, aromatic and polyaromatic compounds can be measured by MUPI-MS[16]. But other substances, even the tensid Triton X 100 which contains a phenolic group and should be ionized very easily by the multiphoton activation, resist investigation by mass spectrometry, although the individual polymer parts were separated by SFC and detected by a FID.

Other groups working on SFC or SFI-MS with conventional ionization techniques like CI or EI are able to obtain MS spectra for these compounds. During the expansion process clusters of CO_2-molecules with non-volatile and polar solutes are created, which reduce the intensity of the unclustered solute molecules in favour of a broad smearing to higher masses. The consequence is a dramatic reduction of the detection sensitivity of the sample of interest. The relatively high pressure inside an ion source in the CI mode is responsible for the observation of molecules only, since clusters are removed by collisions with the plasma gas. This is also true for an EI ion source, where the clusters were destroyed at the hot ion source surfaces. Neither hot surfaces nor high pressures are found inside our mass spectrometric system, therefore clusters cannot be broken up and non-volatile compounds cannot be detected. Triton X 100 molecules were measured, when a heating wire was placed in front of the restrictor nozzle and rapidly heated to about 400°C. The resulting MUPI mass spectrum is shown in Figure 6. Clearly all individual molecular masses can be detected. The polymer units are centred at mass 294 Da(n=2), but a complete sequence is observable up to mass 908 Da. This signal consists of 15 monomer units.

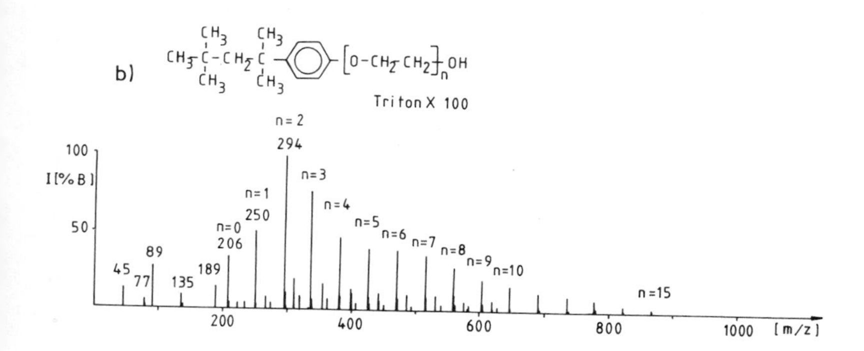

Figure 6. MUPI-mass spectra of *Triton X.*

Liquid introduction combined with laser desorption

To introduce sample molecules from the liquid phase into vacuum, different approaches can be used. As shown in Figure 4c, direct desorption of the liquid phase together with the sample can be performed, or the liquid can be injected through a pulsed valve (Fig. 4d).

In the first case the sample has to be dissolved in isopropanol. The CO_2-laser is focussed on the end of a quartz capillary. Therefore droplets of the solution will be evaporated, because of a high absorption band at 10.6 μm, the CO_2-laser wavelength. For this reason only isopropanol can be used as the solvent, since this molecule absorbs strongly at this laser wavelength.

One disadvantage of this laser desorption assisted liquid introduction is the relatively high pressure rise inside the ion source during the valve pulse, due to the in-line arrangement of the valve nozzle and the ion flight direction. A second disadvantage is the relatively low cooling behaviour of the expanded liquid and the high loss of sample. Nevertheless with this interface it is possible to acquire mass spectra of some smaller molecules in the gas phase. In Figure 7, the mass spectra of *retinol (vitamin A)* are shown. Again soft and hard ionization is possible with the formation of the pure molecular ion as well as the formation of different fragment ions.

To overcome certain problems and to integrate the technique of laser evaporation of intact molecules with liquid injection, an interface was designed whereby gas phase introduction or GC coupling, liquid introduction or LC coupling and solid compound evaporation assisted by laser desorption is possible, without any change in instrumental arrangement.

The developed valve is orientated perpendicular to the flight direction of the desorbed molecules and ions.

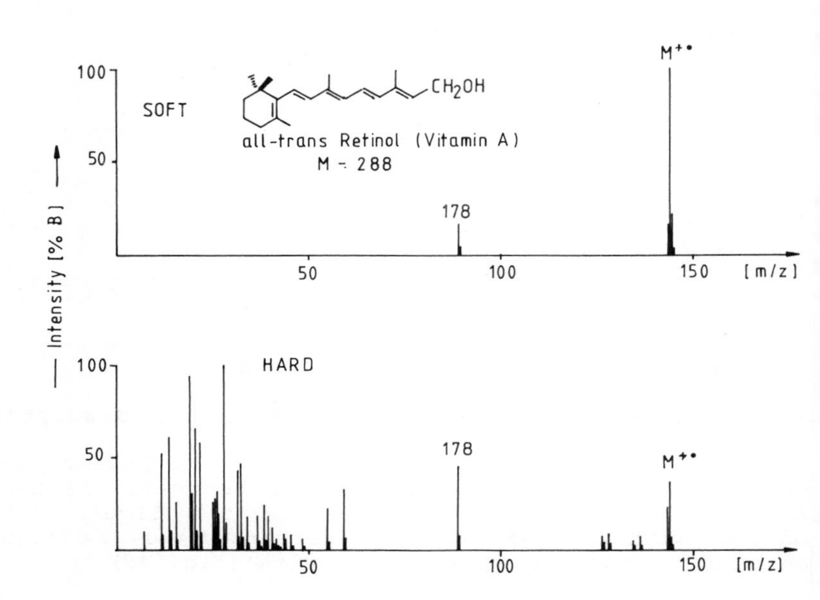

Figure 7 MUPI-mass spectrum of *Vitamin A* obtained by laser assisted evaporation of a methanolic solution.

The solution is pulsed onto a plate directly in front of the nozzle. The plate is transparent to the light of the desorbing CO_2-laser pulse. After some microseconds, when the solvent is evaporated, a CO_2-laser pulse desorbs the precipitated molecules and the gas pulse of a second valve, orientated in line to the ion flight direction and perpendicular to the first valve, transports the desorbed molecules into the ion source. A full description of the valve will be given elsewhere.

Chromatographic techniques combined with MUPI-mass spectrometry

MUPI-MS can be a powerful method in combination with chromatography, because it extends the two dimensional conventional chromatography MS coupling to a four

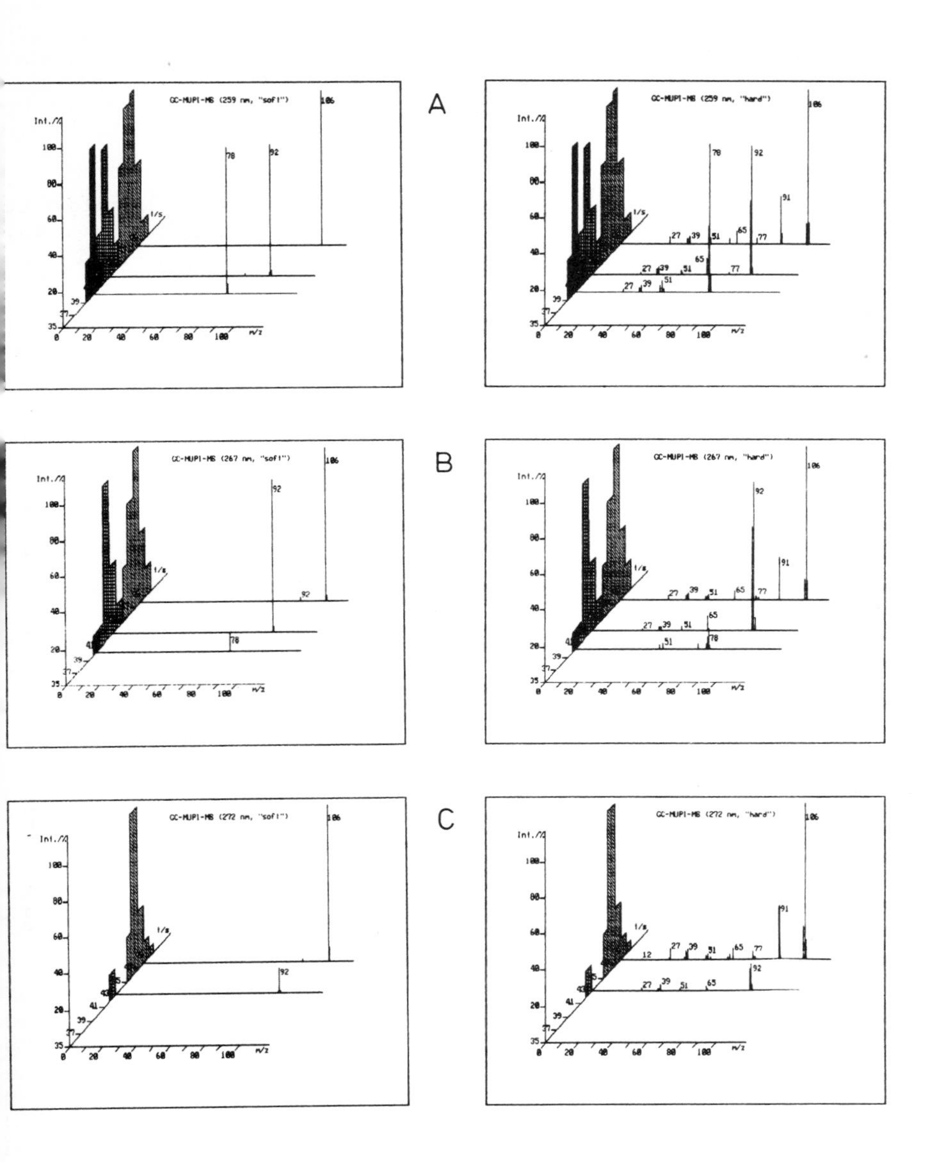

Figure 8 Gas chromatograms and mass spectra from simple aromatic compounds obtained by multi-photon ionization. Left side soft ionization, right side hard ionization.

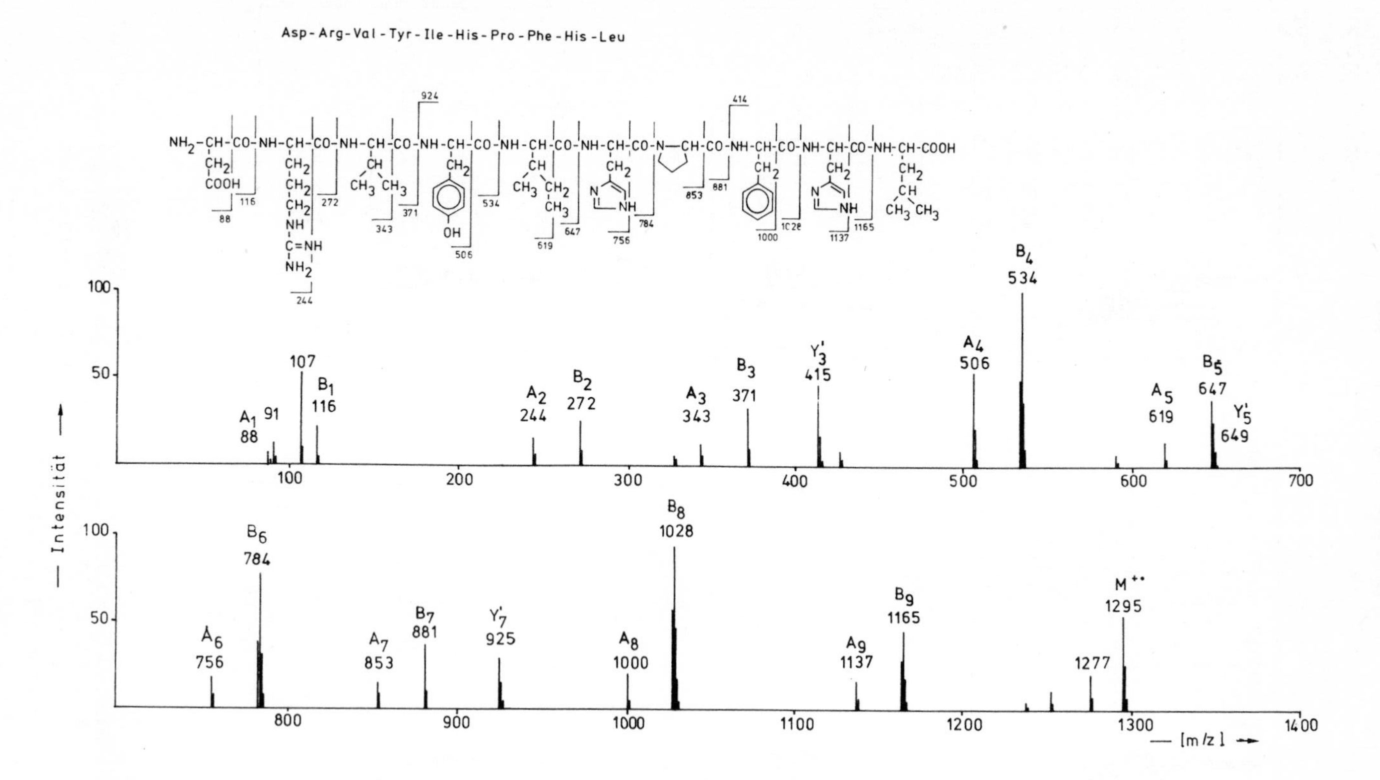

Figure 9 Multiphoton ionization mass spectrum of *Angiotensin I* obtained on the TOF-prototype.

dimensional technique. The additional two dimensions are the ionization wavelength and the photon intensity. The wavelength dependence of the ionization process makes this chromatography-MS coupling possible irrespective of the mobile phases used.

To illustrate the principle of operation, a gas chromatograph was combined with the TOF-MS. A mixture of *benzene, toluene* and *p-xylene* was separated and ionized by three different wavelengths at a given laser intensity, to produce mainly the molecular ions. As shown in Figure 8, it is possible to distinguish between all three samples on the basis of their chromatographic separation. It should be noted, that the measuring system has the potential of acquiring mass spectra in 100 msec. Therefore complete mass spectra can be measured with that repetition rate. This rate is limited only by the electronics and the data system, not by the mass spectrometer. As shown by Figure 8, the ability to scan the ionization laser wavelength allows discrimination of the aromatic compounds on the basis of their absorption. This yields a further dimension in the two-stage GC-MS technique.

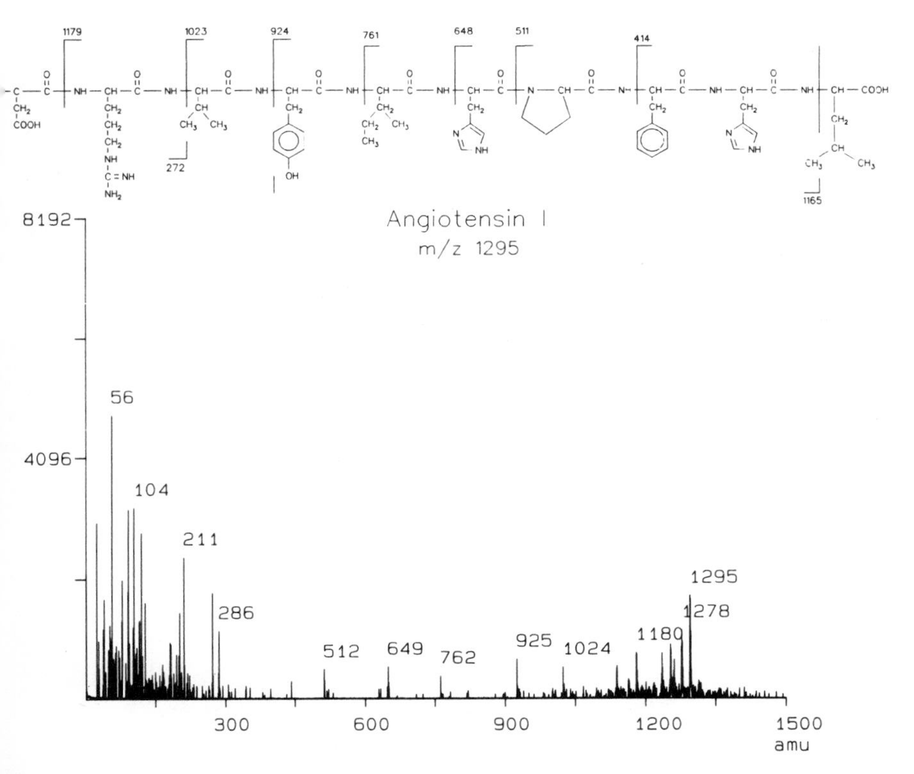

Figure 10 MUPI mass spectrum of *angiotensin I* obtained on the Bruker TOF-1 instrument.

5 SOME EXAMPLES

During recent years MUPI-mass spectra have been published for a large number of substances, such as drugs, peptides, nucleotides, aromatic compounds and others.

As an example of the possibilities of multiphoton ionization applied to smaller biomolecules, the mass spectra of *Angiotensin I* and *Gramicidin D* will be discussed in greater detail. We and others have shown in our investigations that it is possible to sequence peptides in the gas phase. The first mass spectra of *Angiotensin I*[17] published by us in 1986 were taken on our home-build time-of-flight mass spectrometer system. The mass spectrum is shown in Fig. 9. It is dominated by the break-down of the molecular ion in A and B - sequence ions, according to the nomenclature of Fohlman and Roepstorff[18].

By using the Bruker TOF-1 ion source, which gives rise to longer ion lifetimes in the ion source, the mass spectrum is changed drastically. As shown in Figure 10 the Y´-ions now dominate the mass spectrum. This behaviour can be understood by the fact that multiphoton ionization gives rise only to the energetically preferred fragmentation reactions in a given time window. These results have been reported before for smaller aromatic compounds such as chlorobenzenes[19].

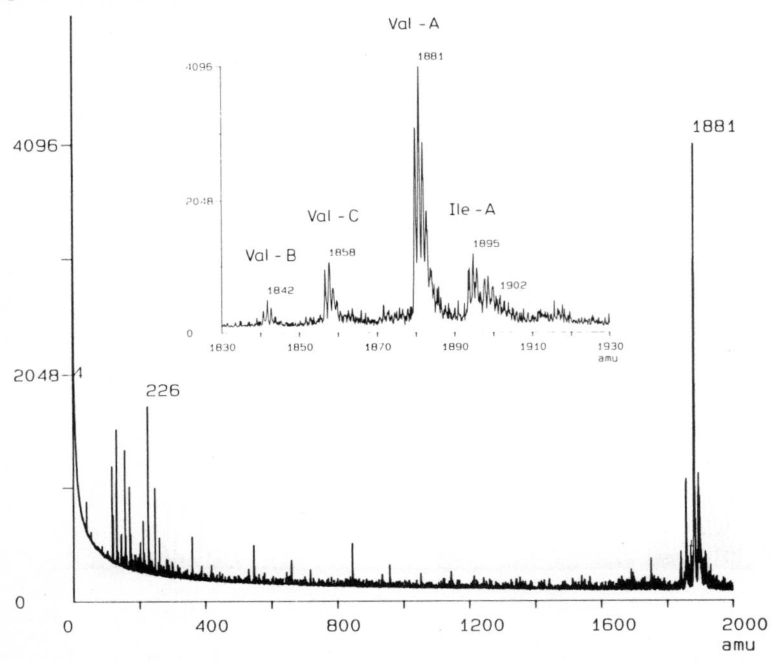

Figure 11 MUPI mass spectrum of *Gramicidin D*.

Figure 11 displays the soft MUPI mass spectrum of the antibiotic *Gramicidin D* (molecular weight = 1880 da), the ionizing wavelength being 255nm. The sample was mixed with maltose hydrate to avoid pyrolytic reaction during the desorption[12]. The pulse length of the desorbing CO_2-laser was 90 ns and the intensity was adjusted to roughly $10^7 W/cm^2$.

The intensity of the ionizing laser was adjusted to $1*10^6$ W/cm^2 to form the molecular ion as the most intensive ion. In this way, the formation of fragment ions is almost completely suppressed. This is very important for the correct interpretation of the molecular ion region of the soft MUPI mass spectrum of *Gramicidin D*, displayed in the insert of Figure 11. By excluding the formation of intense fragment ions, these ion signals may be interpreted as molecular ions from different peptide chains. At least four molecular ions can be distinguished, the molecular weights being 1841 da, 1857 da, 1880 da and 1894 da. In the insert of Figure 11 the mass of the most intense isotopic peak is given for each molecular ion.

It is known from the literature, that *Gramicidin D* is mainly a mixture of the three components *Gramicidin A,*

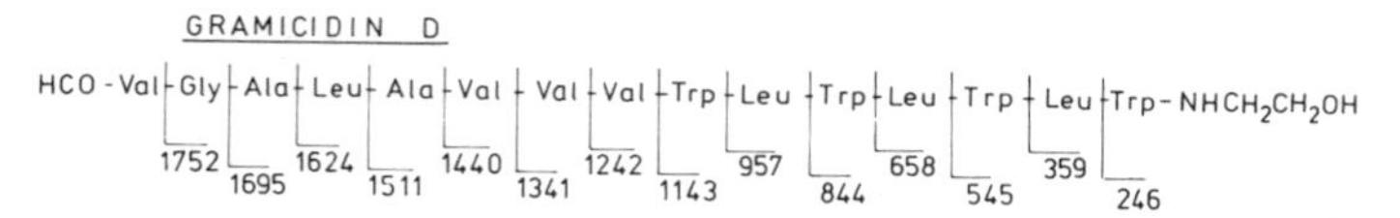

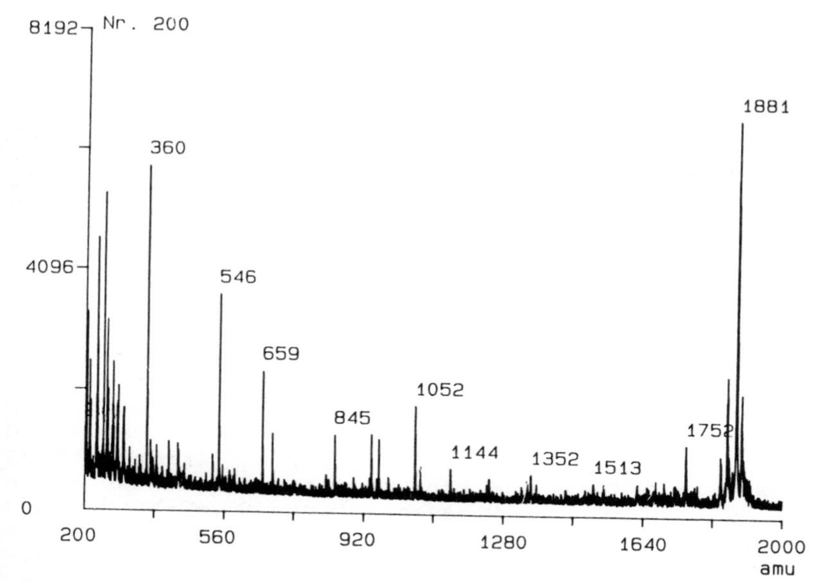

Figure 12 Partial hard ionization mass spectrum of *Gramicidin D*. Note the occurrence of all Y′-ions in mass spectrum.

B, C, which comprise about 87.5, 7.1, 5.1 percent respectively[20]. The structure of these linear pentadecapeptides differs only in the alternative incorporation of *tryptophan* (A), *phenylalanine* (B) and *tyrosine* (C) as aromatic amino acids in position 15 as shown below. The sequence of the main component *Gramicidin Val-A* is N-formyl-Val^{1}-Gly-Ala-Leu-Ala-Val-Val-Val-Trp-Leu-Trap^{15}-Leu-Trap-Leu-Trp-ethanolamine (molecular weight = 1880 da). The N-terminal valine is formulated and the C-terminal tryptophan is peptidically linked to ethanolamine. Each of the components A, B, and C, consists of two chains, one with valine in position 1, comprising 80 - 95% of the component, and the other with isoleucine in position 1.

The molecular ions of the main components of *Gramicidin D* are also clearly observed: the ion signals at m/z 1841, 1857, 1880 are assigned to *Gramicidin B, C* and *A* respectively, with valine in position 1. The peak at 1894 corresponds to the second chain of *Gramicidin A* with isoleucine in position 1.

The partial hard MUPI mass spectrum of Gramicidin D, which is displayed in Figure 12, shows sequence specific fragment ions of the main component valine-gramicidin A. These fragment ions are mainly generated by cleavage of the peptide bond with the charge remaining on the C-terminus containing part of the peptide chain. This fragmentation reaction is accompanied by a hydrogen rearrangement, leading to Y´-type ions. Since all 14 Y´ions of the pentadecapeptide are found in the mass spectrum, the complete sequence of valine-gramicidin is deducible.

Beside the investigations of peptides we have also obtained multiphoton ionization mass spectra of smaller protected nucleotides. In order to develop MUPI mass spectrometry for the structural analysis of fully protected deoxyribonucleoside 3´-phosphates, the isomeric deoxyribo-dinucleotides d(ApTp) and d(TpAp) have been investigated as model compounds. These dinucleotides contain the phosphotriester linkage which is used in oligonucleotide synthesis.

The ionizing wavelength was tuned to give the maximum molecular ion yield, which was found to be at 242.5nm. Figure 13 shows the partial hard MUPI-mass spectrum of d(ApTp) and the short hand structure. The molecular ion is formed with high intensity. As shown previously[21], a characteristic group of fragment ions includes the DMT-residue, the aromatic protective group of the $C_{5'}$-OH- group. Cleavage of the $C_{5'}$-O-bond leads to the formation of the DMT-fragment ion at m/z 303. The fragment ion at m/z 400 is assigned to the DMT-methylenfuran-ether, formed by cleavage of both the $C_{3'}$-O bond and the glycosidic linkage to the nucleic base. One

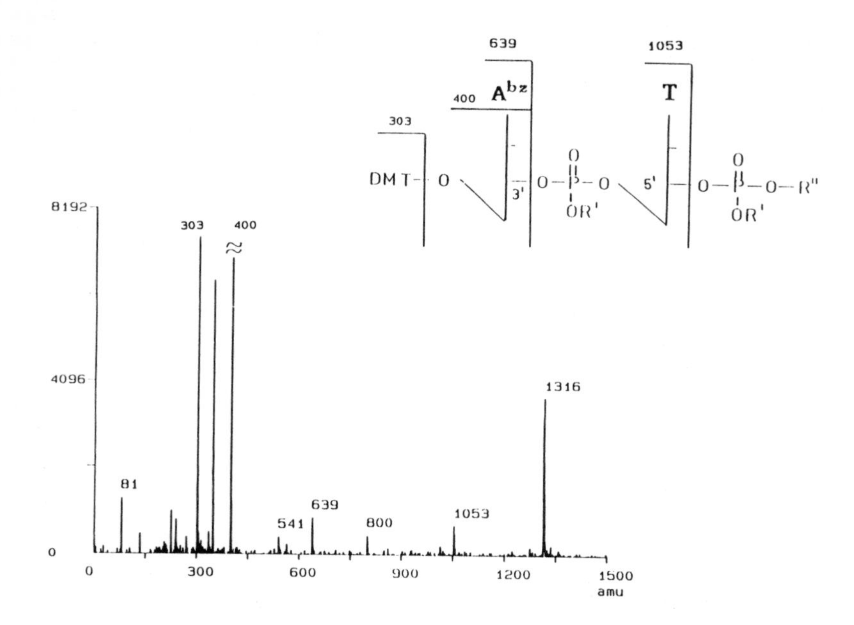

Figure 13 MUPI-mass spectrum of the protected dinucleotide d(ApTp); Molecular weight 1314 Da.

of the most interesting fragment ions is formed by cleavage of the $C_{3'}$-O-bond of the first nucleoside with the charge remaining on the nucleoside. In the mass spectrum of d(ApTp) this sequence specific ion is found at mass 639. Thus the first nucleo-base is unambiguously determined as benzoyladenine. In the mass spectrum of the isomeric deoxy-ribodinucleotide d(TpAp) (Fig. 14) the corresponding fragment ion formed by cleavage of the $C_{3'}$-O-bond appears at m/z 526, thus identifying thymine as the first nucleic base.

For the investigated isomeric deoxy-ribodinucleotides d(ApTp) and d(TpAp) cleavage of the $C_{3'}$-phosphate-ester-bond of the second nucleoside with the charge remaining on the dinucleoside-3′-monophosphate should form a fragment ion at mass 1053. In the MUPI mass spectrum of d(ApTp) this structure specific fragment ion at m/z 1053, which determines the second nucleic base, is displayed with higher intensity compared to the mass spectrum of d(TpAp). Thus, by assigning the fragment ions generated by subsequent cleavage of the $C_{3'}$-O-bond of the second and first nucleoside, it is possible to distinguish the isomeric deoxyribodinucleotides d(ApTp) and d(TpAp) with multiphoton ionization mass spectrometry.

6 CONCLUSION

The combination of multiphoton ionization and mass spectrometry yields clear and easily interpretable mass

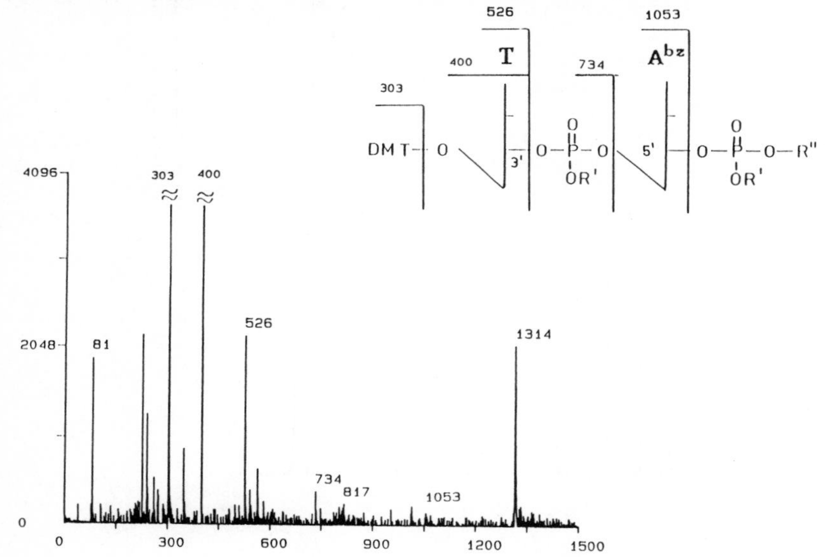

Figure 14 MUPI mass spectrum of the protected dinucleotide d(TpAp); Molecular weight 1314 Da.

spectra of small and medium sized molecules. Although a large number of different molecules have been investigated, the general applicability of this ionization method to all problems in analytical mass spectrometry is still not clear. Further classes of samples have to be investigated. Up to now one can conclude, that aromatic compounds as well as samples containing a large number of aromatic groups are best suited for mass spectrometric investigation by the multiphoton interaction. On the other hand this technique has a great potential for the investigation of basic principles and fragmentation mechanisms in mass spectrometry, because of the relatively exact definition of the internal energies of the ions. The technique of soft ionization allows, in cases of resonant excitation, the nearly exclusive production of molecular ions, while simply increasing the photon density can lead to structure dependent signals.

7 ACKNOWLEDGEMENT

This work is supported by grants from the Deutsche Forschungsgemeinschaft (GR 917/1-2) and the Bundesministerium für Forschung und Technologie. The authors would like to thank Prof. E.W. Schlag for his continuing interest and stimulating discussions about multiphoton ionization mass spectrometry.

8 REFERENCES

1. For a recent overview see: K.L. Busch, G.L. Glish, S.A. McLuckey, "Mass Spectrometry/Mass Spectrometry", VCH Publisher Inc., Weinheim, New York, 1988.
2. F.Hillenkamp, E. Unsöld, R. Kaufmann, R. Nitsche; Appl. Phys. 1975, 8, 341.
3. (a) M. Karas, F. Hillenkamp; Anal. Chem. 1988, 60, 2299.(b) R.C. Beavies, B.T. Chair; Rapid Commun. Mass Spectrom. 1989, 3, 233.
4. (a) D.M. Lubman; Mass Spectrom. Rev. 1988, 7, 535, 559. (b) J. Grotemeyer, E.W. Schlag; Angew.Chem.Int.Ed. Engl. 1988, 27, 447.
5. H.v.Weyssenhoff, H.L. Selzle and E.W. Schlag; Z. Naturforsch. 1985, 40A, 674.
6. I.J. Amster, D.P. Land, J.C. Hemminger and R.T. McIver; Anal.Chem. 1989, 61 184.
7. R.D. Smith, H.T. Kalinoske and H.R. Udseth; Mass Spectrom. Rev. 1987, 6, 445.
8. J. Grotemeyer, J. Lindner, C. Köster, E.W. Schlag; J. Mol. Struc. 1990, 217, 51.
9. C.H. Sin, H.M. Pang, D.M. Lubman, Anal. Chem. 1986, 58, 439.
10. U. Boesl, J. Grotemeyer, K. Walter and E.W. Schlag; Anal. Instrum. 1987, 16, 151.
11. see ref. [4]; J. Grotemeyer, E.W. Schlag; Acc. Chem. Res. 1989, 22, 399.
12. R.C. Beavies, J. Lindner, E.W. Schlag; J. Grotemeyer; Z. Naturforsch. 1988, 43a, 1083.
13. R.C. Beavis, J. Lindner, J. Grotemeyer, E.W. Schlag; Chem. Phys. Lett. 1988, 146, 310.
14. (a) M. Karas, D. Bachmann, F. Hillenkamp; Anal. Chem. 1985, 57, 3925; (b) F. Hellenkamp, M. Karas, D. Holtkamp, P. Klüsener; Int.J.Mass Spectrom. Ion Proc. 1986, 69, 265;(c) M. Karas, D. Bchmann, U. Bahr, F. Hillenkamp; Int.J. Mass Spectrom. Ion Proc. 1987, 78, 53.
15. E.J. Guthrie, H.E. Schwarz; J. Chromatogr. Sci. 1986, 24, 236.
16. (a) C. Köster, J. Grotemeyer, E.R. Rohwer, E.W. Schlag; Adv. Mass Spectrom. 1989, 11b, 1192 (b) J. Grotemeyer, J. Lindner, C. Köster, E.W. Schlag; J. Mol. Struc. 1990, 217, 51.
17. J. Grotemeyer, K. Walter, U. Boesl, E.W. Schlag; Org. Mass Spectrom. 1986, 21, 595.
18. P. Roepstorff and J. Fohlmann; Biomed. Mass Spectrom. 1984, 11, 601.
19. G.R. Kinsel, K.R. Segar, M.V. Johnston; Org. Mass Spectrom. 1987, 22, 627.
20. E. Gross, B.Witkop; Biochemistry 1965, 4, 2495.
21. J. Lindner, J. Grotemeyer and E.W. Schlag, Int.J. Mass Spectrom. Ion Proc.; in print.

ION DISSOCIATION ENERGETICS FROM ANGULAR SCATTER

Colin J. Reid and James A. Ballantine

SERC MASS SPECTROMETRY SERVICE CENTRE, CHEMISTRY DEPT., UNIVERSITY COLLEGE SWANSEA, SINGLETON PARK, SWANSEA

Fast (keV) molecular ions colliding with atoms of a target gas gain internal energy and, as a result, are slowed down and also deflected. The angle of scatter is related to the energy input[1-3] so the (overall) scatter distribution can provide an energy-transfer distribution which reflects the ion's valence-orbital structure[4]. Angular (energy-transfer) distributions can also be obtained for subsets of ions which dissociate into specific fragment ions. These can provide a Breakdown diagram (of fragment-ion relative yield variation with internal energy) for the parent ion[4]. Both types of information are important in, for example , LASER and plasma physics, solution chemistry, analytical mass spectrometry and the chemistry of the upper atmosphere. Data for some positive ions with stable neutral counterparts are given below as energetic data for such can be obtained directly from UV-photoelectron spectroscopy and photoion-photoelectron coincidence methods.

The angle by which a fragment ion is scattered depends not only on the collisional deflection suffered by the parent ion but also on the speed and direction at which the fragment ion is ejected. We have developed a technique, called scatter-profiling[5,6] which allows deconvolution of the collisional-scatter distributions. As shown in Fig. 1, 4 keV energy ions formed in the source of a ZAB-E mass spectrometer are mass selected and focused into a collision cell, and the product ions are translational-energy analysed. Product-ion trajectories are limited to the x-y plane and the incident beam is deflected prior to entering the cell. The cone of fragment ions is sampled on a strip-by-strip basis to give a "scatter profile". The profile which would arise from fragmentational scatter alone is derived from the shape of the relevant peak in the translational-energy (MIKE) spectrum. Numerical deconvolution gives the collisional-scatter profile which is then converted into an angular distribution function.

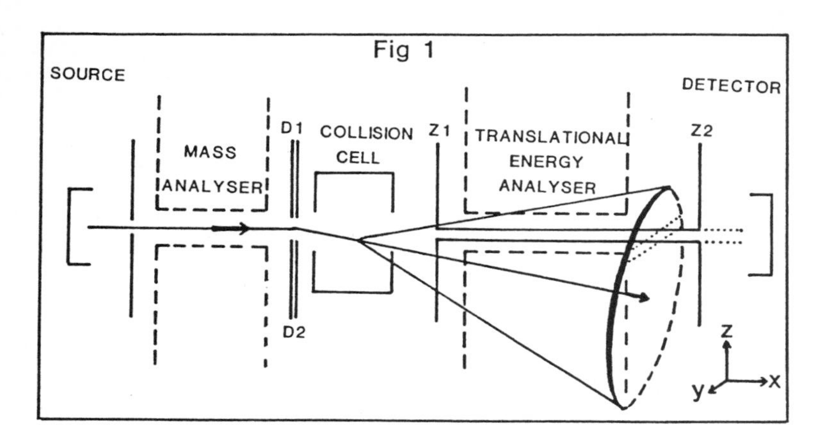

Figure 1 Schematic of the experimental arrangement. By deflecting the incident beam using plates D1 and D2, and limiting detector acceptance with collimators Z1 and Z2, the conically symmetric distributions of fragment ions formed in the collision cell can be measured on a strip-by-strip basis.

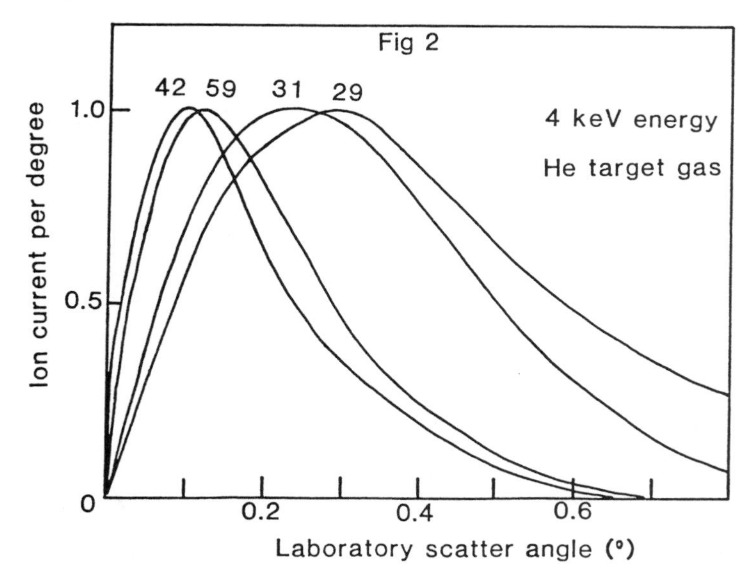

Figure 2. Collisional-scatter distributions for 1-propanol ions which, following 4 keV single collisions with He atoms, dissociate to give ions of the indicated m/z-value.

Angular distribution functions for 1-propanol ions which dissociate specifically to give m/z = 42, 59, 31 or 29 ions, are shown in Fig. 2.

The <u>mean</u> internal energies, ε, gained by 1-propanol ions prior to the above dissociations are about 0.4, 0.7, 3.8 and 5 eV respectively (1 eV ≡ 96 kJ/mol) as determined from peak positions in the translational-energy-loss spectrum. Fig. 2 thus typifies how peak angles and, in particular, root-mean-square angles, θ_{rms}, depend upon mean energy deposition. For 4 keV collision activated decomposition (CAD) reactions of 1-propanol and other ions we find, for He target gas

$$\theta_{rms} = 0.2 \quad (\bar{\varepsilon})^{0.5} \qquad (1)$$

Thus an energy scale can be associated with plots such as Fig. 2. However Eqn. 1 does not imply a one-to-one correspondence between <u>individual</u> angles and energies.

The angular distributions of Fig. 2 are actually broader than the underlying energy-uptake distributions. To modify the curves we studied some <u>non-dissociative</u> collisional excitation reactions whose energy-uptake

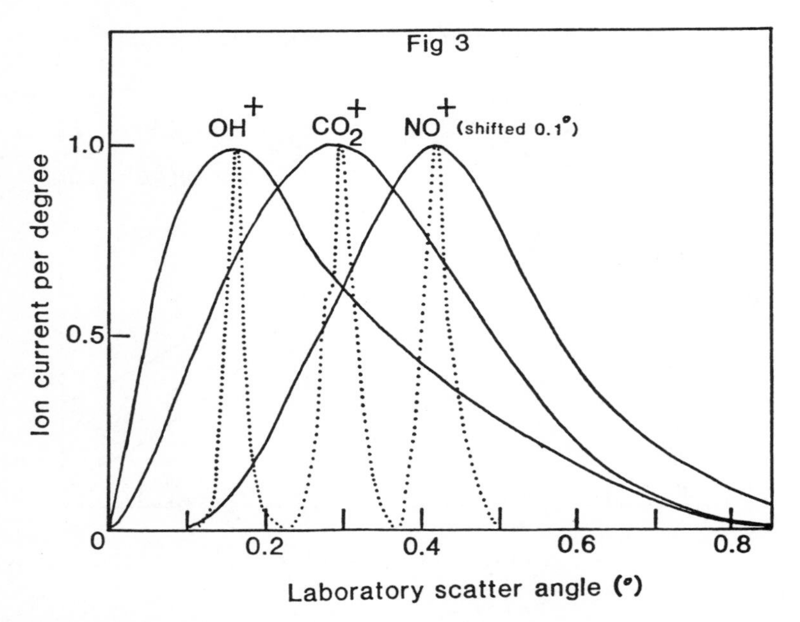

Figure 3 Scatter distributions for non-dissociative excitation rections of some 4 keV molecular ions with He target gas. The dotted curves show the underlying internal-energy-gain distributions, scaled so that peak-maxima coincide.

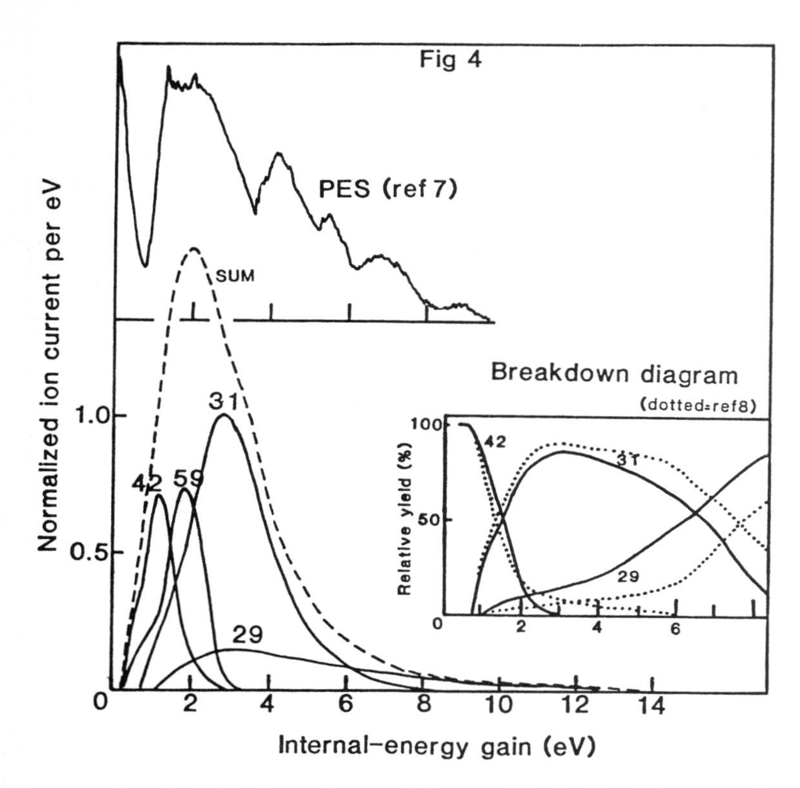

Figure 4 Collisional-scatter distributions for 1-propanol ions, modified to show the underlying internal-energy-gain distributions. The sum of these curves reflects the photoelectron spectrum shown at top of figure, and the relative-yield variations, shown in the right-hand inset, reflect the breakdown curves (shown dotted).

<u>distributions</u> are directly observable (in MIKES) and which are found to be narrow. These reactions involve 4 keV-He excitation of OH^+ (ε = 3.4 eV), CO_2^+(ε = 4.2 eV) and NO^+ (ε = 9.0 eV) and their angular scatter distributions are shown in Fig. 3. The energy functions are shown as dotted lines.

Taking the curves of Fig. 3 as a basis set, CAD angular distributions can be modified to give the energy distributions. Such modified angular functions (MAF's) for the CAD reactions of ionized 1-propanol, are shown in Fig. 4, where areas under curves are proportional to yield. For comparison, the photoelectron spectrum[7] and documented breakdown curve[8] are also given.

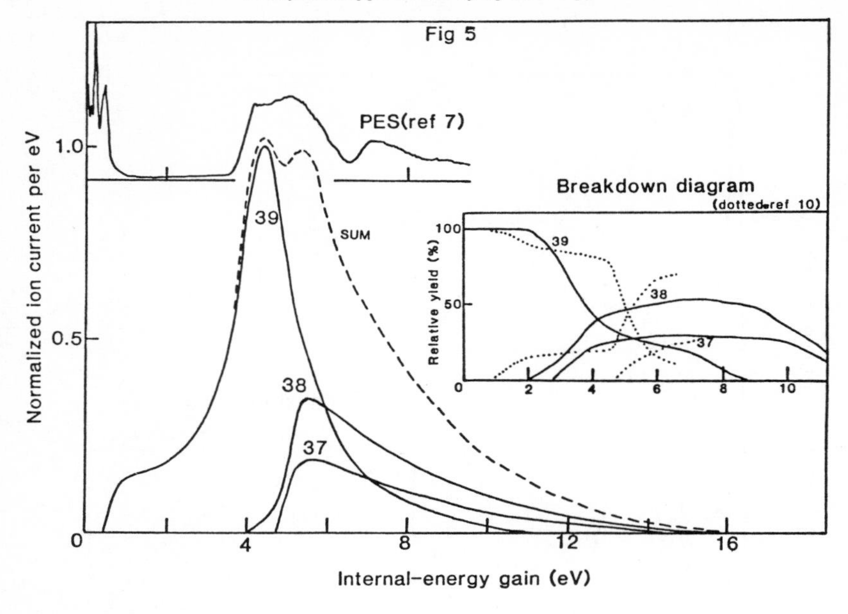

Figure 5 As for Fig. 4, for ionized propyne.

The <u>sum</u> of the individual curves correctly matches with the first band of the photoelectron spectrum. The threshold energies and equi-yield (crossover) points for the m/z = 42, 31 and 29 ions also agree quite well with the known data, as shown in the breakdown diagram.

A similar close match of the sum curve with the photoelectron spectrum is found for ionized ethanol. In this case, however, the present breakdown diagram differs from previous[9] in that formation of stable m/z = 31 ions does not predominate over formation of stable m/z = 45 ions until an internal energy of 2.8 eV is reached. Previous work suggests that this crossover occurs at only 1.0 eV. There is no a-priori reason for expecting angular measurements to give <u>accurate</u> quantitative energy information. Each type of ion behaves differently. For an indirect, semi-empirical technique, however, the data comparisons are quite remarkable as seen for example in the results for ionized propyne shown in Fig. 5. For ionized propyne (m/z = 40) the threshold for dissociation (to m/z = 39) is at an internal energy of 0.8 eV. The MAF curve for this fragment ion, however, shows a most-probable energy transfer of about 4.5 eV. This arises through a lack of suitable states as seen by the gap in the photoelectron spectrum. The sum curve coincides with

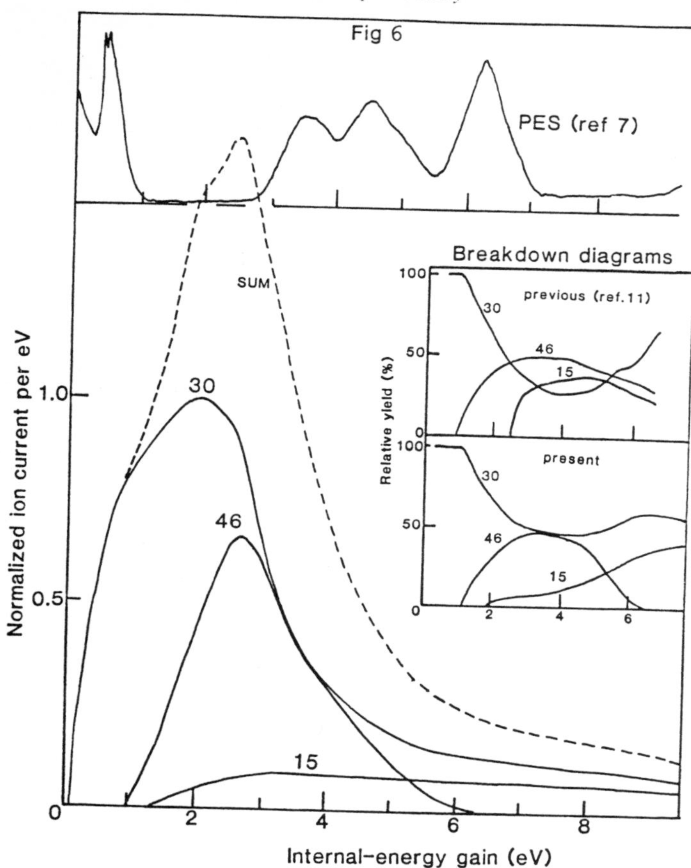

Figure 6 As. Fig. 4, for ionized nitromethane.

the first available PES band. The crossover points in the breakdown curve are also in reasonable agreement with previous work[10].

Finally the results for ionized nitromethane are shown in Fig. 6. This ion shows unusual behaviour, the previous breakdown diagram[11] showing a double crossover of m/z=30 and 46 curves. Present results almost reproduce this behaviour but the m/z = 15 curve is different. Also the sum curve peaks in an empty region of the PES. We tentatively infer from these differences that nitromethane undergoes a change in geometry (and even isomerization) upon ionization.

References

1 J.A. Laramee, P.H. Hemberger & R.G. Cooks, Int.J.Mass Spectrom.Ion Proc. ,1980, 33, 231.

2 C.J. Reid & F.M. Harris, Int.J.Mass Spectrom.Ion Proc. ,1988, 85, 151.

3 C.J. Reid, Int.J.Mass Spectrom.Ion Proc. (accepted May 1990).

4 C.J. Reid, papers to be submitted to Int.J.Mass Spectrom.Ion Proc.

5 C.J. Reid, J.Phys.B (accepted May 1990).

6 J.A. Ballantine & C.J. Reid, Org.Mass.Spectrom 1989, 24, 603.

7 K. Kimura, "Handbook of HeI photoelectron spectra of fundamental organic molecules", Japan Scientific Societies Press, Tokyo 1980.

8 J.J. Zwinselman, S. Nacson & A.G. Harrison, Int.J.Mass Spectrom.Ion Proc.,1985,67, 93.

9 Y. Niwa, T. Nishimura & T. Tsuchiya, Int.J.Mass Spectrom.Ion Proc. ,1982, 42, 91.

10 A.C. Parr, A.J. Jason, R. Stockbauer & K.E. McCulloh, Int.J.Mass Spectrom.Ion Proc.,1979, 30, 319.

11 Y. Niwa, S. Tajima & T. Tsuchiya, Int.J.Mass Spectrom.Ion Proc.,1981, 40, 287.

APPLICATIONS OF SPECTROSCOPY IN COAL CHARACTERIZATION

L. D. Thomas and A. A. Herod

BRITISH COAL CORPORATION, COAL RESEARCH ESTABLISHMENT, STOKE ORCHARD, CHELTENHAM, GL52 4RZ

INTRODUCTION

Despite the extensive use of coal in industry, the structure of coal is still poorly understood. It is generally thought of as a carbonaceous rock, but is a heterogeneous material of great complexity, and contains many elements in addition to carbon and hydrogen. Current theories view coal as a cross-linked polymeric carbon structure, with smaller molecules adsorbed on, or trapped in the "skeleton". The mode of incorporation of inorganic materials in the structure is not understood, but it is thought they may catalyse some conversion processes.

Ultimate and proximate analysis characterise coal used for industrial purposes, but they do not contribute to structural information. Work has been carried out by the Coal Research Establishment (CRE) using spectroscopic techniques such as mass spectrometry, nuclear magnetic resonance and Fourier-Transform infrared spectroscopy. Examples of some of the characterisation work is described.

MASS SPECTROMETRIC TECHNIQUES

Mass spectrometry has provided valuable information concerning the structure of the products of pyrolysis, combustion and liquefaction of coal.

The Study of Volatiles from Bituminous coals, Peat and Brown Coals. Information about the mode of occurrence of chlorine in coal and the manner in which it is released on pyrolysis and combustion is important when considering ways in which its deleterious effects on coal conversion plants, especially combustion equipment, may be countered.

The evolution of HCl and other volatiles during the mild heat treatment of UK coals in helium was followed by mass spectrometry.[1,2] The coals were heated at $2^{0}C\ min^{-1}$ to $300^{0}C$ before being heated isothermally for 24 h. Between 40 and 60% of the chlorine was emitted as HCl; no other chlorine-containing volatiles were identified. The HCl losses

© British Coal Corporation, London 1990

did not correlate with the initial chlorine content of the coals. Volatiles other than HCl included carbon dioxide, sulphur dioxide, water, alkanes, alkyl benzenes, alkyl naphthalenes, alkyl phenanthrenes, and the oxygenates, phenols and dibenzofurans. The volatiles appear to have evaporated from the micropores of the coal rather than having been formed by thermal decomposition. First-order release of HCl from some of the coals suggested that the rate-controlling step may be either diffusion through micropores or breakdown of an ionic complex within the pores.

The study has been extended to include investigation of a Swedish peat and East German brown coals.[3,4] Peat is comprised of plant remains which have undergone an anaerobic decomposition. Brown coals have undergone partial coalification but have not reached the rank of bituminous coal.

Coals are known to be a source rock for methane and are considered to have passed through an oil formation window.[5] The volatiles detected in these experiments correspond to the oil left within the coal matrix. The extent of oil formation is not evident however.

Liquid Chromatographic-Mass Spectrometric Analysis of Coal Tar Fractions. Conventionally, coal-derived tars are separated by solvent fractionation into pentane solubles, asphaltenes (pentane insoluble, benzene soluble) and preasphaltenes (benzene insolubles) with further separation of the pentane solubles into saturated, aromatic and polar fractions by open-column chromatography on silica.

Normal-phase HPLC, using hexane with a gradient to dichloromethane - methanol (95+5) coupled to a mass spectrometer by a moving belt interface, has indicated that the so-called aromatic fraction of a low-temperature hydropyrolysis coal tar contains some saturates (alkanes and cycloalkanes including hopanes, up to C_{64}) and an appreciable amount of polar materials which may represent 50% of the aromatic fraction, in addition to aromatics from alkylbenzenes to alkylrubicenes as extensive homologous series.[6]

Mass spectrometry with electron impact and chemical ionisation using ammonia has given new structural information on the polar materials present in this "aromatic" fraction. Fast atom bombardment of asphaltenes and benzene-insolubles indicated that their structures resemble those of the polar compounds and differ from those of the aromatics. Size-exclusion chromatography confirmed the relative molecular mass ranges detected by mass spectrometry. GC-MS with a capillary column proved inadequate for defining many of the aromatics.

The presence of long alkyl chains is in agreement with the current view of coal structure, but such high relative molecular mass alkanes, cycloalkanes and alkyl aromatics have not been detected previously. This work has extended the upper mass range from previous work and has applied mass spectrometry to the polar and asphaltene materials for the first time; these materials are probably intermediates in the conversion of coal into useful products such as liquid fuels, and their better characterisation should lead to improved processes for the production of low mass products.

Hydrocarbon Type Analysis of Coal Liquefaction Products by Mass Spectrometry. Hydrocarbon type analysis is a mass spectrometry method for analysing complex hydrocarbon oils. Distillate fractions are used with no further fractionation. Although the method was devised for use in the petroleum industry, it has been successfully applied to coal derived fluids, such as coal liquefaction distillate products.[7] CRE was one of the participant laboratories in an exercise organised by the Institute of Petroleum to develop and assess a standard method for the mass spectrometric analysis of petroleum distillate fractions.[8] The aim of the exercise was to promote the use of analytical methods based on mass spectrometry by establishing "off the shelf methods".

A matrix method based on the summation of fragment ion intensities of nine aliphatic and aromatic chemical types, and the molecular ion intensity for naphthalene, found in light gas oils was investigated. The concept of z number is widely used to define chemical structure. It is derived from the general hydrocarbon formula C_nH_{2n+z}.

Results of the round robin exercise showed that different mass spectrometers produced different spectra from each oil and the tuning standard, hexadecane. However, individual laboratories could reproduce their own work very well. To avoid the extensive programme of work necessary for all participants to recalibrate their instruments, it was recommended that any laboratory wishing to use the method should undertake as full a calibration as possible.

Monitoring of Emissions from Coal-Fired Plant using GC-MS. Recently, coal-fired plant have come under increasing scrutiny for the control of emissions such as polynuclear aromatic hydrocarbons (PAH). Methods are being developed at CRE to identify and and quantify these emissions.

An Andersen stack sampling system was used to sample gases from various chimney stacks associated with coal-fired plants, using stainless steel tubes packed with Tenax or XAD-2 resin. The material adsorbed on the resins was extracted with cyclohexane, concentrated and analysed by capillary column GC-MS.[9]

A series of aliphatic and aromatic hydrocarbons was detected. Major peaks were S_8 (cyclooctasulphur) and breakdown products from the trapping resin. The S_8 was thought to be derived from SO_2 attack on the resin, and its subsequent reduction. Other breakdown products were believed to contain nitrogen oxide compounds. The work is being extended to include the evaluation of different resins and control of the temperature and humidity during sampling.

NUCLEAR MAGNETIC RESONANCE SPECTROSCOPY

Analysis of coals and coal products has been extended by the application of solid state and heteroatom NMR.

Solid state characterisation of coals by 13C CP/MAS NMR Spectroscopy. Cross-polarisation (CP) and magic angle spinning (MAS) techniques developed in the mid 1970's overcame the problem of broad, featureless bands in the ^{13} NMR spectra of coals. ^{13}C CP/MAS NMR spectroscopy now permits the direct, non-destructive measurement of carbon aromaticity and functionality in coal on a routine basis with a minimum of sample preparation. A "rank" series of coals and coal precursors, ranging from wood to anthracite has been analysed, as well as some heat-treated coals.[10]

Aromaticity (f_a = ratio of aromatic carbon to total carbon) was seen to increase progressively with rank - wood has mainly aliphatic nature and cellulose residues, through to anthracite with its characteristic aromatic ring systems and very limited aliphatic content.

A linear relationship between aromaticity (f_a) and coal rank (expressed as %C, dmmf) was revealed. Deviations from the relationship can in some cases be correlated with unusual coal properties such as high organic sulphur, which may take the place of C atoms in certain structural units.

Analysis of coal chars shows dealkylation of coals takes place over a relatively narrow temperature range to produce highly aromatic chars, but higher temperature chars (>600^{0}C) do not yield good spectra. It was thought that this may be due to paramagnetic Fe^{3+} species possibly arising from mineral decomposition reactions.

Solid State Heteroatom NMR Analysis of Coals. Some of the problems which occur during coal utilisation are now recognised as being associated with hetero elements such as alkali metals, and phosphorus and boron, and increased attention is being paid to their analysis.[11-14] However, conventional elemental analysis gives little information on their speciation. Solid state NMR analysis of coals directly has become increasingly attractive.

The ^{23}Na spectra of raw coals were interpreted in terms of a fully hydrated Na^+ ion bound either by hydrogen bonding or possibly at one co-ordination site, to a single organic group on the pore surface. The ^{27}Al spectra showed resonances due to both tetrahedrally and octahedrally co-ordinated aluminium, and may be interpreted in terms of differences in the relative proportions of the various clay minerals.

^{31}P MAS-NMR spectroscopy has confirmed that phosphorus is present in coal predominantly as apatite. This mineral is thermally stable under oxidising conditions, and survives largely unaltered in high temperature ashes. However, under the semi-reducing bed conditions of certain stoker-fired boilers, it may be decomposed and the phosphorus volatilised. The ^{31}P MAS-NMR spectra of bonded deposits show phosphorus in a markedly different co-ordination environment to that in apatite, the chemical shift suggesting aluminium phosphate or boron phosphate.

^{11}B MAS-NMR spectra of coals exhibit resonances due to both trigonal and tetrahedrally co-ordinated boron. Trigonal boron is probably present as tourmaline, but the nature of the tetrahedral sites is less certain; it may be held in tetrahedral sites within certain clay minerals. In common with phosphorus, boron may be volatilised during combustion. The ^{11}B MAS-NMR spectra of bonded deposits show a tetrahedral resonance with a chemical shift quite consistent with that of boron phosphate.

A Study of Oxygen Functionality in Coal Tars and Extracts by 29Si NMR Spectroscopy. Oxygen is the most abundant heteroatom in coal. Its levels range from 1-2% in anthracites to more than 20% in brown coals, and even higher in the coal precursors such as peat. Coalification therefore, is accompanied by progressive loss of oxygen from carboxyl and hydroxyl groups, with release of CO_2 and H_2O.

The abundance of oxygen is reflected in the presence of numerous oxygen compounds in the pyrolysis and extraction products. In contrast to the nitrogen compounds, which tend to be basic in character, the oxygen compounds are predominantly acidic, eg phenols and carboxylic acids. Phenols, as tar acids, were once commercially significant, but their presence in coal liquefaction products causes problems in the refined products eg reduction in stability, gum formation, and viscosity increase.

Existing methods of analysis (enthalpimetric, fractionation and GC analysis) tend to encounter problems with selectivity and accuracy. However, silylation of hydroxyl groups followed by ^{29}Si NMR spectroscopic analysis of the derivatives has proved sensitive, highly specific, and quantitative for phenols and carboxylic acids.[15] The samples analysed in this study were derivatised using hexamethyldisilazane.

^{29}Si NMR chemical shifts show little overlap for various hydroxyl types and correlate with pK_a, thereby facilitating assignment of resonances. Tars showed trends in OH type distribution which were consistent with previous studies, and support those models of coal structure in which the OH groups in coal (other than in carboxylic acids) arise almost entirely from aromatic compounds.

Characterisation of Coal Liquefaction Products by 13C NMR Spectroscopy. ^{13}C NMR spectroscopy has played an important role in the structural characterisation of coal liquefaction products at CRE. British Coal have developed a liquefaction process at CRE, which is now undergoing further evaluation on a 2.5 tonne/day pilot plant at Point of Ayr, North Wales. ^{13}C NMR data, in conjunction with ^{1}H NMR, has been used when calculating average structural properties for coal extracts and hydrogenation residues, and concentrations of donatable hydrogen in coal solvents.[7,16]

A part-coupled spin-echo (PCSE) technique for assigning peaks in the spectra of a range of coals was used.[17-19] The combination of PCSE with gated decoupling to estimate the concentrations of CH_3, CH_2, CH and C groups directly, together with resultant improvements to the structural characterisation of these materials, are briefly described. Peaks can be unambiguously assigned to CH_3, CH_2, CH and C groups using the PCSE method.

The peak assignments for the solvent, extract and pitch fractions examined generally confirm those previously reported on the basis of data for model compounds. Concentrations of the various groups can be made. Estimates of aromatic C and CH groups and aliphatic H/C ratios were found to be in good agreement with those obtained indirectly from elemental, ^{1}H NMR and ^{13}C NMR data. The direct estimation of carbon types led to improvements in the structural characterisation of solvents and extracts.

FOURIER-TRANSFORM INFRARED SPECTROSCOPY

The development of Fourier-transform instruments has widened the field of application of infrared spectroscopy enormously, and it has been extended to direct analysis of coals.

<u>Analysis of UK Coals by Diffuse Reflectance FT-IR Spectroscopy.</u> A rank series of UK coals was examined by FT-IR using diffuse reflectance, to assess any relationship between infrared spectral properties and traditionally-derived physical properties, such as volatile matter content of the coals.[20]

Correlations were drawn between FTIR-derived aromaticities and volatile matter contents, reflectances, atomic C/H ratios and swelling properties for a rank series of coals. Subtraction routines were employed to yield spectra equivalent to coal organic matter. Subtraction of the spectrum of mineral matter in the coal (prepared by low temperature or plasma ashing of the coal) from that of the raw coal yielded a spectrum equivalent to the coal organic matter. Conversely, subtraction of the demineralised coal spectrum from that of the raw coal gave a spectrum equivalent to the mineral matter.

Problems were encountered with carbonyl peaks using the DRIFTS technique. This was demonstrated to be a genuine, but surface-specific feature due to oxidation. Transmission techniques would possibly avoid this problem.

<u>Use of Factor Analysis of FT-IR Spectral Data to Predict the Properties of UK Coals.</u> The ability to predict properties of coals using FT-IR spectral data has the potential to provide a rapid analytical method for virtual "on-line" use. This has become increasingly important recently as British Coal's customers are demanding improved coal quality. The ability to

detect "rogue" coals is also important in the new market place created by the privatisation of the electricity supply industry.

A spectral database of UK coals is being created using routine, standardised conditions. Factor analysis will be used to correlate spectral properties with properties of coal, such as carbon content, volatile matter content, ash content and calorific value. This will be used to predict properties of "unknown" coals. It is hoped to extend the database to include foreign coals at a later stage.

REFERENCES

1. N.J. Hodges, W.R. Ladner and T.E, Martin, J.Inst. Energy, 1983, LVI, 428, 158.
2. A.A. Herod, N.J. Hodges, E. Pritchard and C.A. Smith, Fuel, 1983, 62, 1331.
3. A.A. Herod, D. Radek and B.J. Stokes, Paper to be presented to the 25th British Mass Spectroscopy Society Meeting, London, September, 1990.
4. A.A. Herod, N.J. Hodges and B.J. Stokes, Proc. 14th British Mass Spectroscopy Society Meeting, Heriot-Watt University, 1984, 193.
5. T. Chakravorty, W. Windig, G.R. Hill, H.L.C. Muezelaar and M.R. Khan, Energy and Fuels, 1988, 2, 400.
6. A.A. Herod, B.J. Stokes, H.J. Major and A.E. Fairbrother, Analyst, 1988, 113, 797.
7. A.A. Herod, W.R. Ladner and C.E. Snape, Phil. Trans. R. Soc. London, 1981 A300, 3.
8. A.A. Herod, Inst.Petroleum, Quart. J. Technical Papers, April-June 1989, 15.
9. A.A. Herod, Pres. to Chrom. Soc., Warrington, 1988.
10. C.E. Snape, Analytical NMR, Ed. Sternhell and Field, Wiley and Sons, 1989, Chap. 4.
11. O.W. Howarth, G.S. Ratcliffe and P. Burchill, Fuel, 1987, 66, 34.
12. P. Burchill, O.W. Howarth, D.G. Richards and B.J. Sword, Fuel, 1990, 69, 421.
13. P. Burchill, O.W. Howarth and B.J. Sword, Fuel Proc. Tech., 1990, 24, 375.
14. P. Burchill and B.J. Sword, Paper to 9th International Meeting on NMR Spectroscopy, Coventry, July, 1989.
15. O.W. Howarth, G.S. Ratcliffe and P. Burchill, Fuel, 1990,69, 297.
16. J.W. Clarke, T.D. Rantell and C.E. Snape, Fuel, 1982, 61, 707.
17. C.E. Snape, Fuel, 1982, 61, 775.
18. C.E. Snape, Fuel, 1982, 61, 1165.
19. C.E. Snape, Fuel, 1983, 62, 621.
20. P. Burchill, B. Dunstan, C.A. Mitchell and T.G. Martin, Paper to Round Table Meeting, Chemical and Physical Valorization of Coal, CEC, Brussels, June, 1987.

Section 5

COMBINED TECHNIQUES

COMBINED CAPILLARY ELECTROPHORESIS AND ELECTROSPRAY IONIZATION MASS SPECTROMETRY

Richard D. Smith, Joseph A. Loo, Charles G. Edmonds and Harold R. Udseth

CHEMICAL METHODS AND SEPARATIONS GROUP, CHEMICAL SCIENCES DEPARTMENT, PACIFIC NORTHWEST LABORATORY, RICHLAND, WASHINGTON 99352, USA

1 INTRODUCTION

The development of new capillary electrophoresis (CE) methods provides a basis for the efficient manipulation and separation of subpicomole quantities of polypeptides and proteins. Recent advances in microscale methods, such as the demonstration of the tryptic digestion of low picomole quantities of proteins using the immobilized enzyme in a small diameter packed reactor column,[1] provide the basis for such further developments. The use of capillary (free solution) zone electrophoresis (CZE) for separation of proteins,[2] and recent demonstrations of restriction mapping of large deoxyribonucleotides,[3] has propelled potential CE applications into the realm of conventional electrophoresis, while adding the attributes of speed, relatively simple on-line detection, automation, and reduced sample requirements (10^{-17} - 10^{-13} mole). A literal explosion of ancillary methods for sample manipulation, derivatization, and detection as well as new methods of obtaining separation selectivity are being reported. Additionally, other CE formats are attracting increased interest, with the aim of exploiting the unique features of capillary isotachophoresis (CITP),[4] capillary isoelectric focusing (CIEF),[5] capillary electrokinetic chromatography (CEC),[6] and, most recently, capillary polyacrylamide gel electrophoresis (CGE).[7] As a result, there are concomitant and increasing demands upon detector sensitivity and information density.

Mass spectrometry is potentially the ideal detector for CE. At present, CE-MS interfacing methods are based upon either flowing (or dynamic) fast atom bombardment (FAB)[8-10] or electrospray ionization (ESI).[11-15] Our efforts have involved the development of CE-MS interfacing methods that allow operation over an essentially unlimited range of flow rates and buffer compositions without degrading CZE separations.[13] These developments have allowed the first on-line combination of capillary isotachophoresis with MS (CITP-MS),[16] which provides an attractive complement to CZE-MS where (among other situations) greater sample sizes are required. These developments have been augmented by the unique features of electrospray ionization, which include efficient ionization of higher molecular weight compounds and the production of multiply charged ions.[17-19] Of particular interest to us has been the tandem mass spectrometry of multiply charged biopolymers where, most readily, distinctive sequence-specific fingerprinting (or "mapping") may be accomplished. The potential also exists for obtaining substantial sequence related information during the course of a single CE-MS/MS analysis.

2 EXPERIMENTAL

The instrumentation developed at our laboratory has been described elsewhere in detail.[11-13] Figure 1 shows a schematic of the interface and mass spectrometer. Our interface employs a flowing liquid sheath interface which

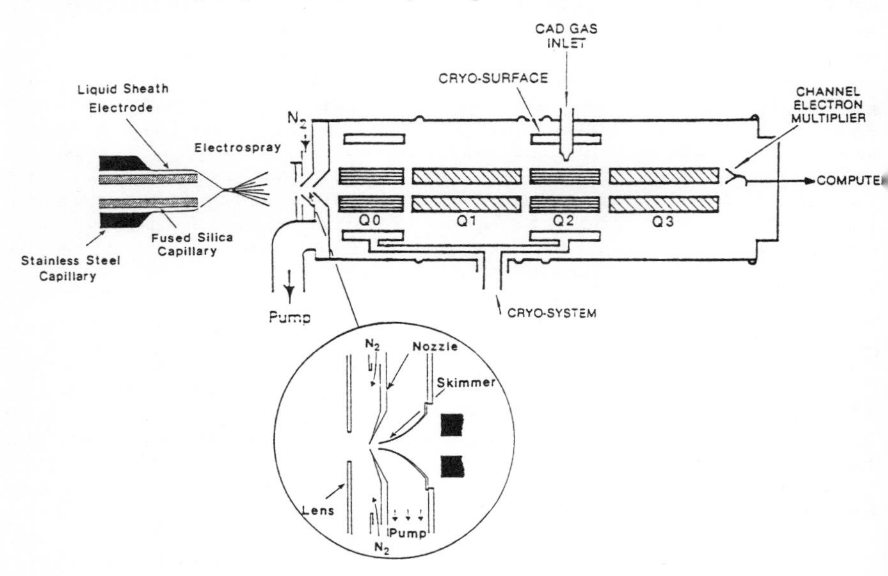

Figure 1. Schematic of the atmospheric pressure electrospray ionization interface and mass spectrometer. The electrospray capillary and sheath electrode assembly are enlarged for clarity.

allows the composition and flow rate of the electrosprayed liquid to be controlled independently of the CZE buffer (which is desirable since the high percentage aqueous and high ionic strength buffers generally useful for CZE are not directly compatible with ESI).[13] The electrical contact is also established through the conductive liquid sheath (typically methanol, acetonitrile, acetone or isopropanol, often including small fractions of water and acetic acid). With this arrangement no significant additional mixing volume (< 10 nL) is produced and analyte contact with metal surfaces is avoided. This interface provides greatly improved performance and flexibility and is adaptable to other forms of CE.[13,16] For direct ESI-MS experiments, syringe pumps control the flow of analyte solution and liquid sheath at ~ 0.5 μL/min and 3 μL/min, respectively. CZE-ESI/MS experiments were conducted in untreated fused silica capillaries using methods that have been described previously.[11-13]

The electrospray ionization source consists of a 50 or 100 μm ID fused silica capillary (which is generally the CZE capillary) that protrudes 0.2 to 0.4 mm from a cylindrical stainless steel electrode. High voltage, generally +4 to 6 kV for positive ions or -5 kV for negative ions, is applied to this electrode. The ESI source (capillary) tip is mounted approximately 1.5 cm from the ion sampling orifice (nozzle) of the quadrupole mass spectrometer. A 3 to 6 L/min countercurrent flow of warm (80°C) nitrogen gas is introduced between the nozzle and source to aid desolvation of the highly charged electrospray droplets and to minimize any solvent cluster formation during expansion into the vacuum chamber. Ions enter through the 1 mm diameter orifice and are focused efficiently into a 2 mm diameter skimmer orifice directly in front of the radio frequency (rf) focusing quadrupole lens (Figure 1). Typically, +350 V to +1000 V is applied to the focusing lens and +200 V to the nozzle (V_n), while the skimmer is at ground potential.

Biochemical samples were purchased from Sigma Chemical Co. (St. Louis, MO, U.S.A.) except bovine apotransferrin (Calbiochem, San Diego, CA, U.S.A.) and were used without further purification.

3 RESULTS AND DISCUSSION

Electrospray Ionization

As originally reported by Fenn and coworkers, proteins can be effectively ionized by ESI yielding a distribution of charge states.[17,18] The ESI mass spectra of eight peptides and proteins of increasing molecular weight are given in Figure 2.

ESI mass spectra of proteins composed of noncovalently bound subunits obtained under our typical operating conditions generally contain only multiply charged ions characteristic of the subunit M_r. Lactate dehydrogenase from rabbit muscle (M_r 140 kDa) is an enzyme composed of four identical subunits of M_r 36 kDa. Molecular ions with over 50 positive charges indicative of the subunit molecular weight are observed.[19] Similarly, human hemoglobin is a tetrameric protein consisting of 2

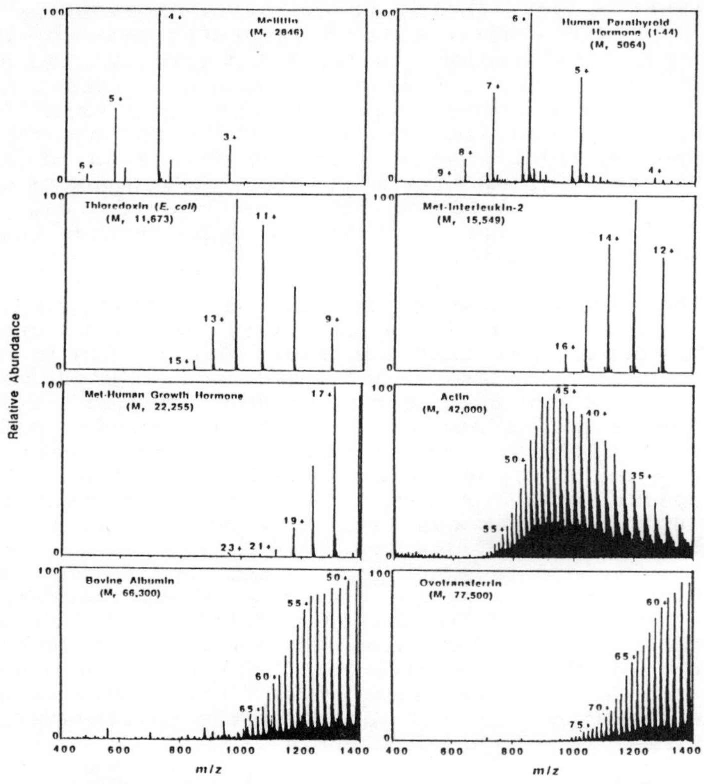

Figure 2. ESI mass spectra of eight peptides and proteins with M_r ranging from 2846 to 77,500.

alpha and 2 beta chains; its spectrum shows multiply charged ions for both the alpha (M_r 15,126) and the beta (M_r 15,865) chain units. Analogously, the ESI mass spectra of heme-containing proteins do not exhibit molecular ions incorporating the heme, unless it is covalently attached. An iron-porphyrin group is associated with hemoglobin; however, peaks characteristic of the subunit with loss of the heme are observed. In contrast, the iron-porphyrin moiety for cytochrome c is covalently linked through thioether bonds to two cysteine residues. Thus, the M_r calculated from its ESI mass spectrum is consistent with the combined polypeptide-heme species.[19]

Increasing the extent of charging for a given molecule is especially important for mass spectrometers with a limited *m/z* range, allowing detection of large molecular species. In addition, mass spectrometer performance is generally improved at lower *m/z*. A substantial fraction of all the proteins attempted to date by electrospray ionization-MS have provided adequate charging to be detected by mass spectrometers with m/z limits of 1400 to 2400 although exceptions have been noted.[18,20]

For larger proteins the contributions ascribed to protein microheterogeneity also become increasingly evident, as shown in Figure 2. Such contributions degrade the ability to resolve individual charge states and ultimately limit the ability to make measurements for species of high molecular weight.[18,19]

Capillary Electrophoresis with ESI-MS

Capillary electrophoresis (CE) in its various manifestations (free solution, isotachophoresis, isoelectric focusing, polyacrylamide gel, micellar electrokinetic "chromatography") is capable of providing rapid, high resolution separations of very small sample volumes. The correspondence between CE and ESI flow rates, and the fact that both address most effectively ionic species in solution, provide the basis for an extremely attractive combination.

Small peptides are easily amenable to CZE-MS analysis with good (femtomole) sensitivity. High efficiency separations of a series of dynorphin and enkephalin peptides have been demonstrated, with over 250,000 theoretical plates obtained by CZE-ion spray-MS.[14] Complex mixtures of peptides generated from tryptic digestion of large proteins are well suited to CZE-MS analysis, as shown by Henion and coworkers.[21] Since trypsin specifically cleaves peptide bonds C-terminally at lysine and arginine, the resulting peptides will tend to form doubly charged, as well as singly charged, molecular ions.[20] Such doubly charged tryptic digest peptides fall within the m/z range of most quadrupole mass spectrometers. The CZE-ESI-MS methods has been extended to the direct separation and detection of proteins,[15] as shown in Figure 3 for separation of three myoglobin species. In this example,approximately 1 pmole of each protein was injected. The horse (HH) and sheep (SS) myoglobins differ only sightly in molecular weight and are not resolved mass spectrometrically. The on-line CZE separation allows the two components to be distinguished, and more accurate molecular weight measurements obtained.

Since ESI-MS detection of proteins often requires a mildly acidic solution, the 80/20 methanol/water solution generally used for the sheath liquid, was augmented by addition of 5% acetic acid. This capability highlights one of the unique features of the sheath flow interface: buffer conditions otherwise inappropriate for ESI-MS can be used and modified "post-column" through the sheath liquid.

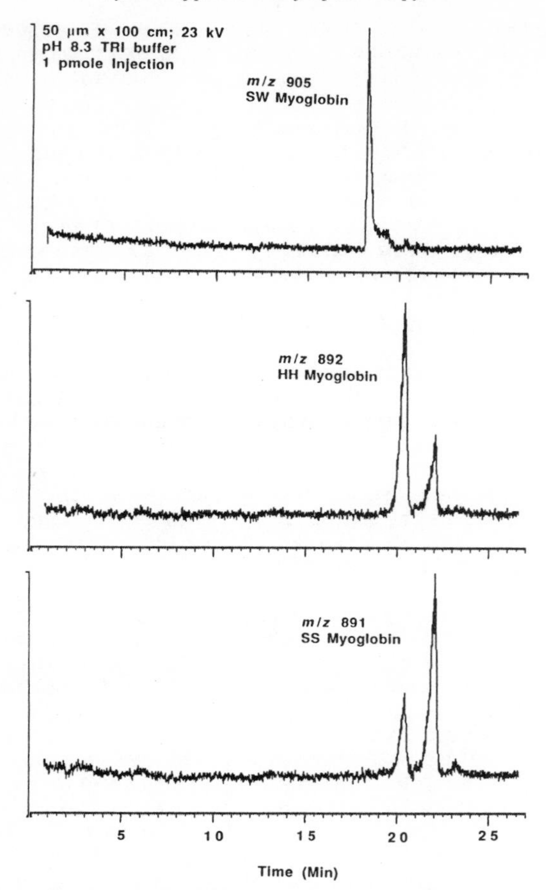

Figure 3. CZE-MS single ion electropherograms from an equimolar mixture of three (sperm whale (SW), horse heart (HH), and sheep skeletal (SS)), myoglobins separated in a pH 8.3 TRIS/HCl buffer in an untreated 100 cm x 50 μm capillary.

Quite high plate counts can be obtained in such separations even without capillary surface treatment, as shown in Figure 4, for a mixture of leucine enkephalin and horse myoglobin (0.1 mM each). The injection volume was ~ 10 nL corresponding to ~ 1 pmole/component and the separation yielded ~ 125,000 theoretical plates for both peaks. Significant improvements in detection limits may result from more efficient transmission of ions generated by the electrospray process.

Because CE relies on analyte charge in solution, and the ESI process appears to function most effectively for ionic species, the CE/ESI-MS combination is highly complementary. We have reported the analysis of CZE/MS of a mixture of quaternary ammonium compounds[11] obtaining over 330,000 plates, an order of magnitude better than obtainable by liquid chromatography (LC) in similar time. In this case sample sizes were ~300 femtomoles; however, detection limits of ≤ 10 attomoles are obtainable using single ion detection.[12]

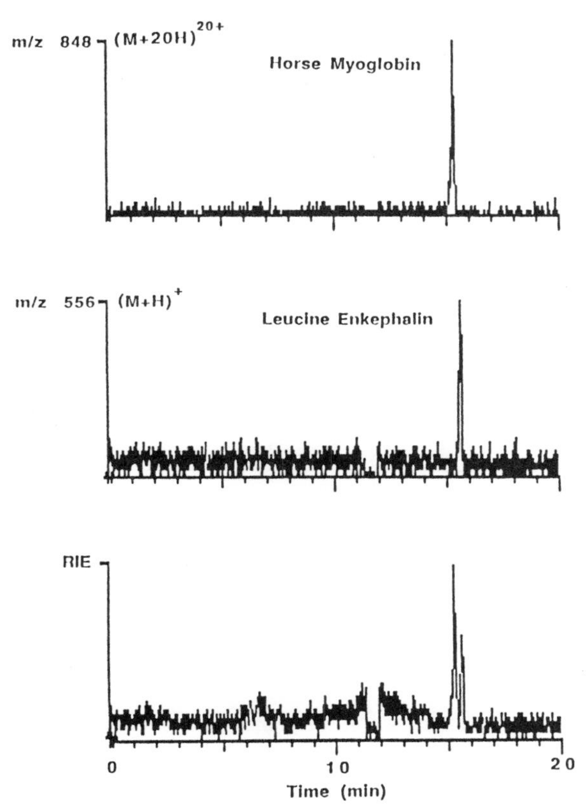

Figure 4. CZE/ESI-MS separation of horse myoglobin (M_r 16,950) and leu-enkephalin (Tyr-Gly-Gly-Phe-Leu, M_r 555) at pH 8.25 in a 125-cm x 50 μm fused silica capillary, at 30 kV (17 uA current).

The steady state concentration of the analyte anion for capillary isotachophoresis (CITP) separations is determined by the lead anion concentration and the relative mobilities of the solute, lead anion and common cation.[22] Accordingly, if the ionic strength of the analyte sample is lower than the lead anion concentration, the analytes will be concentrated as they separate. In a fully developed separation the concentration of each band is similar, and the relative abundance for each component is proportional to the length of the band. Thus, capillary isotachophoresis offers the potential for higher sample loading (and increased molar sensitivity), high resolution separations and actual concentration (in many cases) of separated sample bands.

Figure 5 shows a typical result of a CITP separation of a simple two component peptide mixture. In such a case the electrophoretic mobilities of the two components are quite different and a rapid separation is obtained, yielding nearly ideal bands. Figure 6 illustrates a separation obtained for an enzymatic digestion of cytochrome c. In the case of cytochrome c at least 16 peptides are expected, reflected by the complex nature of the isotachopherogram.

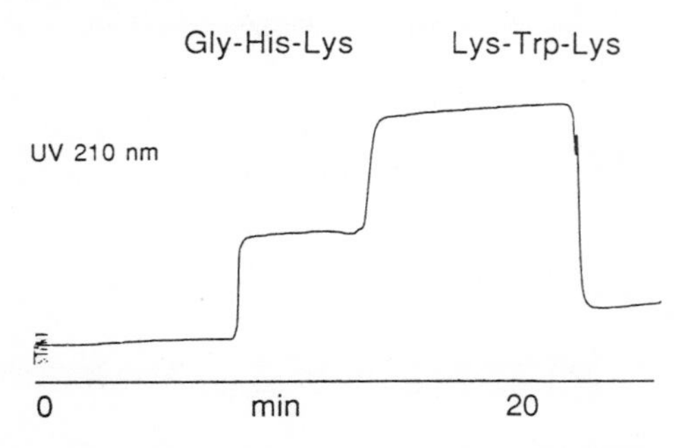

Figure 5. CITP separation of two tripeptides.

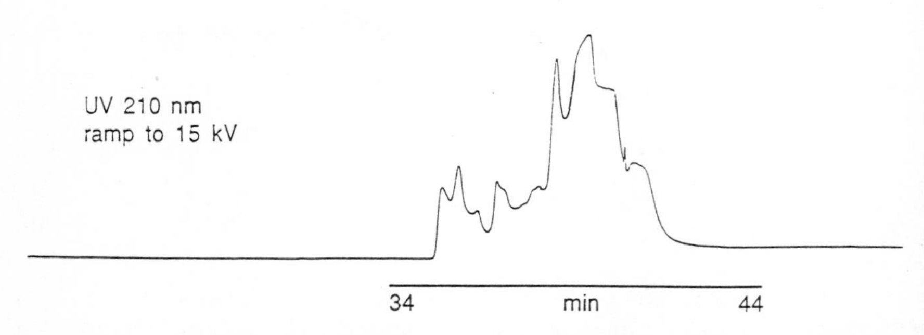

Figure 6. CITP separation of a trypsin enzymatic digest of cytochrome c.

We have previously demonstrated the feasibility of the CITP separation of quaternary phosphonium and ammonium salts, amino acids, and catcholamines[16] and reference peptides[23] with detection by ESI-MS. Detection limits of approximately 10^{-11} M have been demonstrated for quaternary phosphonium salts and substantial improvements appears feasible. Sample sizes which can be addressed by CITP are much greater (> 100 fold) than CZE. Samples eluting in CITP are often flat-topped bands (the length of the analyte band provides information regarding analyte concentration), well suited to MS as the scan speed need not be challenged by the dynamic nature of "sharp" peaks. Most important, however, is that CITP provides a relatively pure analyte band to the ESI source, without the large concentration of supporting electrolyte demanded by CZE. Thus, CITP/MS has the potential of allowing much greater sensitivities in terms of analyte sample concentration.

For our initial attempt at CITP-MS of enzymatic digests we selected glucagon, a polypeptide of molecular weight (MW) 3483. An enzymatic (trytic) digestion of glucagon would be expected to produce four trytic fragments. Figure 7 shows the total ionization current

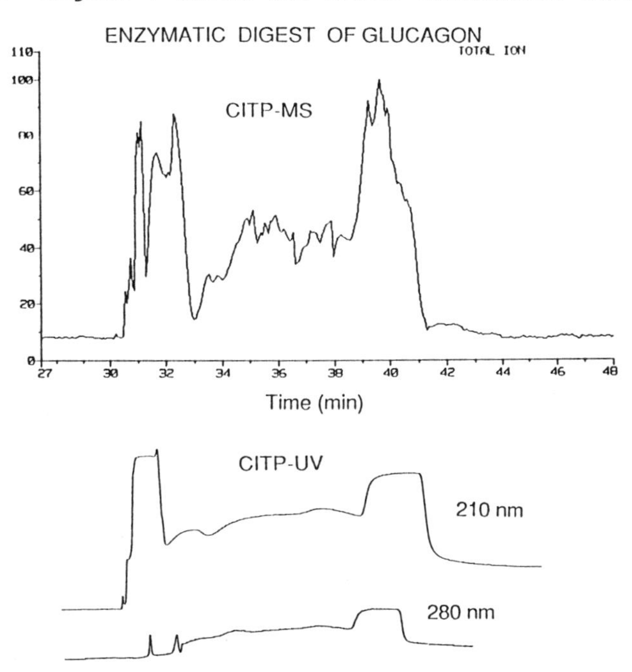

Figure 7. Electrospray total ion current isotachopherogram and corresponding UV absorbance (210 and 280 nm) traces for the CITP/MS analysis of a tryptic digest of glucagon. 2 m x 100 micron I.D. capillary column, leading buffer 0.01 M NH_4OAc, pH 4.9; sample concentration 0.001 M, trailing buffer 0.01 M acetic acid, pH 3.3.

and UV absorbance (210 and 280 nm) isotachopherograms for the CITP/ESI-MS analysis of a mixture of peptides derived from the tryptic digestion of glucagon. The time scale shown on the axis refers to the beginning of the separation. Separation development in CITP may be accomplished before detection starts and, as shown here, the actual time for CITP/MS analysis is determined from the time of application of a "hydrostatic head" to the trailing electrolyte reservoir to cause elution of the focused bands (here at 27 minutes). Strong absorbance at 280 nm indicates the presence of tryptophan in the final band eluting. The total reconstructed ion current for MS detection follows the general features of the UV traces, with better characterized bands apparent at the beginning and the end of the separation. Other differences arise due to the more efficient detection of some components (e.g., arginine) by the mass spectrometer. Figure 8 further gives selected ion isotachopherograms for the

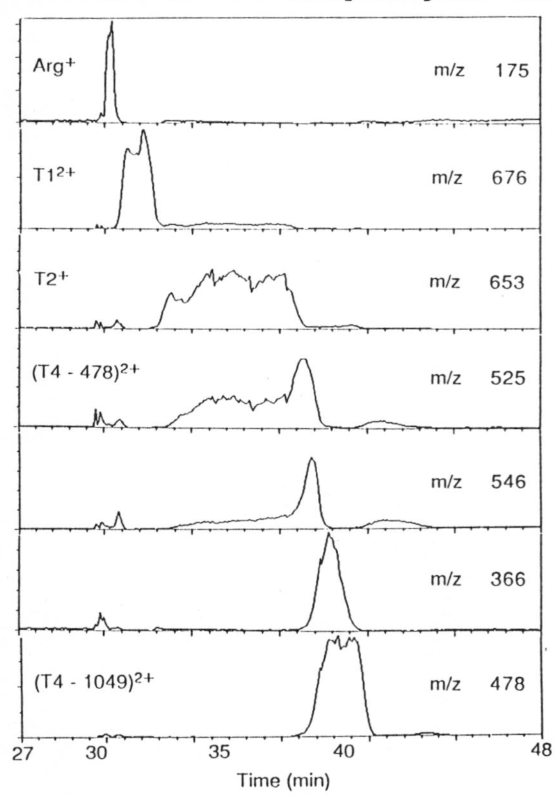

Figure 8. Selected electrospray ion current isotachopherograms for the analysis of an enzymatic digest with trypsin (and contaminating chymotrypsin) of glucagon.

tryptic products arginine (Arg^+, *m/z* 175), $T1^{2+}$ (*m/z* 676) and $T2^+$ (*m/z* 653) for this analysis. Also detected are additional peptide fragments which arise from the action of contaminating chymotrypsin on the predicted tryptic fragments [$(T4-478)^{2+}$ and $(T4-1049)^+$] and on intact glucagon (at *m/z* 546 and 366). The mass spectrum for the T1 tryptic fragment of glucagon is given in Figure 9. The ESI mass spectrum of the T1 fragment contains intense 3+ and 2+ peaks, while the singly charged molecular ion is observed to have much lower intensity.

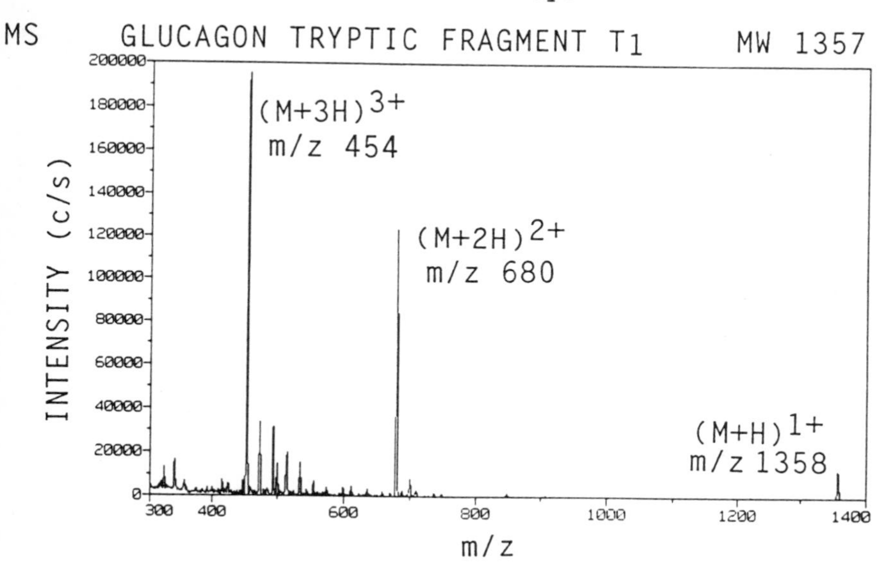

Figure 9. ESI-mass spectrum of the T1 fragment of glucagon (see Figure 8).

An important aim of our CZE-MS and CITP-MS efforts is to develop methods that will yield primary structural information (i.e., sequence) for polypeptides and small proteins. The ESI-MS method affords unique opportunities in this regard since ionization efficiencies are high and good sensitivity can be obtained even for large proteins.[15] The fact that ESI mass spectra generally consist of only intact multiply charged molecular ions is sometimes cited as a disadvantage of this method, since it is claimed that structural information cannot be obtained. However, as we have shown recently, effective dissociation of molecular ions can be induced in the nozzle-skimmer region of the ESI interface.[20,23] A more powerful approach is to apply tandem mass spectrometry of the CID of molecular ions from several of the major charge states.[20,24]

UNIVERSITY OF GREENWICH LIBRARY

The relatively long and stable period of elution of separated bands in CITP facilitates MS/MS experiments requiring longer integration, allowing signal averaging or providing more concentrated samples than provided by CZE. Such an analysis is illustrated in Figure 10 for the various charge states (1+ to 3+) for the T1 fragment of glucagon. While our current data system prevents obtaining MS/MS spectra of more than one charge state from a single separation, future modifications will allow more flexible methods for selecting multiple peaks for MS/MS study.

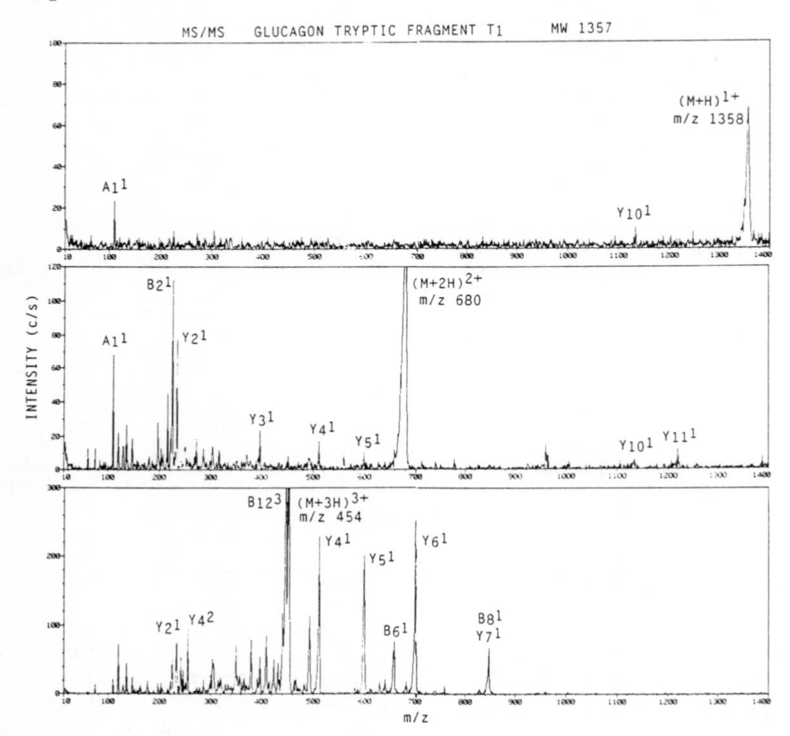

Figure 10. Daughter ion tandem (e.g., MS/MS) spectra for the 1+ to 3+ charge states of the glucagon T1 tryptic fragment.

It is also feasible to use the CID process to directly study intact large molecules. For unknown materials both the absence of product ion charge state information in the CID spectrum and the practical limitations upon mass spectrometer resolution make interpretation of such spectra extremely difficult. However, the fact that significant CID fragment intensities are observed has prompted us to examine the MS/MS spectra of proteins with closely related primary structures.

Figure 11 gives CID mass spectra obtained for the $(M+15H)^{15+}$ charge state of cytochrome c from nine different species. The spectra show the m/z 600 - 1000 region where most product ions are observed.[21] The mass spectra were obtained at collision energies of ~1500 eV laboratory frame and an argon collision gas thickness of 5 x 10^{13} molecules cm^{-2}. These spectra are signal averaged from five scans obtained over a twenty minute period, and were reproducible in terms of relative product ion intensities to better than ±20% on a day-to-day basis. The molecular weights for the various species span a moderate range (M_r ~12,040 to 12,700), and can also be quite readily differentiated by careful M_r measurement. However, of greater significance is the observation of both the strong similarities and the important differences among the spectra. In particular the bovine, rabbit and dog proteins are highly similar (differing by less than four amino acid residues in the sequence) and show the expected similarities in CID products. However, significant differences are also observed. For example, while both the bovine and rabbit cytochrome c show products at m/z 778 ±1, such a peak is absent for the dog cytochrome c which instead shows a distinctive fragment at m/z 766 ±1. The prominent m/z ~904 peak for bovine cytochrome c is absent for rabbit cytochrome c. As expected, cytochrome c proteins with greater differences in sequence show more substantial differences as expected.

These results clearly suggest the potential of tandem mass spectrometry for "fingerprinting" of proteins, even in the absence of the ability to obtain and assign sequence information. Although nanomole quantities of protein were utilized in the present studies, greatly improved sensitivity should be obtainable. An even more attractive approach to such "fingerprinting" applications involves CID in the atmosphere/vacuum interface region. We have recently shown that such CID processes are efficient and conversion to detected products is essentially quantitative. This spectra can be obtained with a few picomoles of material, orders of magnitude less than required for our conventional MS/MS methods. These methods have recently be extended to albumin species of over 66 kDa.[18]

4 CONCLUSIONS

Qualitative investigations of complex systems generally proceed in combined studies with other techniques, e.g., nuclear magnetic resonance, crystallography, etc. Of these, mass spectrometry is frequently the initial experiment by virtue of its relatively high sensitivity. The mass spectrometric study is generally guided by an array of ancillary information which is considered in the course of data analysis--seldom are truly "unknown" proteins considered. Capillary electrophoresis in

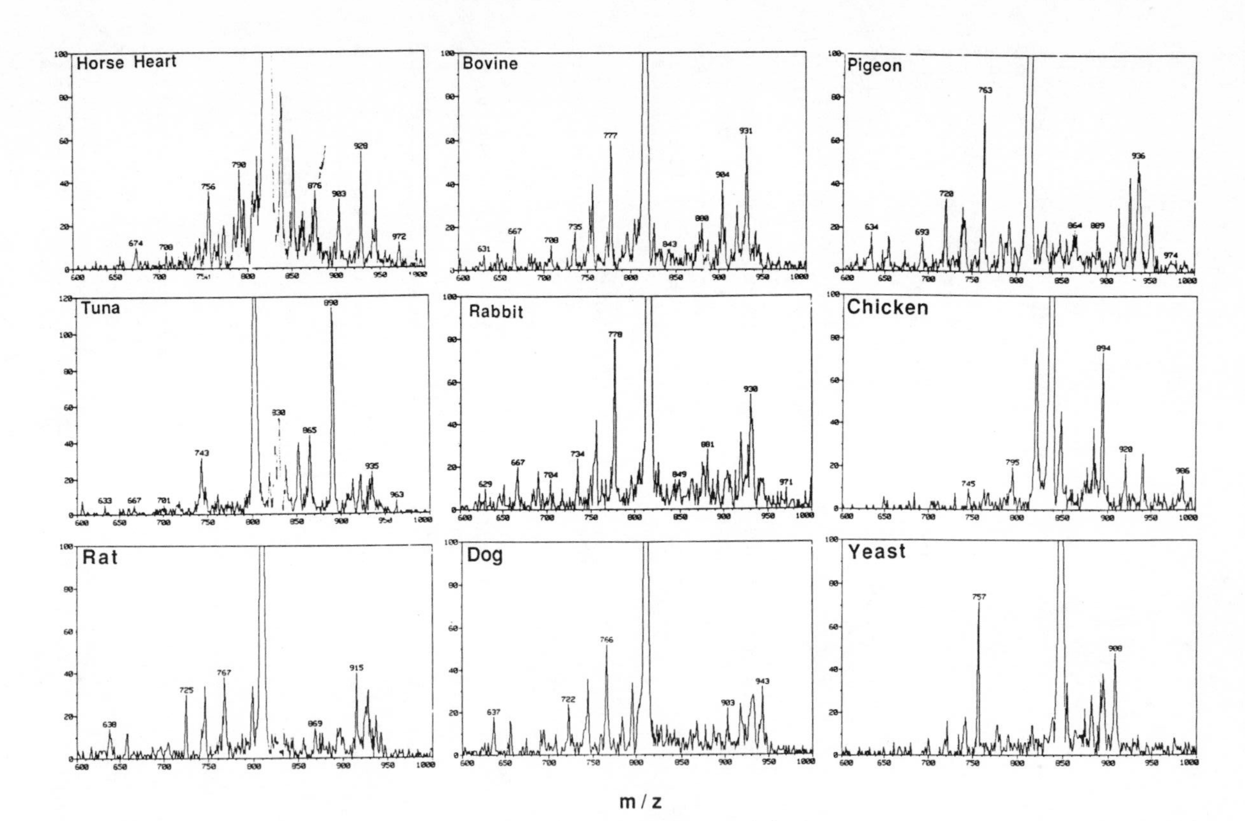

Figure 11. Collision-activated dissociation "fingerprints" obtained for the $(M + 15H)^{15+}$ molecular ion (indicated by *) of cytochrome c from nine different species which show distinctive, but sometimes subtle differences.

combination with mass spectrometry is making increasing contributions to the study of biopolymers. The concurrent developments in both mass spectrometry and capillary electrophoresis are providing complementary tools for manipulation and separation of very small sample volumes. The capability of sampling sub-picoliter volumes from single cells for CE separations is being developed in several laboratories. The mass spectrometric measurement, the determination of mass via the measurement of the ratio of mass to charge, is qualitatively very important, but (except for very small species) insufficient for full characterization. MS/MS studies of large molecules have been shown to yield sequence-related information, but interpretation is complicated due to uncertainties in product charge states.[18,24]

The continued developments leading to improved mass spectrometer performance, and particularly tandem MS, may provide greatly enhanced structural information from such small samples. The recent developments with tandem MS suggest that effective fingerprinting, and perhaps even partial sequence information, may be obtainable for proteins at the attomole level.[18,25]

ACKNOWLEDGMENTS

We acknowledge support of the U. S. Department of Energy, Office of Health and Environmental Research (Contract DE-AC06-76RLO 1830) and the National Science Foundation Instrumentation and Instrument Development Program. Pacific Northwest Laboratory is operated by Battelle Memorial Institute.

REFERENCES

1. K. A. Cobb, J. Liu and M. Novotny, Presented at Pittsburgh Conference on Analytical Chemistry and Applied Spectroscopy, March 6-10, 1989, Atlanta, GA, Abstract No. 1422.
2. (a) S. W. Compton and R. G. Brownlee, BioTechniques, 1988, 6, 432; (b) M. J. Gordon, X. Huang, S. L. Pentoney, Jr., and R. N. Zare, Science, 1988, 242, 224.
3. A. S. Cohen, D. Najarian, J. A. Smith and B. L. Karger, J. Chromatogr., 1988, 458, 323.
4. (a) F. M. Everaerts, J.L. Beckers and Th. P.E.M. Verheggen, 'Isotachophoresis: Theory, Instrumentation and Applications' (Vol. 6, J. Chromatogr. Libr.), Elsevier, Amsterdam, 1976; (b) F. M. Everaerts and Th. P.E.M. Verheggen, 'New Directions in Electrophoretic Methods', J. W. Jorgensen and M. Phillips, American Chemical Society, Washington, D.C., 1987, p.199.
5. (a) S. Hjertén and M.-D. Zhu, J. Chromatogr., 1985, 346, 265; (b) S. Hjertén, K. Elenbring, F. Kilár, and J.-L. Liao, J. Chromatogr., 1987, 403, 47.

6. (a) S. Terabe, K. Otsuka, I. Ichikawa, A. Tsuchiya and T. Ando, Anal. Chem., 1984, 56, 111; (b) S. Terabe, K. Otsuka, and T. Ando, Anal. Chem., 1985, 57, 834.
7. A. S. Cohen, S. Terabe, J. A. Smith and B. L. Karger, Proc. Natl. Acad. Sci. U.S.A., in press.
8. R. D. Minard, D. Chin-Fatt, P. Curry Jr. and A. G. Ewing,.Presented at the 36th ASMS Conference on Mass Spectrom. and Allied Topics, June 5-10, 1988, San Francisco, CA, p. 950.
9. M. A. Moseley, L. J. Detering, K. B. Tomer and J. W. Jorgenson, J. Chromatogr., 1989, 480, 187.
10. R. M. Caprioli, W. T. Moore, M. Martin, B. Dague, K. J. Wilson and S. E. Moring, J. Chromatogr., 1989, 480, 247.
11. J. A. Olivares, N. T. Nguyen, C. R. Yonker and R. D. Smith, Anal. Chem., 1987, 59, 1230.
12. R. D. Smith, J. A. Olivares, N. T. Nguyen, and H. R. Udseth, Anal. Chem., 1988, 60, 436.
13. R. D. Smith, C. J. Barinaga and H. R. Udseth, Anal. Chem., 1988, 60, 1948.
14. E. D. Lee, W. Mück, J. D. Henion and T. R. Covey, J. Chromatogr., 1988, 458, 313.
15. J. A. Loo, H. K. Jones, H. R. Udseth and R. D. Smith, J. Microcolumn Sep., 1989, 1, 223.
16. H. R. Udseth, J. A. Loo and R. D. Smith, Anal. chem., 1989, 61, 228.
17. M. K. Meng, M. Mann and J. B. Fenn, Z. Phys. D - Atoms, Molecules and Clusters, 1988, 10, 361.
18. R. D. Smith, J. A. Loo, C. G. Edmonds, C. J. Barinaga and H. R. Udseth, Anal. Chem. 1990, 62, 882.
19. J. A. Loo, H. R. Udseth and R. D. Smith, Anal. Chem., 1989, 179, 404.
20. J. A. Loo, C. G. Edmonds, R. D. Smith, M. P. Lacey and T. Keough, Biomed. Environ. Mass Spectrom., 1990, 19, 286.
21. E. D. Lee, W. Mück, J. D. Henion and T. R. Covey, Biomed. Environ. Mass Spectrom., 1989, 18, 844.
22. V. Dolnik, M. Deml. and P. Bocek, 'Analytical and Preparative Isotachophoresis', C. J. Holloway, Ed., Walter de Gruyter:Berlin, 1984, p. 55.
23. R. D. Smith, J. A. Loo, C. J. Barinaga, C. G. Edmonds, and H. R. Udseth, J. Chromatogr., 1989, 480, 211.
24. R. D. Smith, J. A. Loo, C. J. Barinaga, C. G. Edmonds and H. R. Udseth, J. Amer. Soc. Mass Spectrom., 1990, 1, 53.
25. J. A. Loo, C. G. Edmonds and R. D. Smith, Science, 1990, 248, 201.

PLASMA ATOMIC EMISSION SPECTROSCOPY FOR ELEMENT SPECIFIC CHROMATOGRAPHIC DETECTION

P. C. Uden

DEPARTMENT OF CHEMISTRY, LEDERLE GRADUATE RESEARCH TOWERS, UNIVERSITY OF MASSACHUSETTS, AMHERST, MA 01003, USA

1 INTRODUCTION

Chromatographic procedures involve transformation of a complex multi-component sample into a time-resolved separated analyte stream, monitored by an analog differential signal. Chromatographic analytes usually change very quickly in nature and with time. Therefore different demands are placed on the analytical 'detection' system than for some other sample inputs.

Selective Chromatographic Detection

Essential to chromatographic instrumentation is a detection device for qualitative and quantitative determination of the resolved components, which responds immediately and predictably to the presence of solute in the mobile phase. An important class of detectors is that giving selective, or specific information for eluates. Spectral property detectors such as the mass spectrometer, the infrared spectrometer and the atomic emission spectrometer fall into this class, showing element selective, structure or functionality selective, or property selective capacities. The chromatographer may consider the spectrometer as a sophisticated chromatographic detector, and the spectroscopist may view the chromatograph as a component-resolving sample introduction device. These observations emphasize the need for optimization of both the separation and the detection process, together with the interface between them.

Element Selective Detection. Here the objective is to obtain qualitative and quantitative information on analyte elemental constitution. Element selective GC detectors, limited to a small number of elements which are in frequent use, are the alkali thermionic detector, for nitrogen and phosphorus; the flame photometric detector for sulfur and phosphorus, and the Hall electrolytic conductivity detector, for halogen, nitrogen and sulfur. In HPLC and SFC, no element specific detectors are in common use. Multi-element chromatographic detection is a worthwhile objective to complement the molecular and structural specific detection given by mass spectroscopy and Fourier transform infra-red spectroscopy. Atomic emission spectroscopy is a natural choice because of its ability to monitor all elements.

Atomic Emission Spectroscopic Detection (AESD)

Atomic absorption (AAS), flame emission (FES), atomic fluorescence (AFS), and atomic plasma emission (APES) have been interfaced for chromatographic detection[1,2]. The latter can give simultaneous multielement measurement, while maintaining a wide dynamic measurement range and good sensitivities and selectivities over background elements. Major plasma emission sources which have been used for GC detection have been the microwave induced plasma, at atmospheric or reduced pressure (MIP), and the DC argon plasma (DCP). The inductively coupled argon plasma (ICP) has been little used for GC, but it and the DCP are valuable as HPLC detectors. Other plasmas have also been evaluated.

The principal advantages of chromatography - atomic plasma emission spectroscopy (C-APES) include; a) monitoring for elemental composition with high sensitivity, b) toleration of non-ideal chromatography, the specificity of plasma emission enabling incomplete chromatographic resolution from complex matrixes to be overcome, c) monitoring for specific molecular functionality by derivative element tagging, d) simultaneous multi-element detection for empirical and molecular formula determination, and e) compatibility with different mobile phases and conditions.

The Microwave Induced Electrical Discharge Plasma (MIP) has been the most widely used atom reservoir plasma system for GC. An argon or helium plasma is sustained in a microwave 'cavity' which serves to focus or couple power from a microwave source, usually operated at 2.45 GHz, into a quartz or other discharge cell. The cavity may be operated at atmospheric or under reduced pressure[3,4]. MIP power levels are usually much lower (ca.50-100W) than for the DCP or the ICP, making their operation simpler. Power densities are similar however due to the smaller size of the MIP. Although MIP temperatures are lower than for some other plasmas, typically in the 3,500 - 4,000°C range, high electron temperatures occur, especially for the helium plasmas, to give intense spectral emission for many elements, including non-metals. A comparison of microwave cavities was made by Risby and Talmi in their general review of GC-MIP[5]. The cavities which have been most widely applied for GC are based on the Beenakker TM_{010} cylindrical resonance cavity[6]. These can sustain argon or helium atmospheric pressure discharges at low power levels and light emitted is viewed axially, an advantage over transverse viewing through the cavity walls whose transmission properties change with time and usage The instrumental advantages of atmospheric pressure operation greatly simplify GC detection.

Another class of atmospheric pressure cavity used successfully in GC is the Surfatron which operates by surface microwave propagation along a plasma column[7]. Surfatrons can sustain a discharge over a wide pressure range.

The Inductively Coupled Plasma (ICP) Discharge[8] is the most widely used analytical spectral emission source. It results from interaction of a radio-frequency field (usually 27 or 41 MHz) on argon gas, flowing through a quartz tube within a copper coil. A stable, intense plasma results at temperatures up to more than 9,000°K. The ICP is a natural complement for liquid chromatography,since it is normally configured for a liquid inlet stream. The Direct-Current Plasma (DCP) is a discharge maintained by a continuous DC arc and stabilized by flowing inert gas[9]. A cathode jet is set above two symmetrically placed anode jets in an inverted 'Y' configuration [10]. Typical power levels are between 500W and 700W, at 40-50V. Solutions are introduced from a nebulizer-spray chamber, or vapor phase samples are directly channelled into the junction of the two columns The DCP has been interfaced with both HPLC and GC, to give an elemental operating range paralleling that of the ICP.

A 60Hz Alternating-Current Helium Plasma (ACP) Discharge has been used for GC[11]. It is a stable, self-seeding emission source, requiring no external initiation and it does not extinguish under high solvent loads. Construction is simple and sensitivity and selectivity for lead and mercury are similar to those obtained for other plasma emission detectors. Capacitively Coupled Atmospheric pressure Plasma (CCP) discharges have been used[12,13] for capillary GC without appreciable band broadening. They can sustain a stable plasma from 10-500W and from 200Kz to 30MHz[14]. High voltage electrodeless discharges in argon ,nitrogen or helium,have been used for GC, utilizing emission from the atmospheric pressure afterglow. The metastable energy carriers in the helium system have the highest collisional energy transfer and the best ability to excite other elements to emission[15]. Low picogram detection limits have been obtained for carbon, halogens, mercury and arsenic.

2 CHROMATOGRAPHIC SAMPLE INTRODUCTION

Since eluent from GC columns is normally at atmospheric pressure, simple interfacing is possible for atmospheric pressure plasmas. For cavities such as the TM_{010}, capillary GC columns can be terminated within a few mm of the plasma, to give minimal 'dead volume'. Heating is needed to prevent analyte condensation along the interface. Helium make-up gas or other reactant gases can be introduced within the transfer line to optimize plasma performance and minimize peak broadening. A sophisticated re-entrant cavity interface is illustrated (Figure 1)[16] in which the plasma is produced in a thin-walled silica discharge tube within a circularly symmetric cavity whose center portion is narrower than the periphery. Water cooling reduces the inside wall temperature from above 1000°C to around 350°C to reduce wall-induced interferences.A threaded tangential flow torch (TFT)[17] has been used to give a self-centering plasma with enhanced emission and better stability, but a large volume (L per minute) of helium flow gas is required. A concentric dual flow torch (CDFT)[18] utilizes two alumina tubes held concentric by a thin wire wrapped around the inner tube, and two plasma gas flows. The gas passing

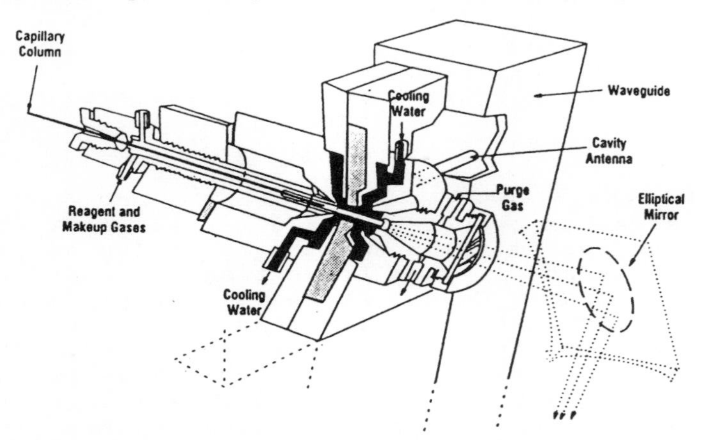

Figure 1.Cutaway view of re-entrant microwave cavity for GC-MIP. (Reproduced by permission from Ref.16, copyright 1990, Dr. Alfred Huethig Publishers, Heidelberg)

between the tubes flows around the wire in a helical path and out along the walls of the outer tube. This arrangement holds the plasma away from the walls and makes it more spatially stable. Another valuable feature is the ability for splitless GC injection of up to 1 µl liquid volumes without extinguishing the plasma.

The microwave induced plasmas (MIP) have found much greater use in GC than in HPLC interfacing although the application of the direct injection nebulizer (DIN) for microbore column flow rates may expand HPLC applications[19].

The HPLC interface for ICP and DCP is simpler than for GC since a heated transfer line is not needed, but post-column peak broadening must be reduced by minimizing the interface tube length and volume[20,21]. The major sensitivity limit is the poor (1% or less) transfer efficiency of eluent into the plasma excitation region due to ineffective nebulization and desolvation. A better approach for HPLC-plasma interfacing uses the DIN[19] to transfer mobile phase flows as high as 0.5 ml/min into the plasma without appreciable peak broadening and high efficiency.

Supercritical fluid chromatographic (SFC) interfacing to plasmas in general is intermediate in complexity between GC and HPLC requirements. For capillary SFC mobile phase flow rates in particular, analyte transport may be quite efficient and precision of measurement is good[22].

3 ANALYTICAL INFORMATION FROM C-AED

Interelement Selectivity

This refers to detection of the desired element without interference or contribution from signals of other elements present simultaneously in the eluate. It depends on emission properties of the element, interferences, and ón resolution and general performance of the spectroscopic system. Inter-element selectivity, is often defined as the peak area response per mole of analyte element divided by the peak area response of the 'background' element per mole of that element. Good selectivity against carbon is most desirable, but performance against other background elements is also important. Selectivities vary greatly among elements, between plasmas and with instrumental conditions so calibration is always necessary. Carbon selective detection provides 'universal' detection for organic compounds.

Elemental Sensitivity and Limits of Detection

Sensitivity depends on the emission intensity at the measured wavelength. Each element emits at various wavelengths and the optimum must be chosen, considering both sensitivity and selectivity. Detection limits for different elements differ by two or three orders of magnitude, and this will affect inter-element selectivity when spectral overlap is present.

Linear Dynamic Measurement Range

Linear dynamic ranges in capillary GC-AED extend from the sample capacity limit of the column (ca.100 ng) to the detection limit of the target element (1 - 100 pg). In HPLC-AED, the upper limit may be higher since more sample can be accomodated, but the lower limit is also raised because of poor transfer of analyte peaks to the plasma and other samples losses. Chemical, gas dopant and plasma-wall interaction effects modify the limits.

Table 1 shows some elemental detection limits, selectivities and linear dynamic ranges for GC-MIP.

4 GAS CHROMATOGRAPHIC APPLICATIONS

GC-AED of Non-metallic Elements - GC-MIP Detection

The helium microwave induced plasmas (MIP) have been the most used for non-metals detection, since for many of these elements, argon metastable energy carriers show insufficient collision energy transfer for adequate excitation.

Reduced Pressure GC-MIP. McCormack et al.[25] and Bache and Lisk[26] first reported practical reduced pressure argon GC-MIP for P, S, F, Cl Br, I, and C, with detection limits between 10^{-7} and 10^{-12} g/s, but selectivities against carbon were generally only between 10 and 100. McLean et al. developed a tunable detection system with detection limits in the 0.03 - 0.09 ng/s range for C, H, D, F, Cl, Br, I and S, and selectivities from 400 to 2300[4]. As and Sb were determined in environmental samples by derivatization at detection limits of 20 and 50 pg[27] A commercial multi-channel instrument using a polychromator and Rowland Circle optics was evaluated by Brenner[28]. For operation at 0.5 - 3 torr, with various scavenge gases, detection limits were from 0.02 ng/s for Br to 4 ng/s for O. Linearity was 3-4 decades and carbon selectivities were 500-1000. Hagen et al.[29] used elemental derivatization

TABLE 1 Selected Detection Limits and Selectivities for Atmospheric Pressure Helium Microwave Plasma GC Detection. (Compiled from refs. 23(a), and 24(b)).

Element	λ (nm)	Det. Limit pg/s (pg)	Sel. vs. C	Dyn. Range
C(b)	247.9	2.7 (12)	1	1,000
H(b)	656.3	7.5 (22)	160	500
D(b)	656.1	7.4 (20)	194	500
B(b)	249.8	3.6 (27)	9,300	500
Cl(a)	479.5	40	3,000	10,000
Br(a)	478.6	60	2,000	1,000
F(a)	685.6	50	20,000	2,000
S(a)	180.7	2	8,000	10,000
P(a)	177.5	1	5,000	1,000
Si(b)	251.6	9.3 (18)	1,600	500
O(a)	777.2	120	10,000	5,000
N(a)	174.2	50	2,000	20,000
Ge(b)	265.1	1.3 (3.9)	7,600	1,000
Sn(b)	284.0	1.6 (6.1)	36,000	1,000
As(b)	228.8	6.5 (155)	47,000	500
Fe(b)	259.9	0.3 (0.9)	280,000	1,000
Pb(b)	283.3	0.17 (0.71)	25,000	1,000
Hg(b)	253.7	0.6 (60)	77,000	1,000

taggants such as chlorofluoroacetic anhydride to permit Cl and F specific detection of acylated amines, while Zeng et al.[30] have optimized oxygen specific detection, an area certain to grow more important with the increasing use of oxygenates in fuel oils.

Sklarew et al.[31] noted that although the low pressure plasma gives optimal excitation energy for non-metals, its sensitivity is poorer than that of Beenakker-type cavities because of its transverse viewing geometry. Olsen et al.[32] compared reduced pressure and atmospheric pressure MIP plasmas for Hg, Se and As detection in shale oil matrices, and found the latter to be superior both in detection limits and in selectivities.

Atmospheric Pressure GC-MIP. Increased efficiency of transfer of microwave power to the discharge using such cavity structures as the Beenakker TM_{010}, allows plasmas to be sustained at atmospheric pressure. Another advantage is that light emitted from the plasma can be viewed axially, rather than transversely through the discharge tube wall,thus avoiding variable response upon extended tube use.

Initial use of the TM_{010} for packed column GC required solvent venting to avoid extinguishing the plasma[33]. A major application was to halogen specific GC detection of purgeable haloorganics in drinking water[34]. Although the MIP is less sensitive than electron capture detection for polyhalogenated compounds, it responds much more uniformly to the level of each halogen independently of analyte molecular structure. Sub-ppb detection was easily achieved by extraction and 'purge and trap' techniques. A clear advantage is specificity for individual halogens. Chiba et al. considered some problems of inconsistent element responses for different molecular structures, and performed pre-pyrolysis between chromatograph and cavity to give more reproducible element ratios[35].

Flexible fused silica capillary (FSOT) GC columns can be terminated within a few mm of the plasma, and they have been widely used in capillary GC-MIP. Detection limits and selectivities from such systems are listed in Table 1 and widespread adoption of the technique for more routine use is now possible with the introduction of a commercial instrumental system[36]. A rapid-scanning spectrometer has been applied for multi-element detection, elemental response per mole for C, Cl and Br being independent of molecular structure despite the low power (50-60 W) used[37]. Near infra-red (NIR) atomic emission has been evaluated for GC-MIP, a cooled TM_{010} cavity and TFT being used at 370 watts with a Fourier transform NIR spectrometer[38].

Oxygen-selective detection has been reported by Bradley and Carnahan[39] with a TM_{010} cavity in a polychromator system. Background oxygen spectral emission from plasma gas impurities, leaks or back-diffusion into the plasma was minimized to give sensitivities between 2 and 500 ppm in different complex petroleum distillates. The selective detection of phenols in a light coal liquid distillate is seen in Figure 2, in which 2a shows carbon emission, 2b oxygen emission and 2c flame ionization detection.for a phenolics concentrate of the same distillate. Peak A is phenol, peak B o-chlorophenol, peak C o-chlorocresol, peak D m- and p-cresols and peak E C_2-phenols

An example of multiple element detection with the diode array detection system described earlier[16] is shown in Figure 3, carbon, chlorine, sulfur and nitrogen detection being obtained in a chromatogram of a chemical waste dump sediment extract.

TM_{010} cavities have been used for a number of

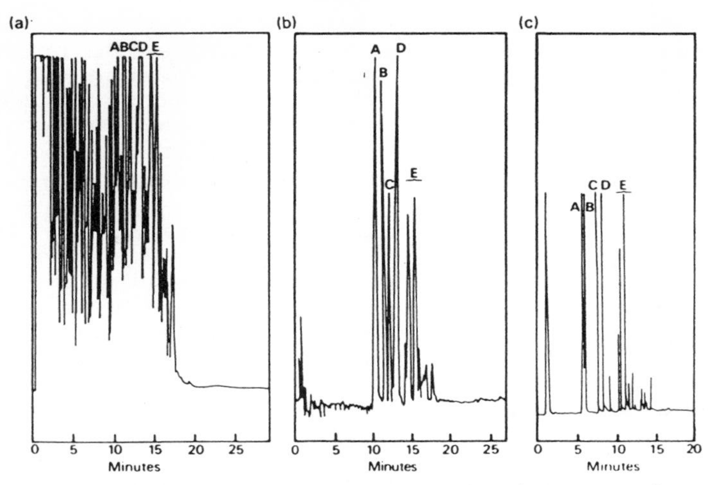

Figure 2. Gas chromatograms of light coal distillates: a) C emission, b) O emission and c) FID trace of phenolics concentrate. Reproduced by permission from Ref.39.

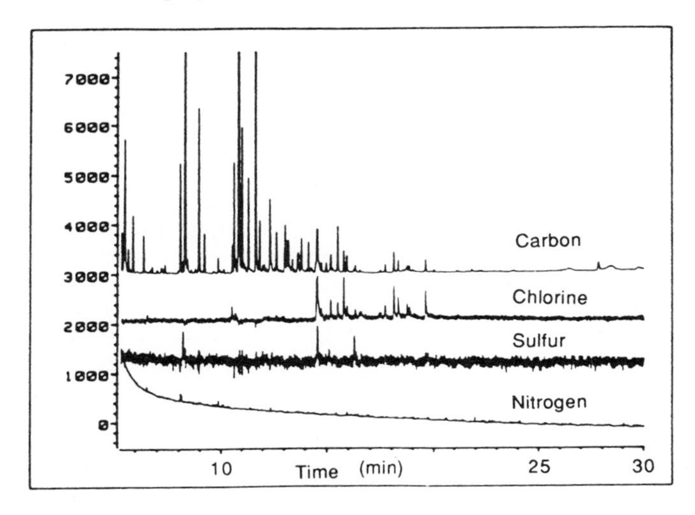

Figure 3. GC-AED analysis of a chemical waste dump sediment extract. Reproduced by permission from Ref. 16.

detailed capillary GC-MIP elemental investigations, including boron specific detection for diol boronate esters and other compounds[40]. The Surfatron-MIP has proved to be of value in determination of P, S, Cl and Br in pesticides, with detection limits from 3 - 60 pg/s[41].

GC-DCP Detection Argon DCP has been used to detect non-metallic elements such as boron and silicon; for the latter element the absence of interfering spectral response from the quartz discharge tube often used in the MIP is an added advantage. A selectivity for silicon over carbon of 2×10^6 with a detection limit of 25 pg/s was reported[10].

GC-AED Detection of Metallic Elements

Many volatile organometallic and metal chelate compounds can be effectively gas chromatographed [42], and GC-AED detection is valuable in confirming elution and acquiring sensitive analytical data.

GC-MIP Detection As shown in Table 1, GC-MIP data has been obtained for many metals, and some such as lead, and mercury have been the subject of a number of studies. Each of these elements is determinable with TM_{010} cavities to sub-pg/second detection limits. In their comparison of reduced and atmospheric pressure MIP systems, Olsen et al. found for the latter system a one pg detection limit for mercury, with selectivity over carbon of 10,000[32]. GC-MIP of volatile elemental hydrides of germanium, selenium and tin gave sub-ng detection[43] and there is considerable potential for the determination of these elements in environmental matrixes.

The study of metal chelates of sufficient volatility and thermal stability for GC has received much attention, most applications being for complexing ligands of 2,4-pentanedione (acetylacetone) and its analogs[42]. Among examples of application have been GC-MIP analysis for chromium as its trifluoracetylacetonate in blood plasma, with excellent quantitation and precision[44]. Trace determinations of beryllium, copper and aluminum have been reported and ligand redistribution and reaction kinetics of gallium, indium and aluminum chelates have been followed[45]. Figure 4 illustrates the element specific detection of copper, nickel, palladium and vanadyl chelates of N,N'-propylene-bis(trifluoroacetylacetoneimine) (H_2pnTFA_2) which forms stable complexes with divalent transition metals.The peak shape of the vanadyl species results from on-column interconversion of different isomeric forms.

Some π-bonded organometallics such as metallocene derivatives are well-behaved in capillary GC; a TM_{010} cavity gave excellent detection of iron, cobalt, nickel, chromium, manganese and vanadium compounds, confirming elution of some previously

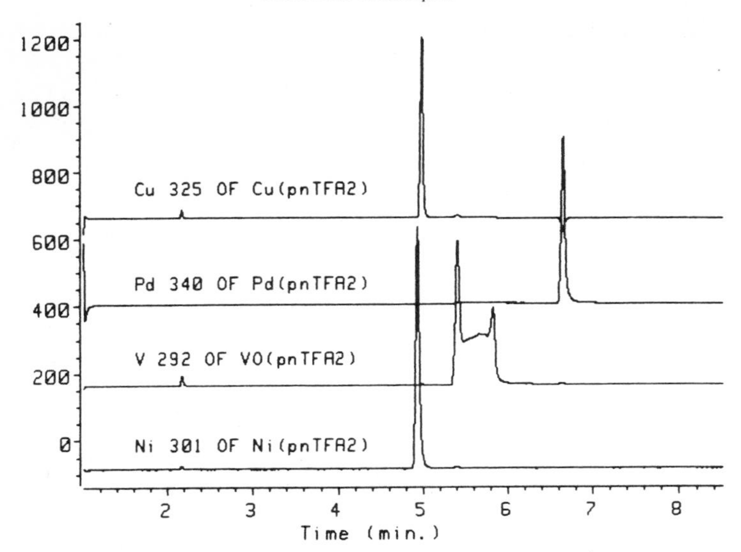

Figure 4. GC-AED of copper, palladium, vanadium and nickel chelates. Column 6 m x 0.25 mm DB 5 Temperature program 150-210°C at 10°C/min.

unchromatographed compounds[46]. In Figure 5 is illustrated the high selectivity of the GC-AED system over carbon. The compounds cymantrene (cyclopentadienylmanganese tricarbonyl), and MMT (methylcyclopentadienylmanganesetricarbonyl), typical organomanganese petrol additives, were detected selectively at the 10 ppm level.

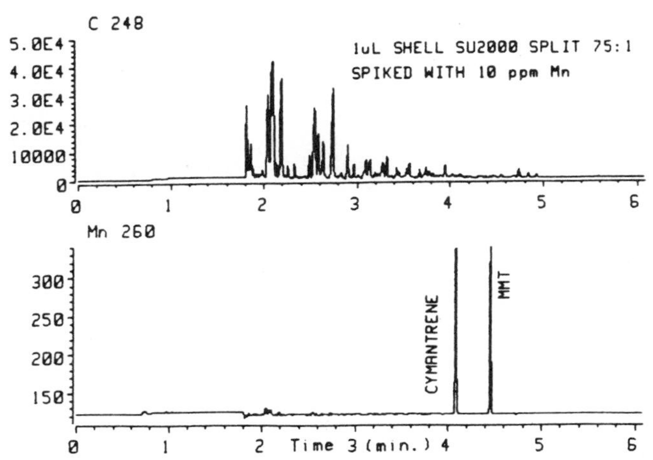

Figure 5. GC-AED of the organomanganese compounds. cymantrene and MMT, spiked into gasoline at 10 ppm. Column 25 m x 0.32 m HP1. Temperature Program from 70°C at 20°C/min.

GC-DCP Detection. GC-DCP was also used for the determination of MMT, direct injection of 5 μL samples of gasoline allowing determination at a few ppm with a precision of 0.8-3.4% rsd[47]. Redistribution reactions of silicon, germanium, tin and lead compounds were examined[48]. Detection limits (pg/s) and carbon selectivities using a three-electrode plasma jet were Cr (4, $4x10^8$), Sn (60, $2.5x10^6$), Pb (100, $5x10^5$), and B (3, $3x10^5$)[10]. The versatility of GC-DCP has been emphasized in a simple system dedicated to the specific determination of methylmercury compounds in fish[49].

5 LIQUID CHROMATOGRAPHIC APPLICATIONS

Most development in HPLC-AED detection has been with the ICP and to some extent with the DCP. Metal specific detection is predominant and will probably remain so until interface systems to remove HPLC mobile phases, preferentially while transfering eluate peaks to a plasma optimized for non-metals, can be devised. Such an interface may incorporate a moving band eluate transport device, or it may be based on thermospray or particle beam technology. The major problem in HPLC-plasma interfacing is poor plasma compatibility with analytical flow rates of mobile phases. All specific element atomic spectroscopic detectors except for graphite furnace atomic absorption , employ on-line nebulization and excitation of small volumes (5 - 200 μL) of liquid which are converted into an aerosol and introduced into the atomization-excitation cell. The major reason for relatively poor detection limits is the relatively ineffective conversion of effluent flow into aerosol and its transport to the plasma; typically only 1 - 5% of the sample reaches the plasma. Poor tolerance of the plasma to common solvents used in HPLC, particularly in reverse phase ion pairing and size exclusion chromatography, is typical. Solution to these problems necessitates more quantitative nebulization, atomization and excitation of HPLC samples as well as better transport systems.

HPLC-ICP Detection Numerous applications of HPLC-ICP have appeared since 1979[2], but detection limits for many elements in real samples at levels of environmental significance has only been moderate. Current developments suggest that substantial enhancement in working sensitivities is feasible.

MDLs for aqueous mobile phases, whose characteristics are familiar from standard sampling procedures, are usually two or more orders of magnitude worse than in continuous flow ICP-AES. Moreover normal phase HPLC using organic solvents such as hexane or methyl isobutyl ketone presents further problems, since ICP behavior is less well defined and spectral background interference is greater. One approach has been to use microbore HPLC and lower mobile phase flow rates. For samples of copper and zinc diketonates and dithiocarbamates, peak broadening was minimised by optimal design of the interface, connecting tubing, nebulizer, spray chamber and plasma torch to give virtually constant peak width ratios for ICP and UV detection[50]. At above 15 mL/min flow rates, sensitivity was independent of flow rate and the ICP operated as a mass flow sensitive detector; below that flow rate however, it acts as a concentration sensitive detector.

HPLC-ICP has been effective for metalloids; a 130 ng/mL detection limit for arsenic in organo-arsenic acids being obtained for 100 µL samples[51]. The ICP spectrometer utilized 48 channels operating at 1.2 kW, allowing sampled chromatograms for As, Se and P to be displayed on-line.

Arsenic and cadmium compounds showed detection limits for 50µL injections at 3.1 ng/mL for As as arsenite and 0.12 ng/mL for Cd as the nitrilo-triacetate[52]. Reverse phase determination of organo phosphorus and sulfur anions has also been carried out[53]. Size Exclusion (SEC) - ICP has been applied to molecular size distribution of sulfur, vanadium and nickel compounds in petroleum crudes and residua[21]. Ferritin, an iron-containing protein which exists in various forms, has been analyzed using aqueous SEC good repeatability being found at ng levels[54]. To overcome transfer problems of HPLC eluate to the ICP, a useful device is the Direct Injection Nebulizer (DIN)[19], a total injection microconcentric nebulizer which can achieve almost 100% nebulization and transport efficiency. Detection limits from 164 ng/mL for sulfur to 4 ng/mL for zinc were reported.

<u>HPLC-ICP-Mass Spectrometry</u>. Among the most sophisticated developments in HPLC-ICP is that of interfaced HPLC-ICP-Mass Spectrometry[55] for which detection limits of 500 pg/peak or below have been found for many elements. Ion exchange and ion pair chromatography were used for triorganotin species[56], and arsenic speciation has been the subject of a number of studies also using ion exchange[57,58]. This technique shows much potential in biomedical and clinical studies where analyte levels are usually

below the capabilities of ICP emission detection. Thus the technique has been used to monitor patient treatment with gold-based drugs used to treat rheumatoid arthritis[59].Sensitivity to gold containing drugs in blood or urine at levels down to 5 ppb was achieved.

HPLC-DCP Detection. The good tolerance of the DCP to a wide range of solvents has enhanced its value as an LC detector. The first procedure described for HPLC-DCP used standard nebulization for reverse phase chromatography, but an impact device proved better for normal phase hydrocarbon and halocarbon eluents. Metal chelates were determined with mass flow detection limits of 0.3 ng/s for copper and 1.25 ng/s for chromium[20]. Applications for the inorganic anions sulfate, nitrate and acetate as their cadmium salts were reported, but with minimum detectable levels only in the 100 ppm range[60]. The determination of Cr(III) and Cr(VI) (as chromate) by reverse phase ion pairing gave detection in the 5-15 ppb range[61]. Applications included biological samples from ocean floor drillings, chemical dump site, surface well water and waste water samples.

HPLC-MIP Detection. The low powered MIP cannot be directly interfaced to conventional HPLC columns, since the discharge is quenched by mL/min liquid flow streams Some approaches to this problem have been explored. The only direct introduction of liquid into an MIP involved flowing the effluent over a heated wire and vaporizing it by a cross-stream of helium into the discharge. This system shows some potential for reversed phase separations[62]. A mixed gas oxygen-argon MIP was applied successfully for HPLC of mercury compounds. Methanol/ water mixtures with up to 90% of the former solvent being tolerated; detection limits for organically-bound Hg were in the ng range, but response was found to be dependent upon molecular structure[63]. A high-power (kW) discharge, operating in the radio frequency or microwave range, accomodated HPLC solvent-flows provided that nebulization was adequate[64]. A moving-wheel sample transport-desolvation system was developed in which aqueous solvent is evaporated by a flow of hot nitrogen, leaving dry analyte to be transported into the plasma, where it is volatilized, atomized and excited. The plasma used was a small volume helium MIP, operated at 100W and detection limits were in the range 0.4 -20 mg of halogen[65]. The top and side views of the LC-MIP interface are shown in Figure 6. 1 is a friction wheel drive, 2 a guided wheel gearing, 3 the interface chamber housing, 4 the

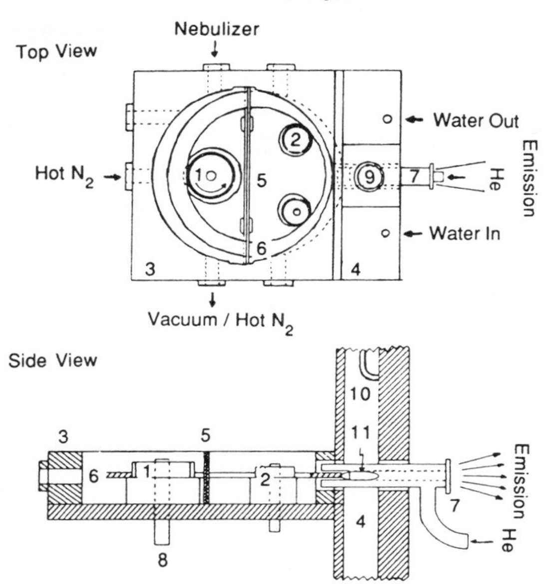

Figure 6. LC-MIP Interface. ((Reproduced by permission from Ref 65.)

TM_{010} resonator, 5 the separation plate, 6 a stainless steel wheel, 7 the fused silica plasma torch with face plate, 8 the driver wheel shaft, 9 an N connector, 10 the coupling loop and 11 the plasma region.

The direct injection nebulizer may prove useful for HPLC-MIP Alternatively the removal of the solvent may be possible with the 'Thermospray' and 'Particle-beam' approaches now used in HPLC-MS; the interfacing problems for lower-powered MIPs are parallel to those in HPLC-MS and HPLC-FTIR. Another possibility is cryo-focusing as also used in HPLC-FTIR. Investigations are also underway on new plasma cavities which may be able to sustain the helium MIP under conventional HPLC flow conditions. Capillary HPLC columns with mobile phase flow rates of a few ml/min provide an interesting possibility for helium MIP interfacing, but sample capacity is a limiting factor.

<u>HPLC-MIP-MS</u>. Heitkemper and co-workers have coupled HPLC to MIP-MS for the determination of halogenated organic compounds[66]. Bromine and iodine were determined at low pg levels but chlorine determination was hindered by large backgrounds.

6 PLASMA EMISSION DETECTION FOR SFC

Analytical SFC has become viable in recent years through the availability of high resolution packed and capillary SFC columns and instrumentation. High resolution SFC along with supercritical extraction (SFE), promises to allow separations in areas where neither GC or HPLC are possible. Adoption of detectors for SFC has proceeded in two distinct directions; where methodology has derived from GC, the flame ionization detector has been favored. For development related to HPLC, the UV/-visible spectrophotometric detector has been adopted. Plasma emission is a natural choice because of its use in GC and HPLC. An initial report described an ICP interface with close to 100% atomization efficiency[67], and a practical interface has been devised for SFC-ICP with application to organosilicon compounds[68].

A surfatron MIP sustained in helium was used for SFC, giving sulfur-specific detection at 921.3 nm with a 25 pg/s limit for thiophene[22]. Modification of plasma excitation by SFC solvents appears to be less troublesome than for typical organic HPLC solvents, and it seems likely that as SFC becomes more widely adopted, element specific detection by atomic plasma emission will become a useful option.

ACKNOWLEDGEMENT

Thanks are due to the Hewlett Packard Company for interest in and support of this work.

REFERENCES

1. L. Ebdon, S. Hill and R.W. Ward, Analyst, 1986, 111, 1113.
2. L. Ebdon, S. Hill and R.W. Ward, Analyst, 1987, 112, 1.
3. C.A. Bache and D.J. Lisk, Anal.Chem., 1977, 39, 786.
4. W.R. McLean, D.L Stanton and G.E.Penketh, Analyst, 1973, 98, 432.
5. T.H. Risby and Y Talmi, CRC Crit.Rev. in Anal.Chem.. 1983, 14(3), 231.
6. C.I.M. Beenakker, Spectrochim.Acta 1976, 31B, 483.
7. M.H. Abdellah, S Coulombe and J.M. Mermet, Spectrochim.Acta, 1982, 37B, 583.
8. R.M. Barnes, Crit.Rev.in Anal.Chem., 1978. 7, 203.
9. R.J. Decker, Spectrochim.Acta, 1980, 35B, 19.
10. J.O. Beyer, Ph.D. Dissertation, University of Massachusetts (1984).
11. R.B. Costanzo and E.F. Barry, Anal.Chem., 1988, 60, 826.
12. B.M. Patel, E. Heithmar and J.D. Winefordner, Anal.Chem. 1987, 59, 2374.
13. D.C. Liang and M.W. Blades, Anal.Chem., 1988, 60, 27.

14. D.Huang, D.C. Liang and M.W. Blades, J.Anal.Atomic Spec.., 1989, 4, 789.
15. G.W. Rice, A.P. D'Silva and V.A. Fassel, Spectrochim.Acta, 1985, 40B, 1573.
16. J.J. Sullivan and B.D. Quimby, J.High Res.Chrom., 1989, 12, 282.
17. A. Bollo-Kamara and E.G. Codding, Spectrochim Acta. 1981, 36B, 973.
18. T.M. Dowling, J. A. Seeley, H. Feuerbacher and P.C. Uden, in 'Element Specific Chromatographic Detection by Atomic Emission Spectroscopy', P.C. Uden ed. American Chemical Society Symposium Series, Washington D.C., to be published 1991.
19. K.E. LeFreniere, V. A.Fassel, and D.E. Eckel, Anal Chem., 1986, 59, 879.
20. P.C. Uden, B.D. Quimby, R.M. Barnes and W.G. Elliott,. Anal Chim. Acta, 1978, 101, 99.
21. D.W. Hausler, Spectrochim. Acta., 1985, 40B, 389..
22. D.R. Luffer, L.J. Galante, P.A. David, M. Novotny and G.M. Hieftje, Anal.Chem., 1988, 60, 1365.
23. R.L. Firor, American Lab., 1989, 21(5), 40.
24. S.A. Estes, P.C. Uden, and R.M. Barnes, Anal.Chem., 1981, 53, 1829.
25. A.J. McCormack, S.C. Tong and W.D. Cooke, Anal.Chem, 1965, 37, 1470.
26. C.A. Bache and D.J. Lisk, Anal.Chem.,.1965, 37, 1477.
27. Y.Talmi and V.E. Norvall, Anal.Chem., 1975, 47, 1510.
28. K.S. Brenner, J.Chromatogr, 1978, 167, 365.
29. D.F. Hagen, J.S. Marhevka and L.C. Haddad, Spectrochim Acta, 1985, 40B, 335.
30. K.Zeng, Q.Gu, G.Wang and W.Yu, Spectrochim Acta,.1985, 40B, 349.
31. D.S. Sklarew, K.B. Olsen and J.C. Evans, Chromatographia, 1989, 27, 44.
32. K.B. Olsen, D.S. Sklarew and J.C. Evans, Spectrochim Acta, 1985, 40B, 357.
33. B.D. Quimby, P.C. Uden and R.M. Barnes, Anal.Chem., 1978, 50, 2112.
34. B.D. Quimby, M.F. Delaney, P.C. Uden and R.M. Barnes, Anal. Chem., 1979, 51, 875.
35. K. Chiba and H. Haraguchi, Anal.Chem., 1983, 55, 1504.
36. B.D. Quimby and J. J. Sullivan, Anal.Chem., 1990, 62, 1027.
37. M. Zerezghi, K.J. Mulligan and J.A. Caruso, J.Chromatog.Sci., 1984, 22, 348.
38. D.E. Pivonka, W.G. Fateley and R.C. Fry, Appl.Spectroscopy, 1986, 40, 291.
39. C. Bradley and J.W. Carnahan, Anal.Chem., 1988, 60, 858.
40. L.G. Sarto Jr., S.A. Estes, P.C. Uden, S. Siggia and R.M. Barnes, Anal.Letters, 1981, 14, 205.
41. B. Riviere, J-M Mermet and D. Deruaz, J.Anal.Atomic Spec., 1987, 2, 705.
42. P.C. Uden, J. Chromatogr., 1984, 313, 3.

43. R.B. Robbins and J.A. Caruso, J.Chromatog.Sci., 1979, 17, 360.
44. M.S. Black and R.E. Sievers, Anal.Chem., 1976, 48, 1872.
45. P.C. Uden and T. Wang, J.Anal.Atomic Spec., 1988, 3, 919.
46. S.A. Estes, P.C. Uden, M.D. Rausch and R.M. Barnes,J High.Res.Chrom., 1980, 3(9), 471.
47. P.C. Uden, R.M. Barnes and F.P. DiSanzo, Anal.Chem., 1978, 50, 852.
48. S.A. Estes, C.A. Poirier, P.C. Uden and R.M. Barnes, J.Chromatogr., 1980, 196, 265.
49. K.W. Panaro, D. Erikson and I.S. Krull, Analyst, 1987, 112, 1097.
50. K Jinno, H Tsuchida, S. Nakanishi, Y. Hirata and C. Fujimoto, Appl.Spectroscopy, 1983, 37, 258.
51. K.J. Irgolic, R.A. Stockton, D. Chakraborti and W. Beyer, Spectrochim.Acta, 1983, 38B, 437.
52. W. Nisamaneepong, M. Ibrahim, T. W. Gilbert and J.A. Caruso, J.Chromatog.Sci., 1984, 22, 473.
53. D.R. Heine, M.B. Denton and T.D. Schlabach, J.Chromatog.Sci.,1985, 23, 454.
54. F.La Torre, N,.Violante, O.Senofonte, C. D'Arpino and S. Caroli, Spectroscopy, 1989, 4(1), 48.
55. J.J. Thompson and R.S. Houk, Anal.Chem., 1986, 58, 2541.
56. H. Suyani, J. Creed, T. Davidson and J. Caruso, J.Chromatog.Sci.,1989, 27, 139.
57. D. Beauchemin, M.E. Bednas, S.S. Berman, J.W. McLaren, K.W.M. Siu and R.E. Sturgeon, Anal.Chem., 1988, 60, 2209.
58. D. Heitkemper, J. Creed, J. Caruso and F.J. Fricke, J.Anal.Atomic.Spect.1989, 4, 279.
59. S.G Matz, R.C. Elder and K.J. Tepperman, J.Anal.Atomic Spect.1989, 4, 767.
60. I.S. Krull, in 'Liquid Chromatography in Environmental Analysis'(J. F. Lawrence, ed.), 1983, chapter 5, Humana Press,Inc. Clifton, N.J., USA.
61. I.S. Krull, K.W. Panaro and L.L. Gershman, J.Chromatog.Sci., 1983, 21, 460.
62. H.A.H. Billiet, J.P.J van Dalen, P.J. Schoemakers, and L deGalen, Anal.Chem., 1983, 55, 847.
63. D. Kollotzek, D. Oechsle, G, Kaiser, P. Tschopel and G. Tolg, Fresenius Z.Anal.Chem., 1984, 318, 485.
64. D.L. Haas, J.W. Carnahan and J.A. Caruso, Appl.Spectroscopy, 1983, 37, 82.
65. L. Zhang, J.W. Carnahan, R.E. Winans and P.H. Neill, Anal.Chem., 1989, 61, 895.
66. D. Heitkemper, J. Creed and J.A. Caruso, J.Chromatog.Sci.,1990, 28, 175.
67. J.W. Olesik and S.V Olesik, Anal.Chem., 1987, 59, 796.
68. K.A,. Forbes, J.F Vecchiarelli, P.C. Uden and R.M. Barnes, Anal.Chem., 1990, 62, 2033.

CHLORINE AND SULPHUR ANALYSIS IN POLYMERS BY INDUCTIVELY COUPLED PLASMA EMISSION SPECTROMETRY (ICP-ES)

M. J. Hepher, C. L. R. Barnard, and D. Fortune

DEPARTMENT OF PHYSICAL SCIENCES, GLASGOW COLLEGE, GLASGOW G4 0BA

1 INTRODUCTION

Chlorine is a main constituent in such polymers as polyvinyl chloride and is often used as an intermediate functional group en route to the synthesis of other surface activated polymeric preparations. Such syntheses as those of diene-divinylbenzene styrene copolymers [1], used in piezoelectric gas sensors research [2], may be achieved using chloromethylated divinylbenzene styrene as the anchor polymer for the diene. The diene receptor molecule may react stereospecifically with sulphur dioxide by a reversible cycloaddition reaction [3] important for the development of sulphur dioxide specific sensors. It is important to know the degree of chloromethylation of the anchor polymer prior to diene substitution to help understand and so improve the synthesis reaction conditions for control of diene substitution and choice of 1,3 diene type. A secondary indicator is the degree of sulphur binding and release to the diene product.

2 EXPERIMENTAL

A novel thermal decomposition utilising selective volatile trapping was investigated for chlorine and sulphur prior to determination by ICP-ES. Polymer samples ranging in size from 0.25g to 1.00g were decomposed at approximately 600^{0}C and the resulting volatiles trapped using reaction specific reagents, Figure 1. The chlorine trap used 1% silver nitrate in 0.36M nitric acid solution while the sulphur trap used 1% aqueous zinc acetate. The volume of trapping reagent was kept to the minimum practicable, 2 to 5 cm^{3}, while the pathlength for passage of volatiles through the trap was maximised. The bubble size of these evolved volatiles was kept low by using a small (1mm internal diameter) outlet tube below the surface of the reagent.

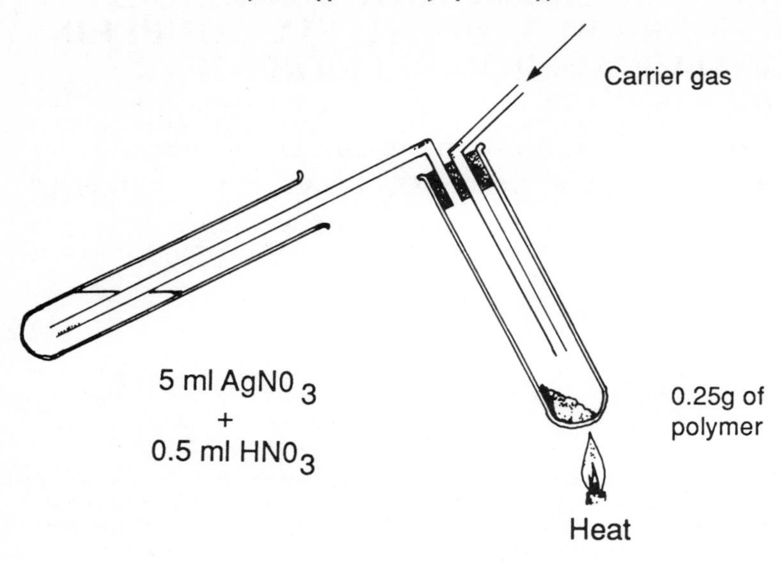

FIGURE 1 Decomposition and Trapping Apparatus

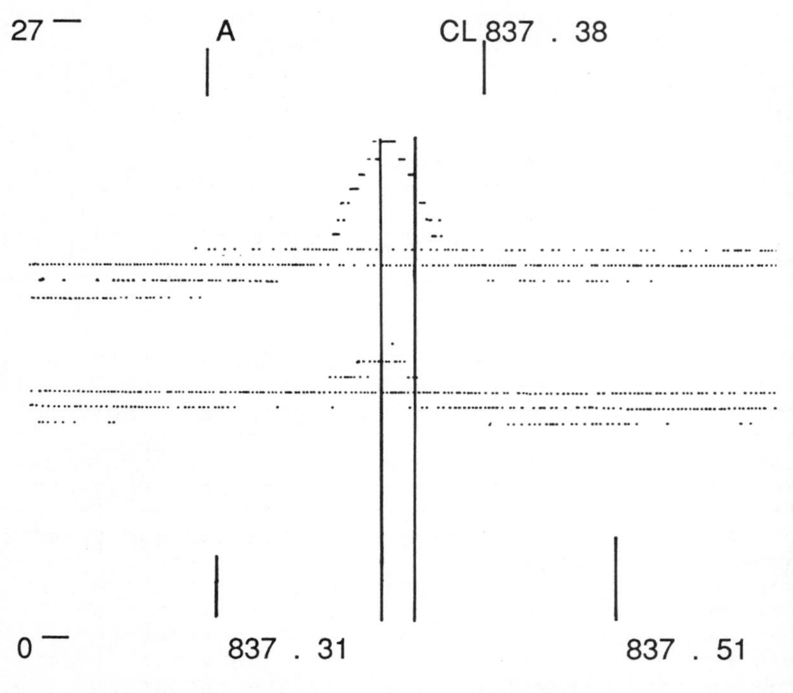

FIGURE 2 Emission Profile for Chlorine; Top peak Aqueous 6066ppm Cl, Top middle peak Aqueous baseline, Bottom Middle peak Chlorinated Polymer sample Volatiles in Ammonium Hydroxide, Bottom peak Ammonium Hydroxide baseline.

Evolved chlorine was determined as silver chloride, which was collected by centrifugation, the supernatant liquid removed by Pasteur pipette and the precipitate washed several times with water. The resultant clean product was dissolved in 1% ammonium hydroxide solution and made up to volume prior to chlorine analysis. This analysis was carried out directly on the ICP using the chlorine line at 837.38nm, Figure 2, and also indirectly by the analysis of the silver in the silver chloride using the silver emission line at 338.29nm, Figure 3.

Sulphur was determined directly on the zinc acetate trapping solution using the emission line at 182.04nm, Figure 4.

The method development may be considered as a two stage operation. The first of these stages is the optimization of conditions for the trapping of volatiles, while the second is the optimization of the ICP-ES operating conditions. The latter conditions include the effect of emission line selection, nebuliser pressure, torch power, sample flow rate and monochromator viewing height.

Figures 2, 3 and 4 also illustrate the capabilities of the Jarrel Ash Plasma 300 data display system. The numbers in the top left hand corner of these figures are scaled photon counts, the letters A or B signifying the visual and ultraviolet wavelength channels respectively, the slit window is marked by two short lines at the top of each profile with the centre of the window marked with a full line while the second full line marks the displacement from the slit centre for the first sample run in each case.

The preliminary work described here indicates that the zinc acetate trapping procedure for sulphur produces a small emission interference, but this only marginally effects the published limits of detection.[4] The overall accuracy, limit of detection and the percentage recoveries for each of the procedures may be evaluated using standard chloride and sulphate spikes added to the silver nitrate and zinc acetate trapping solutions. Table 1 shows such test data obtained to check the silver nitrate trapping procedure and the indirect chlorine method using silver chloride standards. It has been previously stated that optimization of the plasma operating conditions is a prime objective of this work to yield the best limits of detection, accuracy and reproducability. However there is often a compromise between solvent choice for optimum sample preparation and for optimum plasma operation. This dichotomy is well illustrated by the three dimensional plots shown in Figure 5, in which the drift in anode current indicates

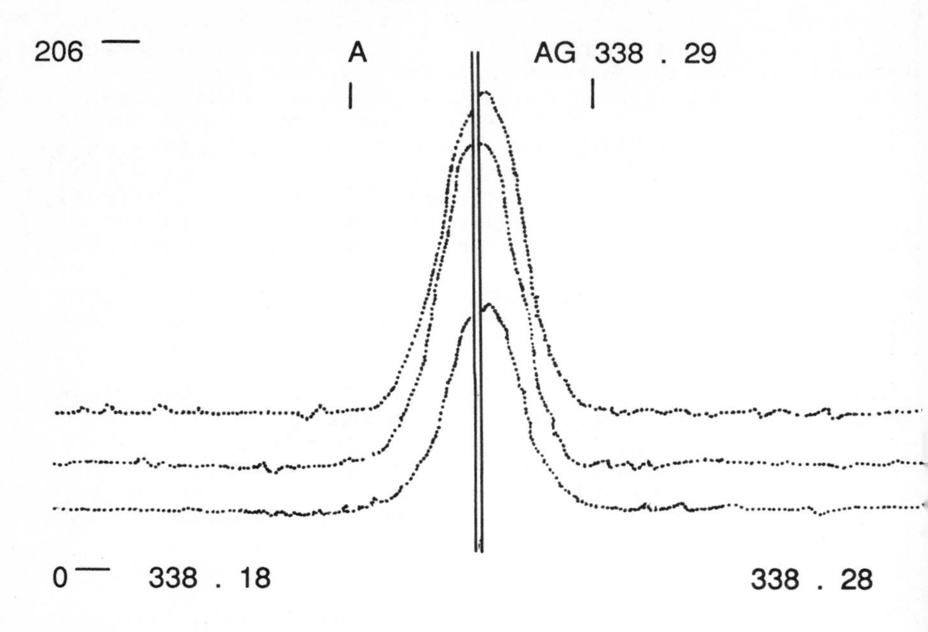

FIGURE 3 Emission Profiles for 6.35ppm Silver Standard Illustrating the Effect of Nebuliser Pressure; Top peak 20psi, Middle peak 30psi, Bottom peak 40psi.

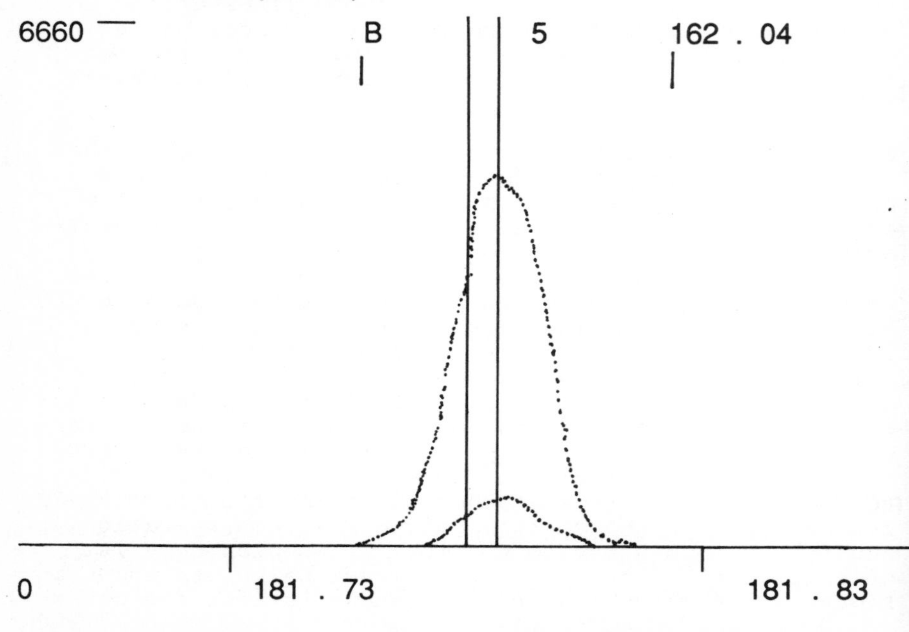

FIGURE 4 Emission Profiles for Sulphur Standards; Top peak 1000ppm Sulphur, Bottom peak 100ppm Sulphur.

TABLE 1 Test Recoveries for Silver Chloride Precipitation and Washing Procedure.

Theoretical Ag Conc (ppm)	Determined Ag Conc (ppm)	Recovery (% Ag)
151.9	146.0	96
30.4	27.0	89
15.2	13.0	86
3.0	2.0	67
1.5	2.0	133

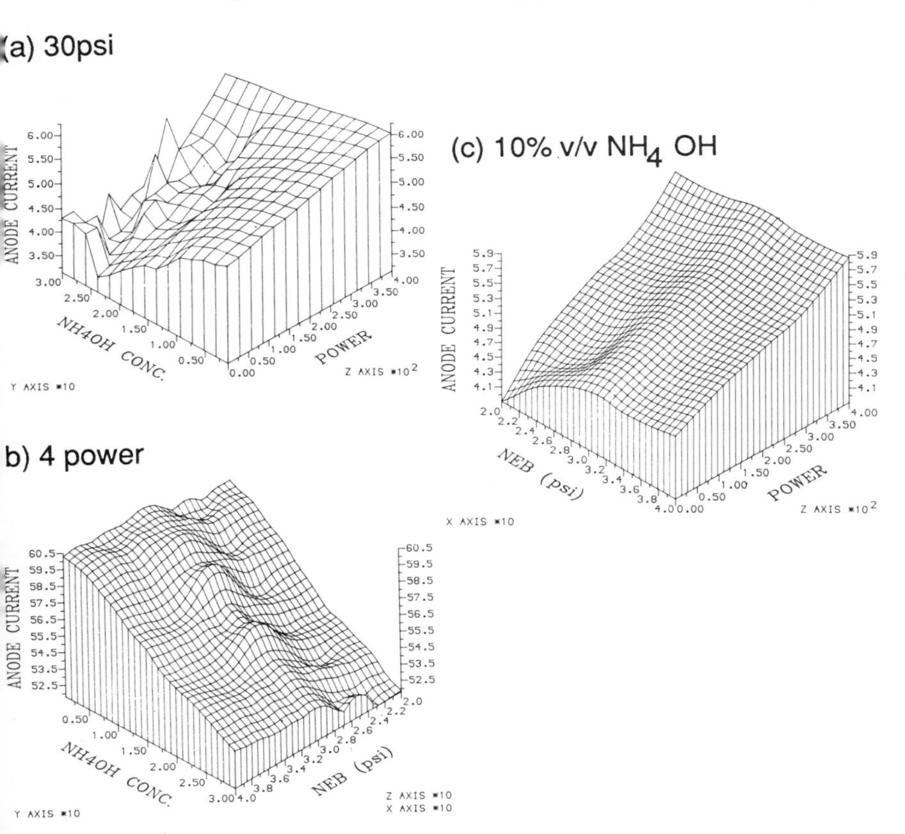

FIGURE 5 Example of the effect on Anode current of (a) Varying Power and Ammonium Hydroxide Concentration at Fixed Nebuliser Pressures (20-40 psi), (b) Varying Nebuliser Pressure and Ammonium Hydroxide Concentration at Fixed Power (0 - 4), (c) Varying Nebuliser Pressure and Power at Fixed Ammonium Hydroxide Concentrations in Nanopure water (1-30% NH_4OH).

detuning of the radio frequency argon torch.[5] In this case the solvent of choice for sample preparation, ammonium hydroxide, causes anode currents to drift below 360mA which can extinguish the torch.

3 CONCLUSIONS AND DISCUSSION

The precipitation and washing procedure for the novel trapping shows around 90% recovery of chloride down to the 5ppm level, measured indirectly as silver. The trapping procedure as applied to polymeric chlorine containing samples currently shows low recoveries. Such loss is not due to volatile chlorine compounds passing through the trap as a second trap in series with the first showed no chlorine absorption. However a small oil fraction at the surface of the trapping medium may be rich in organic chlorine-containing fragments, and is currently under investigation.

Ammonium hydroxide affects the anode preset drive current but not significantly at the 1% v/v level. Nebuliser pressure and power settings can be used to sustain an argon gas plasma for increased ammonium hydroxide concentration. Solvent matrix-match standards are required for the ammonia solution method. Use of the chlorine line is possible but the detection limit is poor, i.e. around 1000ppm Cl. Considering the plasma alone, use of the silver emission line would give limits of detection of around 4ppb Ag (equivalent to 1.3ppb Cl), but taking into account the sorption technique, the overall limit of detection for the method rises to 5ppm Cl.

REFERENCES

1. T. J. Nieuwstad, A. P. G. Kieboom, A. J. Breijer, J. van der Linden and H. van Bekkum, Rec. Trav. Chim. Pays-Bas, 1976, 95, 225.

2. T. Edmonds, M. J. Hepher and T. S. West, Anal. Chim. Acta., 1986, 187, 293.

3. T. L. Gilchrist and R. C. Storr, Organic Reactions and Orbital Symmetry, Cambridge University Press, 1972.

4. Thermal Jarrel Ash Plasma 300 Operation Handbook.

5. Ginosurf (Surface Plot on Prime), Version 1, Glasgow College, May 1988.

CHROMATOGRAPHY WITH FLUORESCENCE AND LUMINESCENCE DETECTION

H. Lingeman, C. Gooijer, N. H. Velthorst, and U. A. Th. Brinkman

DEPARTMENT OF GENERAL AND ANALYTICAL CHEMISTRY, FREE UNIVERSITY, DE BOELELAAN 1083-1081 HV AMSTERDAM, THE NETHERLANDS

1 INTRODUCTION

One of the most important questions, within the framework of this review is; why do fluorescence and luminescence belong to the most popular detection modes in chromatography? The answer is that these techniques are amongst the most sensitive and selective detection methods. Furthermore, they can provide both qualitative and quantitative information, and can be used for the determination of analytes present in complex matrices (e.g. biological, environmental).[1]

It will be obvious that quite a number of parameters have an influence on the intrinsic fluorescence properties of the analyte. The most important ones, in the present discussion, are the chromatographic mobile phase and the molecular structure of the analyte. The fact that only aromatic and highly conjugated compounds possess native or intrinsic fluorescence is an advantage in terms of selectivity, but at the same time it is a disadvantage in terms of applicability. However, this limitation can be overcome by using a derivatisation procedure.[2]

One of the main interferences in fluorescence detection, is Raman scattering of mobile phase components. This is especially troublesome when the Stokes shift, the difference between excitation and emission wavelengths, is relatively small. In the most popular mode of chromatography - reversed phase column liquid chromatography (RP-CLC) - this is especially a problem. In this mode the mobile phase consists of mixtures of water and methanol, or water and acetonitrile. This means that for analytes with small Stokes shifts there will be considerable overlap between the Raman band of the mobile phase solvents and the emission band of the analyte. However, by changing the excitation wavelength the Raman band will shift, but the emission wavelength will not and so the interference can be minimised.[3]

An example is the determination of dansyl derivatives of amino acids. For the separation RP CLC should be used, but the use of aqueous mobile phases results in unfavourable detection limits. The emission wavelengths in chloroform, 1,2-dichloromethane, acetonitrile, and water are 497 nm, 510 nm, 533 nm, and 578 nm, respectively. In the same order the intensity of the emitted light drops by about 90% in total.[4]

2 FLUORESCENCE DETECTION IN COLUMN LIQUID CHROMATOGRAPHY

One of the numerous examples of using the native fluorescence of solutes for detection purposes, is the determination of the catecholamine-like drug dobutamine in rat plasma.[5]

The bioanalytical procedure is relatively straightforward. The analyte is isolated from the matrix by liquid-liquid extraction and analysed with RP ion-pair CLC (Fig. 1). Because the Stokes shift is rather small,

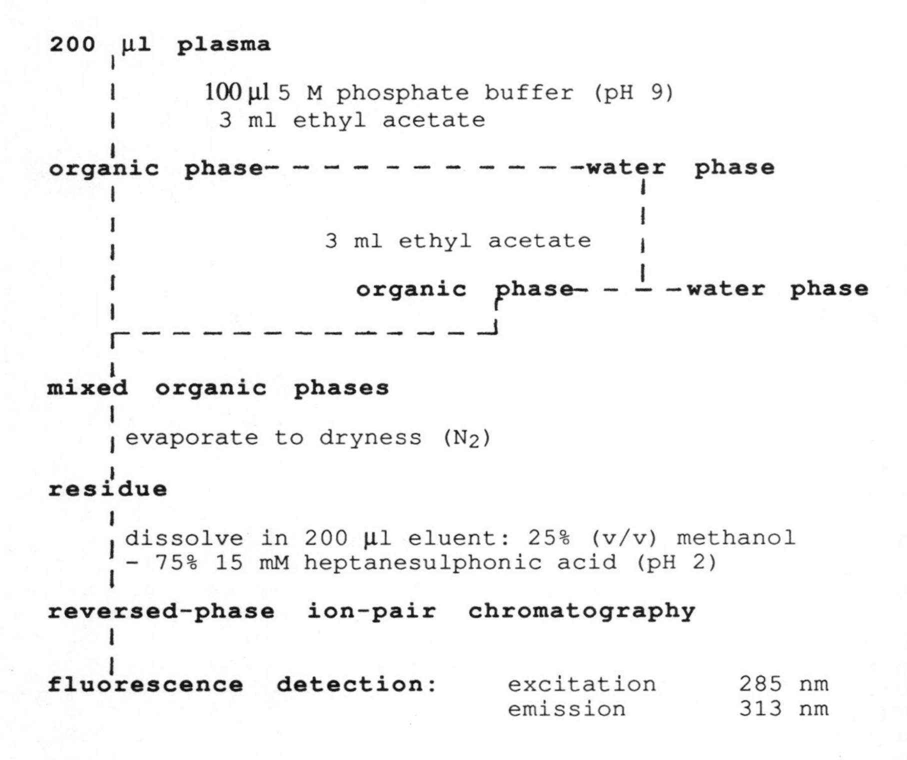

Figure 1. Bioanalytical procedure for determination of dobutamine in plasma.

only 30 nm, a grating detector is advantageous, in spite of the fact the filter detectors are, in principle, more sensitive.[2]

The determination limit of dobutamine is less than 10 ppt, which is not rather sensitive, but for example, the detection limit of the highly fluorescent aflatoxins is less than 10 pg, indicating that indeed fluorescence detection in CLC is a sensitive detection mode.[6]

3 FLUORESCENCE DETECTION IN GAS CHROMATOGRAPHY

In addition to CLC, gas chromatography (GC) is an important separation technique. For the separation of a number of solutes, e.g. polynuclear aromatic hydrocarbons (PAH's), GC is more suitable than CLC. One of the positive features of GC, compared to CLC, is that universal detection devices (i.e. flame-ionisation) and a number of sensitive and selective detectors (e.g., electron-capture, thermionic) are available. However, the use of fluorescence detection may have some additional advantages for the analysis of complex samples

Table 1 Vapour-Phase Fluorescence Excitation and Emission Maxima for some Aromatic Hydrocarbons[7]

Analyte	Maximum	Wavelength (nm)
	Excitation	Emission
3-Dimethylbenzene	265	289
1,3,5-Trimethylbenzene	267	292
1,2,4,5-Tetramethylbenzene	274	296
Naphthalene	264	330
1-Methylnaphthalene	270	338
2-Methylnaphthalene	264	335
Fluorene	252	308

(Reprinted with permission from reference 7).

An example is the determination of aromatic compounds in light hydrocarbon fractions. Interfacing of a GC and a fluorescence detector is relatively simple. The two instruments are connected via a heated transfer line and the chromatographic effluent is passed through a heated flow cell. Using capillary GC, kerosene and petrol samples can be analysed. Selective detection of the two- and three ring aromatic solutes is possible because the vapour-phase excitation maxima, at 160°, of the two-ring and three-ring aromatics are approximately 265 nm and 250 nm, respectively (Table 1). Moreover, the two-ring compounds are eluted before 13 min, and the three-ring analytes after 13 min (Table 2), which means that at 13 min the detection wavelengths can be changed, so allowing adequate selectivity and sensitivity.[7]

Table 2 Concentrations of Aromatics in Light Hydrocarbon Fractions

Sample	Concentration ($\mu g\ ml^{-1}$)			
	Naphthalene	1-Methyl Naphthalene	2-Methyl Naphthalene	Fluorene
White spirit	913	72	149	
White spirit	592	46	70	
White spirit	268	20	28	1
Kerosene	70	99	135	5
Kerosene	506	550	721	13
Petrol	142	274	632	29
Retention time	4.3	6.3	5.9	18.2

(Reprinted with permission from reference 7).

The detection limit for fluorene is 0.3 ppm and for the naphthalenes it is ca. 3 ppm (Table 2). These detection limits are certainly sufficient for the determination of these compounds in light hydrocarbon fractions. The advantage of using fluorescence detection instead of, for example, flame-ionisation detection is that co-eluting aliphatic hydrocarbons are not detected, and so an increased selectivity is obtained. The selectivity of the systems allows the direct injection of these samples without any sample clean-up.

Using special equipment, for example, a silicon intensified target camera in combination with an optical multichannel analyser provides a highly flexible fluorescence detection device in GC. The systems consists of a GC, a heated transfer line, a 50-mm long rectangular quartz gas flow cell with an internal diameter of 3 mm heated to about 250°, a light source, some optics, a heating device, an excitation and emission monochromator, and a 12.5 mm wide silicon intensified target camera - with 500 channels - allowing multiple detection over a selectable spectral interval of 62.5 nm. All these data can be stored in the two 500-word memories of the optical multichannel analyser.[8]

The silicon intensified target/optical multichannel analyser can be used in several ways. In the integration mode data are accumulated over a certain period of time resulting in an increased signal-to-noise ratio. The detection limits, using the integration mode, are about the same as using a fast-scanning photomultiplier tube. The subtraction mode allows background corrections, which is important for both qualitative and quantitative experiments. The most important mode, however, is the real time mode. In this mode entire spectra of the analytes can be recorded on-the-fly resulting in rapid semi-qualitative information of the analytes (Fig. 2).

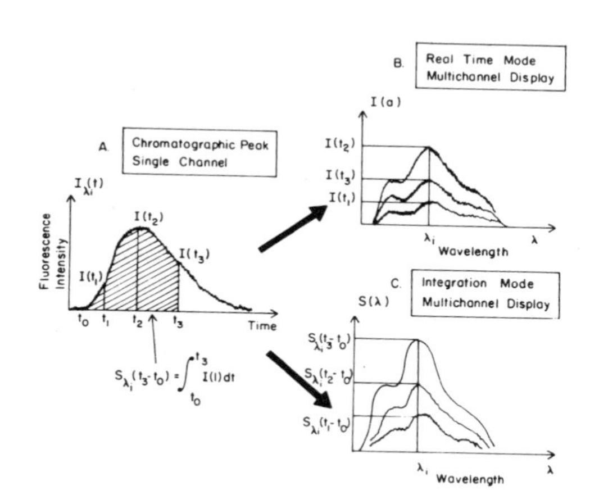

Figure 2. Possibilities of silicon intensified target-optical multichannel analyser device. A, change in fluorescence intensity at one wavelength; B, real-time mode spectra at different times; C, integration mode spectra accumulation over a period of time (Reprinted with permission from ref. 8)

4 FLUORESCENCE DETECTION IN COLUMN LIQUID CHROMATOGRAPHIC - THIN-LAYER CHROMATOGRAPHY COMBINATIONS

Another widely-used chromatographic technique is thin-layer chromatography (TLC). This technique is not so important for quantitative experiments, but in combination with CLC it has some nice features. Although fluorescence detection in TLC is well known, the combination of TLC and CLC is rather new. Storage of the effluent from a RP-CLC separation on a TLC plate offers two advantages: (i) detection principles that are not applicable on-line with flow systems can be used, and (ii) the TLC plate itself allows a second separation to be carried out.

The CLC system is a RP microbore system. This is because only flow rates of about 15 µl/min can be used to spray the column effluent onto the TLC plate. The interface is a modified TLC line-applicator with a heated spray jet assembly. Storage and second separation can be performed on commercially available RP or cellulose acetate HPTLC plates.

As an example, some PAH's are separated with the microbore RP-CLC system and detected with fluorescence spectroscopy. Comparing the chromatograms obtained for a standard mixture of PAH's - known to be present in marine sediment samples - after on-line RP-CLC and conventional fluorescence detection, with the stored chromatogram - recorded by fluorescence scanning of the deposited trace on the TLC plate with a densitometer - showed that there was hardly any loss in resolution after immobilisation of the CLC effluent.

The improved efficiency of this combination of techniques can be seen in Figure 3. In this Figure the results of the TLC separation of a marine sediment sample, after previous CLC separation and storage, is shown. The second separation is performed by developing the TLC plate in a direction perpendicular to the deposition CLC trace. The PAH's which are not separated in the isocratic CLC system are now resolved by using this two-dimensional chromatographic technique.

As discussed before the recording of excitation and emission spectra on-the-fly is still troublesome and rather expensive in CLC, but by using the CLC-TLC combination this can be performed easily at relatively low costs. Furthermore, this system not only allows the recording of spectra at room-temperature, but also at relatively low temperatures providing high resolution spectra which can be used for identification purposes. The general conclusion is that with this combination of relatively inexpensive techniques, significantly more information can be obtained, with respect to sensitivity and selectivity.[9,10]

5 FLUORESCENCE DERIVATISATION IN COLUMN LIQUID CHROMATOGRAPHY

The major problem of using fluorescence detection in chromatography in general, and CLC in particular, is that quite a number of analytes do not show sufficient native fluorescence or possess native fluorescence at unfavourable wavelengths. One way to circumvent this problem is to derivatise the analyte.[11] Derivatisation reactions can be performed in different modes, for example, before and after the CLC separation, and in an on-line or off line mode. Recently, a number of excellent monographs on derivatisation, and especially, on fluorescence derivatisation, have been published.[1,2,6,11] Therefore, only a one more or less generally applicable

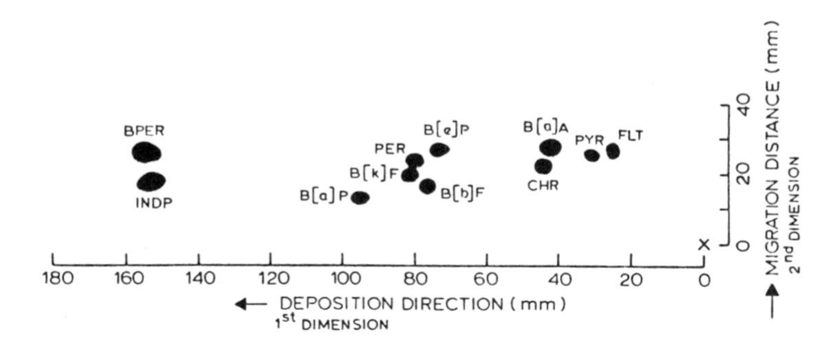

Figure 3. Two-dimensional CLC-TLC of a marine sediment sample. First dimension: isocratic RP-CLC separation; CLC effluent deposited on a 30% acetylated cellulose TLC plate. Second dimension: TLC perpendicular to the deposition trace.B[a]A, benz[a]anthracene; B[b]F, benzo[b]fluoranthene; B[k]F, benzo[k]fluoranthene; B[a]P, benzo[a]pyrene; B[e]P, benzo[e]pyrene; BPER, benzo[g,h,i)perylene; CHR, chrysene; FLT, fluoranthene; INDP, indeno[1,2,3-c,d]pyrene; PER, perylene; PYR, pyrene

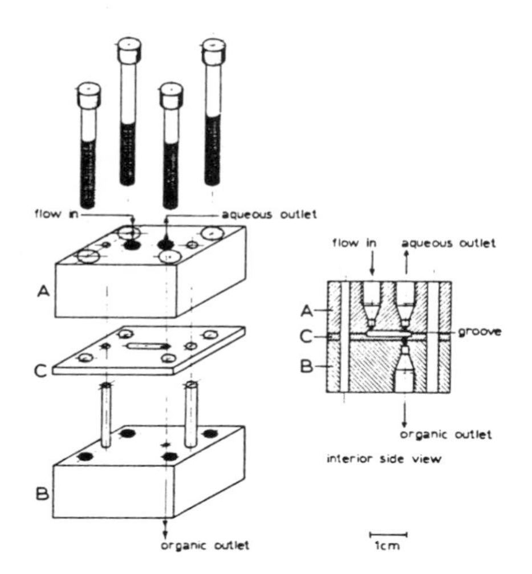

Figure 4. Design of the sandwich phase separator. A mixture of water and a non-miscible organic solvent enters at 'flow in' and after passing the groove, water exits through the 'aqueous outlet' and the organic phase to the 'organic outlet

post-column on-line derivatisation procedure will be discussed: the ion-pair extraction of phenoxy acid herbicides.[12]

The principle of the procedure is as follows: the acids are separated in their neutral form with RP-CLC using a relatively low pH. After separation, the analytes are deprotonated by mixing the column effluent with a hydroxide solution. The ion-pairing agent, a fluorescent quaternary ammonium compound, is also present in this alkaline solution. But because the excess of the ion-pairing agent has the same fluorescence properties as the neutral ion pair formed (ion-pair of the analyte and the quaternary ammonium ion) these two solutes have to be separated before detection can take place. Therefore, using another pump, a non-miscible organic solvent (i.e. chloroform) is added to the column effluent. The neutral ion-pair will be extracted to the organic phase and the excess of the reagent will stay behind in the aqueous phase. By using a sandwich phase separator (Fig. 4) the organic phase can be separated on-line from the aqueous phase, and finally the organic phase can be detected. The linear dynamic range for the phenoxy acid herbicides is about 1 to 100 ng with a repeatability of 0.5% and a detection limit of ca. 400 pg.[12]

Several fluorescent quaternary ammonium compounds are commercially available nowadays, but it is relatively easy to synthesize a new reagent 'fitting' with the polarity of the compound that has to be determined. For example, naphthalenedialdehyde, a well known probe for primary amines, is reacted with aminotrimethylammonium chloride resulting in a highly fluorescent quaternary ammonium compound, which is the probe described in the study above.

The advantage of using the naphthalene derivative instead of the more commonly used ortho-phthaldialdehyde is the higher excitation wavelength, which offers more selectivity and sensitivity.

6 LASER-INDUCED FLUORESCENCE DETECTION IN COLUMN LIQUID CHROMATOGRAPHY

In cases where sensitivity and/or selectivity are the most important goals in developing a CLC procedure, laser-induced fluorescence (LIF) detection can be a valuable technique.

However, until now continuous wave lasers - providing only a few lasing lines - have generally been used and the result is that a considerable number of analytes possessing no native fluorescence, weak native fluorescence, or native fluorescence at unfavourable wavelengths can not be detected with LIF detection.

Therefore, applicability extending techniques such as frequency doubling, variable wavelength dye lasers, and derivatisation techniques are an integral part of LIF detection.[13,14]

Principles, Set-up, and Advantages of Laser-Induced Fluorescence Detection

The acronym LASER stands for Light Amplification by Stimulated Emission of Radiation which means that photons are involved which are identical in phase, direction and amplitude and that the output beam is highly singularly directional, intense, monochromatic, and coherent.[13]

The practical advantages of LIF detection, in addition to the much higher photon flux, are the limited influence of straylight effects, because due to the high monochromacity relatively small Rayleigh and Raman bands are obtained resulting in a favourable optical window. However, in comparison with conventionally-induced fluorescence (CIF) these bands are more intense, resulting in a higher background signal. Furthermore, accurate focusing of the laser beam onto the flow cell is possible resulting in a limited loss of the excitation energy.

At present no commercial equipment is available, but an LIF detection system can easily be constructed in house. Since the laser is only used as the light source, this means that in addition to the CLC system an optical-train must be present. In this optical-train, between the flow-cell and the photon-counting unit, a blue/green sensitive cooled photomultiplier tube (PMT), a condenser lens (to focus the light onto the PMT), a monochromator or an interference filter (which is only transparent for the emitted fluorescence light), a straylight filter to eliminate the interfering Rayleigh scattering, and some recording facilities are placed. In order to achieve a high and reproducible signal all these optical components should be positioned in a straight line. To facilitate the positioning and to increase reproducibility, an optical fibre can be installed between the flow-cell and the PMT.[13,14]

The most critical part of the set-up is the flow-cell. For conventional CLC, the design can be relatively simple. The cell may be placed in a metal box with two holes in a straight line for the incident and the transmitted laser beam. In order to avoid reflections, the inside of the box is coated with graphite. The home-made quartz cell consists of a square block of quartz with an internal circular bore of about 1 mm. The flow-cell housing is mounted directly in front of the PMT.[3]

An important question is which type of laser should be used: a continuous-wave, a pulsed or a dye laser? First of all it should be realised that all lasing systems, except for the variable wavelength dye lasers, emit only a limited number of characteristic wavelengths, and because for selective and sensitive LIF detection only wavelengths between 250 and 500 nm can be used, the choice is limited to the argon-ion, helium-cadmium, and nitrogen gas lasers. This wavelength limitation can be explained by the fact that at wavelengths below 250 nm a relatively high background will be observed and that the number of analytes that can be excited at wavelengths over 500 nm is limited.

Continuous-wave lasers are normally preferred over pulsed lasers because of the instability of the latter systems, and although the advantage of variable wavelength dye lasers is that they can be tuned over a certain wavelength range, these systems have the disadvantage that they are relatively expensive and inefficient.

Frequency Doubling and Laser-Induced Fluorescence Detection in Column Liquid Chromatography

Because the majority of analytes do not possess sufficient native fluorescence at suitable wavelengths, there is a need for applicability extending techniques. The first possibility is the use of frequency doubling techniques.[3]

The use of frequency doubling, which means dividing the wavelength into half, is favourable because molar absorptivities are normally higher in the ultra-violet part of the spectrum and because a wide variety of analytes can be excited at these wavelengths. Moreover, the background signal in LIF detection is governed by Rayleigh scatter, refractions and reflections, but mainly by Raman scatter from the eluent and luminescence from cell walls and impurities. Using an excitation wavelength of 458 nm, the Raman spectrum extends to about 560 nm, which means that fluorescence detection can only be performed in one of the windows or at wavelengths over 560 nm. However, using the frequency-doubled 257 line (doubling of the 514 nm emission line of the argon ion laser) all wavelengths over 290 nm can be used for detection purposes. In other words, the advantage of shorter wavelengths is that Raman scatter of the eluent does not interfere any more with fluorescence of the analyte.

As an example, a standard mixture of PAH's was studied both by LIF and CIF detection. Improvement of the detection limits varied from a factor of 4 to 30, and by using LIF detection, the minimum detectable concentration was about 500 fg. The limited gain in sensitivity can be explained by the fact that the optimum excitation wavelength was not always used and because frequency doubling of the argon-ion laser is not very efficient: an initial laser power of 500 mW (at 514 nm) results in 5 mW of the frequency doubled line (at 257 nm).[3]

Variable Wavelength Dye Lasers and Laser-Induced Fluorescence Detection in Column Liquid Chromatography

The second possibility for extending the applicability range of LIF detection is the use of a variable wavelength dye laser. These lasers consist of two coupled lasing systems. The first one, a neodymium-YAG, nitrogen, or argon-ion laser, is the pump laser for the second one in which an organic dye or mixture of dyes is present. The advantage of variable wavelength dye lasers is that they can be tuned over a certain wavelength range. For example, when a rhodamine dye is used, all wavelengths between 537 nm and 618 nm can be isolated.[13,14]

However, since the number of analytes with sufficient native fluorescence at excitation wavelengths over 500 nm is limited, frequency doubling of the pump laser or derivatisation is still unavoidable in many cases. Another limitation of variable wavelength dye lasers is that they are relatively expensive and not very efficient. The efficiency of the dye laser system, for example, is only about 10%.

A nice application of a nitrogen pumped dye laser is the analysis of nitrobenz-oxadiazole derivatives of primary amines. The sensitivities, obtained with this technique, are compararable with sequentially excited two-photon excitation using the 488 nm emission line of the argon-ion laser.[15]

Derivatisation and Laser-Induced Fluorescence Detection in Column Liquid Chromatography

The final possibility for extending the applicability range of LIF detection is the use of a derivatisation procedure. Nowadays a number of fluorescence probes are available possessing high reactivity for one or more functional groups, and excitation wavelengths which fit with one of the available lasing lines.[13,14]

For example, dopamine, one of the catecholamines, can be derivatised with ortho-phthaldialdehyde. After excitation with the 350-360 nm multi-line of the argon-ion laser a detection limit of 16 pg is obtained.[13]

An example of the use of the inexpensive helium cadmium laser in combination with a derivatisation procedure is the determination of coumarin derivatives of solvolysed plasma steroids. Capillary RP CLC in combination with a capillary flow-cell with a volume of 100 nl, results in a detection limit of 10.[16,17]

The most important conclusions using LIF detection are that the priority order of applicability extending techniques is: (i) the use of frequency doubling, (ii) the incorporation derivatisation procedures and (iii) the application of variable wavelength dye lasers.

Although not many applications have been described so far, the use of a frequency doubled or tripled neodymium-YAG laser or a frequency doubled excimer laser seems to have a high potential in the near future. Especially, since the majority of the analytes fluorescing at higher wavelengths will also emit light at wavelengths between 240 nm and 270 nm and because selectivity enhancing techniques such as two-photon excitation, and time-resolved techniques can be applied.

Diode Laser-Based Laser-Induced Fluorescence Detection in Column Liquid Chromatography

One of the disadvantages of LIF detection is that the available lasing systems are relatively expensive. However, since monochromacity of the laser beam is far more important than the output power, relatively weak and cheap diode lasers can also be used. The limitation is that these lasing systems possess lasing wavelengths over 600 nm, but nowadays a number of dyes are known possessing good fluorescence properties in this wavelength range.

An example is the determination of some serum proteins labelled with indocyanine green. The proteins are separated by gel filtration chromatography and by using a 15 mW diode laser with a lasing wavelength of 780 nm, the detection limit for albumin is about 1 pmol, which is 1-2 orders of magnitude better than by using CIF detection. Another interesting aspect of diode lasers is that the stability is better than 1%, which is significantly better than other lasing systems.[18]

7 IMPROVEMENT OF SELECTIVITY FOR FLUORESCENCE DETECTION IN COLUMN LIQUID CHROMATOGRAPHY

In addition to the laser-based detection systems described some other principles can be used to improve the selectivity of fluorescence detection. The most important ones are two-photon excitation, supersonic jet spectroscopy, and time-resolved detection.

Two-Photon Excitation in Fluorescence Detection

The improved selectivity of two-photon excitation is caused by the fact that contrary to normal fluorescence, the emission wavelength, is shorter than the excitation wavelength, which means that interferences from scattered light are eliminated. For the cytostatic agents adriamycin and daunorubicin a gain in sensitivity of a factor 5 is observed, but far more important is the gain in selectivity, which is especially important for analyses in complex matrices.[16]

Supersonic Jet Fluorescence Detection

An interesting combination of techniques is the coupling of TLC, laser desorption, and supersonic jet fluorescence detection. A combination allowing the simultaneous recording of excitation spectra and chromatograms of a sample developed on a TLC plate.[19]

The sample is deposited on a flexible TLC plate in a straight line, and subsequently the plate, which is mounted on a sliding sheet roller attached to a supersonic jet nozzle, is developed with a suitable solvent. At a certain distance the sample is desorbed and vaporised, by a dye laser beam, into a carrier gas. The molecules entrained in the carrier gas are subsequently expanded into a vacuum to form a supersonic jet and the sample is detected by fluorescence detection, induced by a second dye laser. The excitation spectrum can be measured by scanning the wavelength of the beam of the desorption laser.

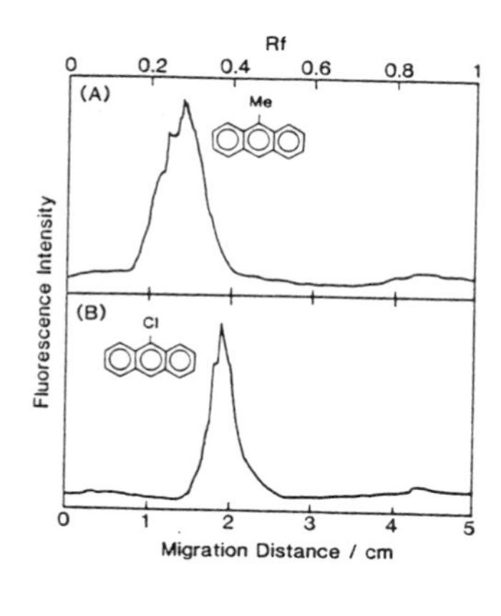

A

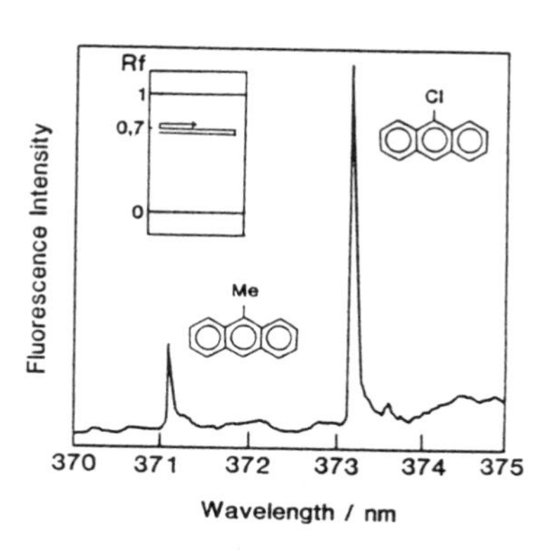

B

Figure 5. Chromatograms of a mixture of 9-chloroanthracene and 9-methylanthracene (A) and supersonic jet excitation spectrum for chemical species at R_f 0.7 (Reprinted with permission of reference 19).

In Figure 5A the thin-layer chromatograms for 9-methylanthracene and 9-chloro-anthracene are shown. As can be seen these components are difficult to separate by TLC, and moreover, the room temperature excitation and emission wavelengths are almost identical for these two solutes. The excitation and fluorescence wavelengths are 371.2 nm and 390.6 nm for 9-methylanthracene and 373.3 nm and 393.5 nm for 9-chloroanthracene, respectively. The detection limit for 9-chloroanthracene is about 10 ng. In Figure 5B the supersonic jet excitation spectrum is given for a species with an R_f value of 0.7. Under the chromatographic conditions used 9-methylanthracene and 9 chloroanthracene can not be separated, and since the room temperature excitation and emission spectra are also identical, they can not be resolved with conventional fluorescence. Using supersonic jet spectrometry, in which the sharp 0-0 transition is observed, they can be resolved.

Time-Resolved Luminescence Detection

The selectivity of luminescence detection in CLC can be improved by using labels with long emission wavelengths (> 450 nm). At these wavelengths fluorescence from interfering compounds, present in real samples, is usually low. By using a pulsed light source and a gated photomultiplier the luminescence of labels with a long luminescence lifetime can be discriminated from background luminescence and scattering. Since luminescence with a long lifetime in aqueous solutions containing oxygen is a rare phenomenon, detection is rather selective. It has, furthermore, an inherent sensitivity since low background luminescence results in a low noise level and therefore, a better signal-to-noise ratio.[20,21]

For example, lanthanide ions like the terbium (III) ion - possessing a relatively long lifetime (ca. 0.4 ms) in aqueous solutions - can be used for this purpose. Tb(III) luminescence can be observed after indirect excitation using a donating ligand. Several donating ligands have been described, but it is also possible to synthesize donating ligands using non-donating analytes and a fluorescence derivatisation reagent.[20]

Thiol-containing solutes, e.g. cysteine, can be labelled with 4-maleimidesalicylic acid in the pre-column CLC mode. In the next analysis step the derivatives are separated with CLC. The Tb(III) ion is added after the CLC separation and as a result of the complexation reaction with the maleimide derivative, luminescence can be detected. The limit of detection using the time-resolved method is about 1.5 x 10^{-7} which is a factor of 5 better than fluorescence detection.

8. CHEMILUMINESCENCE DETECTION IN COLUMN LIQUID CHROMATOGRAPHY

Chemiluminescence detection is one of the most sensitive detection modes in CLC. This is mainly because excitation of the analytes is performed by means of a chemical reaction and, as a result of a strongly decreased background, the signal-to-noise ratio is significantly increased. Various chemiluminescence reactions can be applied for detection purposes in CLC, but the most frequently used are the peroxyoxalate and the luminol system.[22,23]

Isoluminol Chemiluminescence Detection

The mechanism for the chemiluminescence of luminol and isoluminol derivatives is based on the reaction of luminol or isoluminol with hydrogen peroxide and a base catalyst forming an excited solute which emits light

Normally in CLC the hydrogen peroxide and the catalyst are added to the effluent, after the CLC separation, as two separate solutions, because hydrogen peroxide reacts with the catalyst. However, every additional pumping system will increase the background noise. The inconvenience of handling three flowing solutions (eluent, hydrogen peroxide and catalyst) can be circumvented by using electrochemical generation of hydrogen peroxide. This reagent is generated on-line

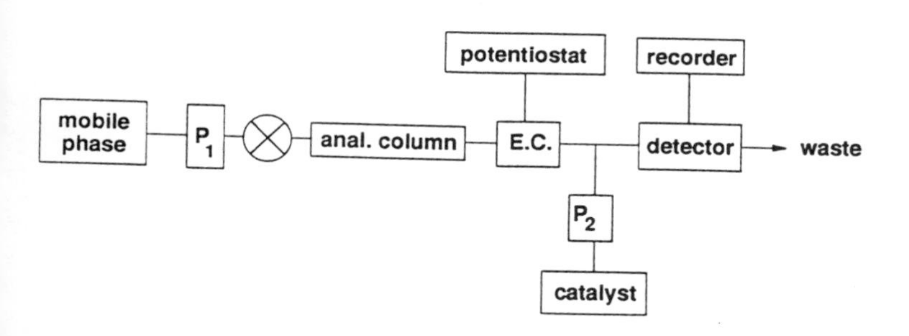

Figure 6. Column liquid chromatographic - chemiluminescence system with on-line electrochemical reagent generation. E.C., electrochemical flow cell (-600 mV); P_1, mobile phase pump; P_2, microperoxidase pump

from oxygen present in the eluent instead of via the addition of hydrogen peroxide by a pump (Fig. 6).

An example is the pre-column derivatisation of the analgesic drug ibuprofen with N-(4-aminobutyl)-N ethylisoluminol (ABEI). The carboxylic acid function of the analyte is first activated with a carbodiimide, and subsequently the activated carboxylic acid function reacts with the primary amine function of the isoluminol derivative. The determination limit for ibuprofen in plasma samples is about 1 ppb.[23]

Peroxyoxalate Chemiluminescence Detection

Another chemiluminescence system is the peroxyoxalate system in which an aryl oxalate reacts with hydrogen peroxide forming a reactive dioxetane dione. This dioxetane dione reacts with a fluorophore resulting in an excited fluorophore which emits light. This means that using this type of chemiluminescence reaction all fluorophores with an excitation energy of less than 105 kcal mol^{-1} can be determined.[22]

The combination of chemiluminescence, providing high sensitivity, and immobilised enzyme reactors, providing high specificity, is one of the most powerful reaction/detection systems in CLC.

An application is the determination of choline and acetylcholine in serum and urine. The system set up consists of an eluent pump, a vessel for the eluent, an injection valve, a cation-exchange column for the separation of choline and acetylcholine, and an enzymatic post-column reactor in which acetylcholine esterase and choline oxidase are immobilised on Sepharose (Fig.7).

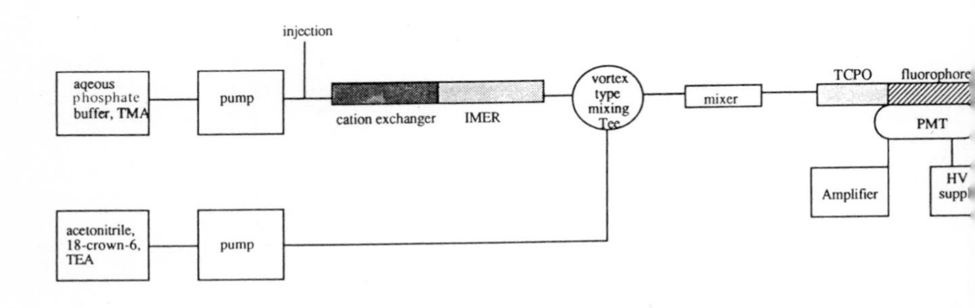

Figure 7. Schematic of acetylcholine/choline immobilised enzyme reactor system.

During the reaction betaine and hydrogen peroxide are formed in the enzyme reactor. The effluent of the reactor is mixed with a make up flow and pumped through a column containing solid trichlorophenyloxalate and an immobilised fluorophore (aminofluoranthene on glass beads). The make-up flow is needed because the enzymatic reaction should be performed in pure water, while the chemiluminescence reaction should be performed in an organic solvent. The determination limits are about 500 fmol for choline and acetylcholine with a repeatability of 3.5%. Using this type of immobilised enzyme reactor over 200 samples can be analysed before the reactor should be renewed.

9. CONCLUSION

Summarising it can be stated that the combination of a chromatographic separation method and a luminescence detection mode offers a high degree of selectivity and sensitivity and both quantitative and qualitative results can be obtained even in complex samples.

REFERENCES

1. J. Goto in H. Lingeman and W.J.M. Underberg (Editors), 'Detection-Oriented Derivatization Techniques in Liquid Chromatography', Dekker, New York, 1990.

2. H. Lingeman, W.J.M. Underberg, A. Takadate and A. Hulshoff, J. Liq. Chromatogr., 1985, 8, 789.

3. R.J. van de Nesse, G.Ph. Hoornweg, C. Gooijer, U.A.Th. Brinkman and N.H. Velthorst, Anal. Chim. Acta, 1989, 227, 173.

4. A.T. Rhys Williams, 'Fluorescence Detection in Liquid Chromatography', Perkin-Elmer, 1981.

5. H. Lingeman, J.A. Haverhals, H.J.E.M. Reeuwijk, U.R. Tjaden and J. van der Greef, Chromatographia, 1987, 24, 886.

6. A. Hulshoff and H. Lingeman in S.G. Schulman (Editor), 'Molecular Luminescence Spectroscopy: Methods and Applications, Part I', Wiley Interscience, New York, 1985.

7. H. Bagheri and C.S. Creaser, Analyst, 1988, 113, 1175.

8. R.P. Cooney, T. Vo-Dinh and J.D. Winefordner, Anal. Chim. Acta, 1977, 89. 9.

9. R.J. van de Nesse, G.J.M. Hoogland, H. de Moel, C. Gooijer, U.A.Th. Brinkman and N.H. Velthorst, submitted for publication.

10. U.A.Th. Brinkman, G.J. de Jong and C. Gooijer, in E. Reid, J.D. Robinson and I.D. Wilson (Editors), 'Bioanalysis of Drugs and Metabolites, Vol. 18', Plenum, New York, 1988.

11 H. Lingeman and W.J.M. Underberg (Editors), 'Detection-Oriented Derivatization Techniques in Liquid Chromatography', Dekker, New York, 1990.

12. C. de Ruiter, W.A. Minnaard, H. Lingeman, E.M. Kirk, U.A.Th. Brinkman and R.R. Otten, Int. J. Environ. Anal. Chem, in press.

13. C.M.B. van den Beld and H. Lingeman, in Baeyens, De Keukeleire and Korkidis (Editors), 'Luminescence Techniques in Chemical and Biomedical Analysis', Dekker, New York, 1990.

14. H. Lingeman, R.J. van de Nesse, U.A.Th. Brinkman, C. Gooijer and N.H. Velthorst, in E. Reid (Editor), 'Methodological Surveys in Biochemistry and Analysis, Vol. 20', Royal Society of Chemistry, London, 1990.

15. P.B. Huff, B.J. Tromberg and M.J. Sepaniak, Anal. Chem., 1982, 54, 946.

16. J.C. Gluckman, D. Shelly and M.V. Novotny, J. Chromatogr., 1984, 317, 443.

17. M.V. Novotny, J. Pharm. Biomed. Anal., 1989, 7, 239.

18. K. Sauda, T. Imasaka and N. Ishibashi, Anal. Chem., 1986, 58, 2649.

19. T. Imasaka, K. Tanaka and N. Ishibashi, Anal. Chem., 1990, 62, 374.

20. M. Schreurs, C. Gooijer and N.H. Velthorst, Poster presented at the 18th ISC, Amsterdam, September 23-28, 1990.

21. M. Schreurs, G.W. Somsen, C. Gooijer, N.H. Velthorst and R.W. Frei, J. Chromatogr., 482 (1989) 351.

22. P. van Zoonen, C. Gooijer, N.H. Velthorst, R.W. Frei, J.H. Wolf, J. Gerrits and F. Flentge, J. Pharm. Biomed. Anal., 1987, 5, 485.

23 O.M. Steijger, G.J. de Jong, J.J.M. Holthuis and U.A.Th. Brinkman, submitted for publication.

FLUORESCENCE SPECTROSCOPIC AND HPLC STUDIES OF INTRINSIC FINGERPRINT RESIDUES

G. A. Johnson, C. S. Creaser and J. R. Sodeau

SCHOOL OF CHEMICAL SCIENCES, UNIVERSITY OF EAST ANGLIA, NORWICH, NORFOLK, NR4 7TJ

1 INTRODUCTION

In the process of touching the various surfaces at, or around the scene of a crime, the criminal frequently leaves behind a series of latent fingerprints which can be utilised for identification purposes. It is of interest to forensic science to visualise latent fingerprints on a number of difficult surfaces using more sensitive techniques.

Typically, a fingerprint weighs about 200 μg (when dry) and consists of 99% water, with the remaining 1% composed of natural skin secretions [1]. Fingerprint residue is known to contain amino acids, peptides and proteins [2,3]. This chemical composition is exploited by the traditional ninhydrin fingerprint test. Current fingerprint development work has concentrated on photographing the luminescence of the latest fingerprint observed following excitation at 488 and 514nm by an Ar^+ laser. However, if the three aromatic amino acids tyrosine, phenyl alanine and tryptophan are present, then fingerprint residue would be expected to fluoresce when excited in the ultraviolet.

These compounds have not previously been studied by forensic workers, since little is known about the *intrinsic* fluorescence of non-chemically derivatised fingerprints. Fluorescence spectroscopy and high performance liquid chromatography (HPLC), using fluorescence spectrometric detection, have therefore been employed to identify, separate and quantify these components.

2 EXPERIMENTAL

Absorbance spectra were recorded on a Hewlett Packard HP8452A diode array spectrophotometer, controlled from a Hewlett Packard Vectra PC compatible computer (running

HP8452A software). Spectra were measured over the range 180 to 800nm.

Fluorescence excitation and emission spectra were recorded on a Perkin Elmer LS50 luminescence spectrometer controlled by Perkin Elmer fluorescence data manager (FLDM) software installed on an Epson PC AX2 computer. Both excitation and emission slit widths on the spectrometer were set to 5nm.

For the HPLC studies, the method of Viell et al was adopted [4]. A Waters model 510 pump was linked to a Rheodyne 7125 manual injection valve (fitted with a 20μl sample loop) and a 25cm X 4.6mm(i.d.) 5um Spherisorb ODS II analytical column (Jones Chromatography). For fluorescence detection, the Perkin Elmer LS50 luminescence spectrometer was fitted with a 20μl flow cell (Hellma). Both excitation and emission slit widths on the spectrometer were set to 10nm. Integration of the chromatographic data was carried out (in both peak height and peak area mode) using Kontron Integration Pack software installed on an Amstrad HD1640 HD20 personal computer. The mobile phase was; 41.5% citric acid (0.05M, BDH Ltd.), 41.5% sodium acetate (0.05M, BDH Ltd.), 17% methanol (HPLC grade, Fisons PLC). The pH was adjusted to approximately 4.6.

Each amino acid has different wavelengths of excitation and emission, so a wavelength program was used to optimise detection on the LS50 (Table 1).

Table 1. Wavelength program for fluorescence detection of aromatic amino acids.

Time (s)	λ_{ex} (nm)	λ_{em} (nm)	Amino acid
0-200	276	313	Tyrosine
200-350	258	290	Phenyl alanine
350-500	276	350	Tryptophan

Linearity and quantitative studies of the amino acids were performed by injecting aliquots (20 μl) of standard solutions of tyrosine, phenyl alanine and tryptophan (Sigma) in 50:50 methanol (HPLC grade, Fisons, PLC):water (Analar, Fisons, PLC) onto the HPLC chromatograph.

Fingerprint residue was taken from four donors (two male, two female) by inverting a vial containing 2ml of a 50:50 methanol :water mixture over the fingers and thumbs of each hand. Aliquots of these were then injected onto the system, and peak response compared to a known concentration standard.

3 RESULTS AND DISCUSSION

Fingerprint residue absorbs strongly with an absorbance maximum at 276nm, a region of the UV spectrum that is characteristic of amino acids, peptides and proteins. When excited at this wavelength intense fluorescence is observed in the range 300 - 400 nm, (Figure 3 ii), also indicating the presence of these components.

The three aromatic amino acids all showed good chromatographic separation (Figure 1) on the Spherisorb ODS column using the wavelength program given in Table 1. Linearity was observed for tyrosine between 0.21 and 20.8µg/ml, phenyl alanine between 1.0 and 200.0µg/ml, and tryptophan between 0.01 to 19µg/ml. Limits of detection were 3ng/ml, 2690ng/ml and 2.7ng/ml respectively.

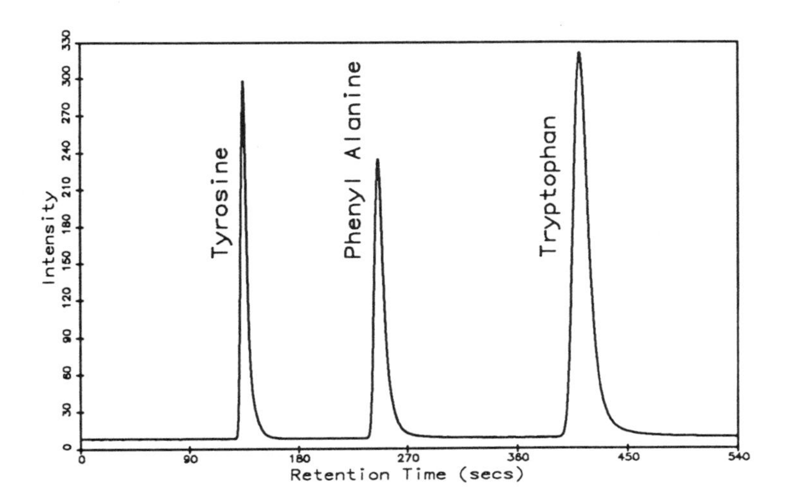

Figure 1. HPLC separation of tyrosine, phenyl alanine and tryptophan using fluorescence detection.

Chromatograms of the fingerprint residues of four donors are shown in Figure 2. The average amount of tyrosine and tryptophan from each fingerprint washing from donors 1-4 are given in Table 2. Average amounts of tyrosine collected from individual fingerprints were between 19.7 and 103.3ng for each individual donor (with an average of 49ng for all donors) and for tryptophan between 9.8 and 35.2ng (with an average of 17.4 ng for all donors). Phenyl alanine was not detected in any of the residues. This is because the limit of detection (LOD) for phenyl alanine is 1000 times higher than for either tyrosine or tryptophan. Quantum yield measurements of phenyl alanine, tyrosine and tryptophan have been made in neutral aqueous solution as 0.02, 0.15

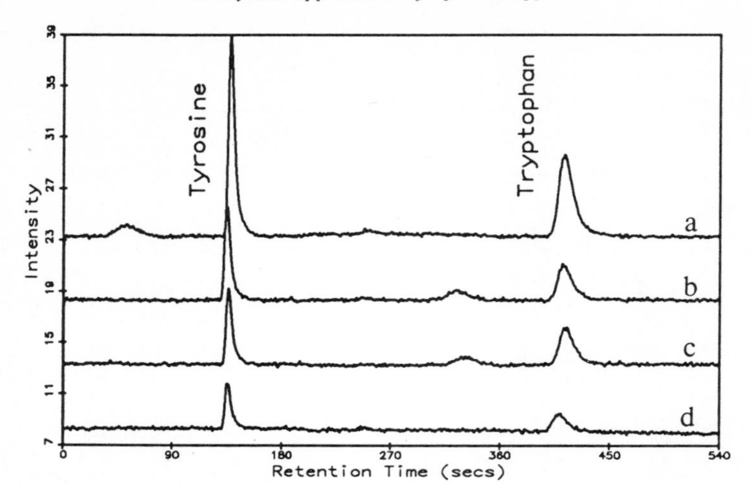

Figure 2. Determinations of tyrosine and tryptophan concentrations in fingerprint residue of; (a) donor 1 (male), (b) donor 2 (male), (c) donor 3 (female) and (d) donor 4 (female).

and 0.13 respectively.[5] It may be seen that there is a difference in quantum yield of only about 7 between phenyl alanine and both tyrosine and tryptophan. However, the larger differences observed for the LOD can be accounted for by quenching in the different solvent system used in this work.

A comparison of fluorescence excitation and emission spectra from synthetic fingerprint (3:1 tyrosine:tryptophan in methanol/water) and real fingerprint residue in methanol/water is shown in Figure 3. The tyrosine and tryptophan 'synthetic' fingerprint

Table 2. Quantitation of tyrosine and tryptophan in fingerprint residue from four donors. (†Average amount of amino acid from each fingerprint washing).

Donor Number	Tyrosine (ng)†	Tryptophan (ng)†	Ratio of Tyr:Try	% Contribution to fluorescence
1	103.3	35.2	2.9	75
2	39.7	9.8	4.1	93
3	33.3	14.7	2.3	79
4	19.7	9.8	2.0	86
Average	49.0	17.4	2.8	83

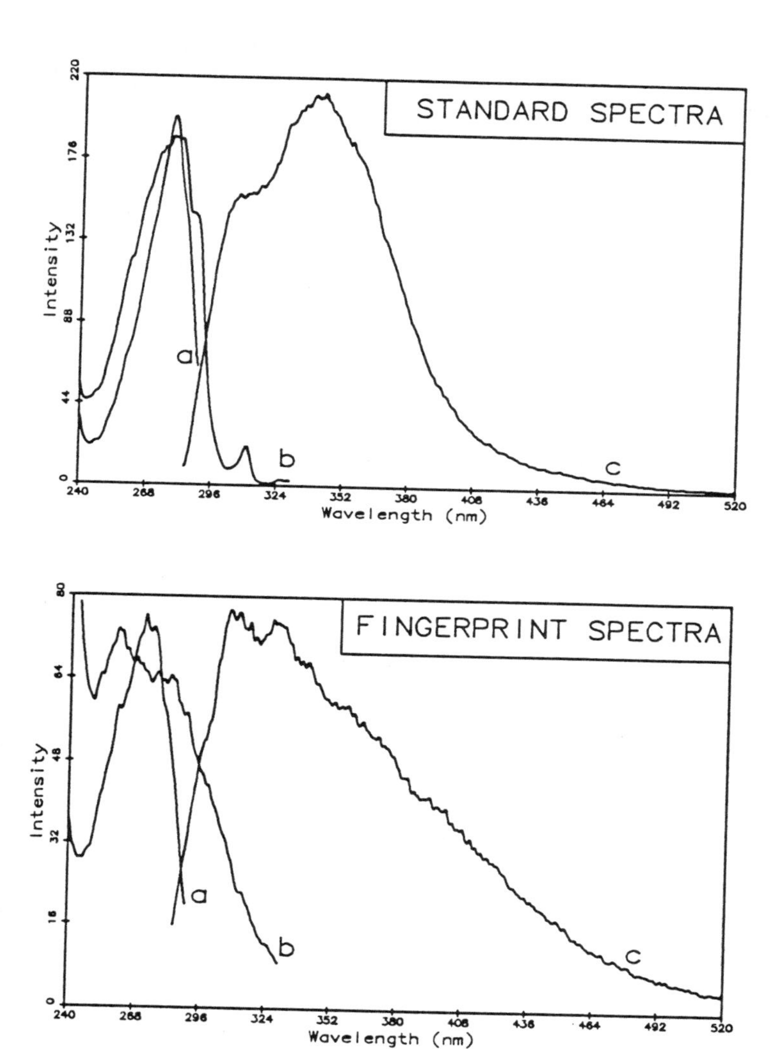

Figure 3. Fluorescence excitation and emission spectra for i) 'synthetic' fingerprint residue (3:1 tyrosine:tryptophan standard in methanol - top spectra), and ii) real fingerprint residue in methanol (bottom spectra). Excitation spectra are labelled a and b with emission at 304 and 342nm respectively. Spectra labelled c are emission spectra with excitation at 276nm.

residue mixture shows a broad emission profile similar to the real fingerprint residue in the region 300-400 nm.

The significance of tyrosine presence in the FPR is supported by the similar excitation spectra (Figure 3, spectra labelled a) for both 'synthetic' and real fingerprints with the emission optimised for tyrosine (λ_{em} = 304 nm). However, the excitation spectrum of tryptophan (λ_{em} = 342 nm) is less clearly defined for the real fingerprint residue (Figure 3, spectra labelled b) indicating interference from other components.

Quantitative fluorescence measurement, made on the basis of the peak heights of the emission spectra at 304 nm (tyrosine) and 342 nm (tryptophan) for both synthetic and real fingerprint residue account for between 75 and 93% of the intrinsic fingerprint fluorescence (Table 2). This represents an average of 83% of the ultraviolet fluorescence emission attributable to tyrosine and tryptophan for the four donors.

REFERENCES

1. G.C. Goode and J.R. Morris,"Latent fingerprints: A review of their origin, composition, and methods for detection", AWRE report 022/83, Aldermaston, Bucks., October 1983.

2. P.B. Hamilton, Nature, 1965, 205, 285

3. P. Hadorn, F.Hanimann, P.Anders, H.C. Curtius and R. Halverston, Nature, 1967, 215, 416.

4. B. Viell, B. Weidler, B. Krause and K.H. Vestweber, J. Pharm. and Bio. Anal., 1988, 6, 933.

5. "Excited States Biol. Mol., Proc. Int. Conf.", Ed. J.B. Birks, Wiley, Chichester, England, 1974, p. 375-87.

THE DETERMINATION OF LINDANE AND *p*-CHLORO-*m*-XYLENOL: A COMPARISON BETWEEN INFRARED AND CHROMATOGRAPHIC METHODS

Colin Peacock

CHEMICAL ANALYSIS SERVICE, SCHOOL OF PHYSICS AND MATERIALS, LANCASTER UNIVERSITY, BAILRIGG, LANCASTER LA1 4YA

1. THE PROBLEM:

The analysis by gas-chromatography with electron capture detection of an insecticidal lotion for cattle and horses containing nominally 0.11%w/w Lindane (γ-1,2,3,4,5,6-hexachlorocyclohexane), 0.7%w/w PCMX (p-chloro-m-xylenol) plus some acetone in a base of liquid paraffin always gave results for Lindane which were low compared with the amounts of material thought to have been added during manufacture.

Over a period the mean value for Lindane was analyzed as 0.10% rather than the target 0.11%. However the PCMX results were in the right region. The specification for the active ingredients was that they should be within 10% of the nominal values so there was difficulty attaining this requirement.

Why was there this discrepancy?

2. THE CAUSE:

Gas-chromatography/ mass spectrometry showed that Lindane eluted in the midst of the paraffin whereas PCMX elutes far earlier (Fig 1,2).

Batch to batch differences in paraffin meant that the ratio of Lindane to paraffin passing the detector varied from sample to sample (Fig.3). It is known that whilst the ECD does not detect hydrocarbons its response towards chlorine is partially quenched in their presence. It would be necessary to make up fresh standards each time a new batch of liquid paraffin was used.

Could a simpler method be found?

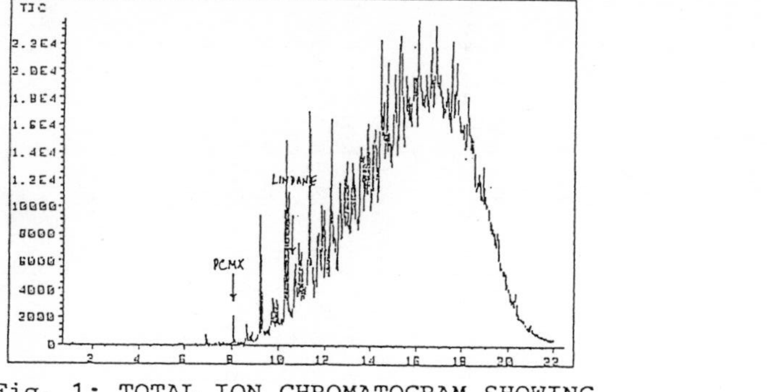

Fig. 1: TOTAL ION CHROMATOGRAM SHOWING WHERE PCMX AND LINDANE ELUTE COMPARED WITH THE PARAFFIN.

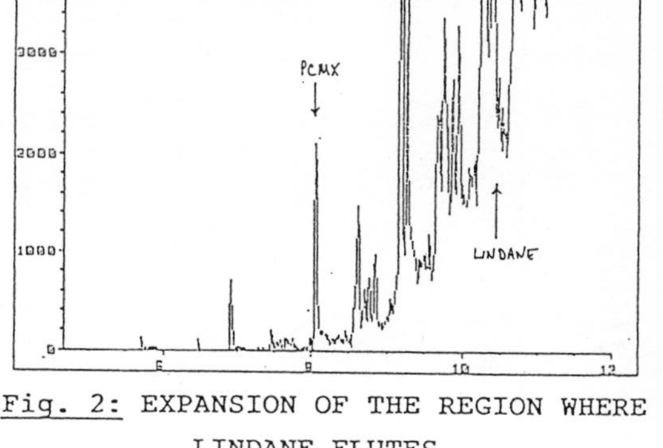

Fig. 2: EXPANSION OF THE REGION WHERE LINDANE ELUTES.

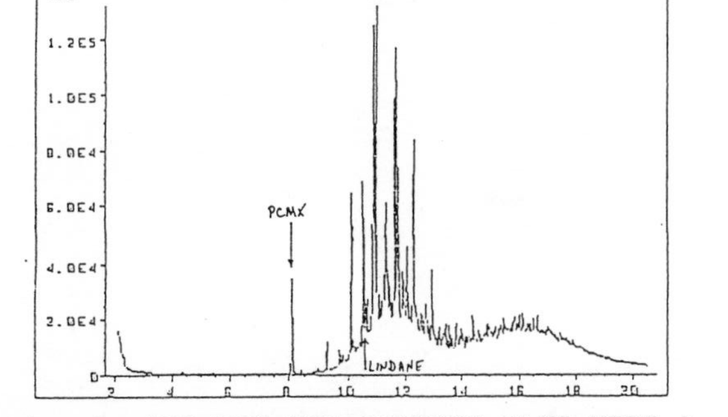

Fig. 3: TIC FOR THE PRODUCT MADE WITH A DIFFERENT BATCH OF LIQUID PARAFFIN.

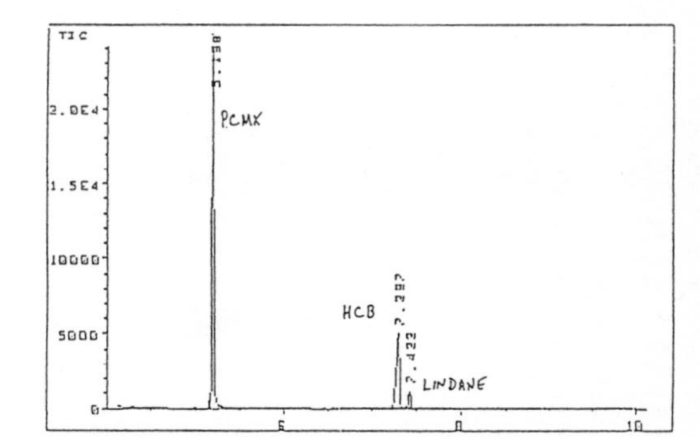

Fig. 4: ANALYTICAL (SINGLE ION MONITORING) CHROMATOGRAM.

3. FIRST ATTEMPT AT A SOLUTION:

GAS-CHROMATOGRAPHY/ MASS SPECTROMETRY:

At first it was feared that GC/MS would be prone to the same interference: that in the presence of so much co-eluting hydrocarbon there would be a chemical ionization effect altering the ion intensities from batch to batch. In fact this was not observed using hexa-chlorobenzene as internal standard (Fig.4). Standard solutions prepared without any liquid paraffin gave results much more in keeping with the composition expected from the quantities used in manufacture.
A problem arose however with PCMX. This molecule is very polar and in the instrument available using a non-polar capillary column, splitless injection and glass liners there was a marked non-specific absorption of analyte resulting in the response varying with the amount injected. It was not practicable to cure this problem with the instrumentation available. The other approach of adding a second, chemically similar internal standard was rejected as adding extra complexity to the analysis. Furthermore the reproducibility of the method was not very good : about ± 3% even for Lindane.

Was there another solution?

4.SECOND ATTEMPT AT A SOLUTION:

INFRA-RED SPECTROPHOTOMETRY:

The advent of the ratio-recording and fourier-transform infra-red spectrometers has given us a generation of instruments capable of good photometric accuracy. IR spectroscopy is, however, not normally used for quantitative determination of components below the % level in mixtures. In this case the matrix is relatively simple, liquid paraffin being a much beloved medium for IR spectroscopy (Fig.5). To get sufficient absorbance by Lindane it was necessary to use a 1mm pathlength cell: at this thickness paraffin has not such a simple spectrum (Fig.6). However there are considerable windows which can be used if the paraffin spectrum is subtracted out (Fig.7).
Two bands each for acetone and PCMX and one band for Lindane were identified as being free from mutual interferences and useful for quantitation (Figs 8-12).

Typical results have been (in % w/w):

	Acetone		PCMX		Lindane	
	530cm-1	1090cm-1	640cm-1	1030cm-1	530cm-1	GC/MS
Sample 1:	1.27	1.25	0.73	0.74	0.109	0.109
Sample 2:	1.30	1.32	0.65	0.64	0.102	0.103
Sample 3:	1.32	1.29	0.71	0.71	0.106	0.104
Specification:	1.1-1.4		0.63-0.77		0.100-0.120	

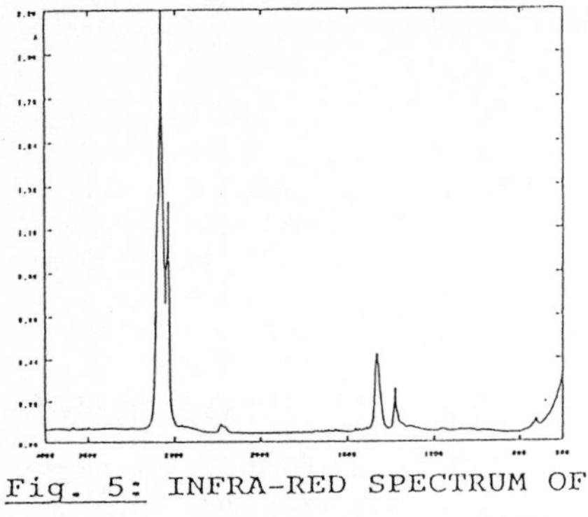

Fig. 5: INFRA-RED SPECTRUM OF A CAPILLARY FILM OF LIQUID PARAFFIN.

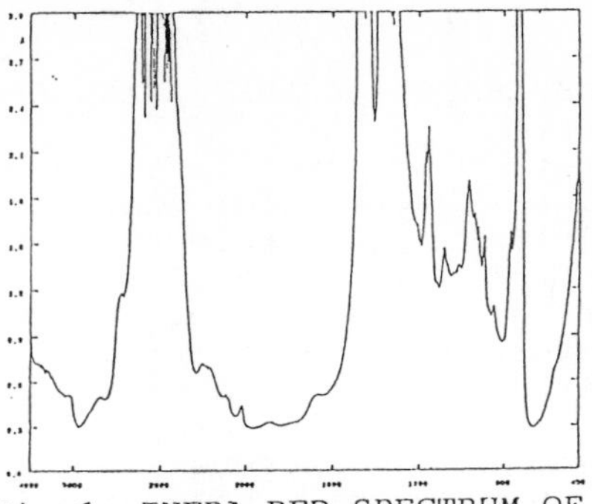

Fig.6: INFRA-RED SPECTRUM OF A 1mm FILM OF LIQUID PARAFFIN.

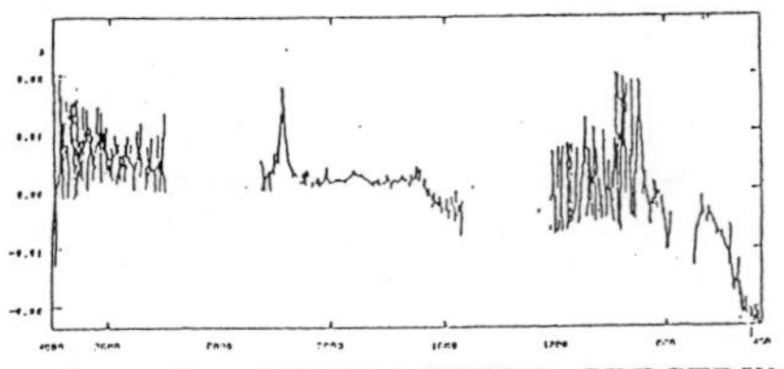

Fig. 7: SUBTRACTION SPECTRUM OF LIQUID PARAFFIN FROM LIQUID PARAFFIN SHOWING AVAILABLE WINDOWS AND BACKGROUND NOISE.

5.REMAINING PROBLEMS:

The analyses for acetone and PCMX can be done on two bands giving a check in case some impurity has an absorption underneath one of the analyte absorbances. The above method gives no such check for lindane. It is therefore necessary to use more of the spectral data, including bands in regions where absorbances overlap. This could be done using a full "number-crunching" data analysis but the number of data points in such a complex spectrum, with many bands only a few cm-1 apart, makes that rather over elaborate. Instead the peak heights for all the major bands were measured from the overall baseline for a set of four standards and factorized into the contributions from each component manually using a simple word-processor spreadsheet. This calibration set then only needs modifying for changes in cell pathlength prior to each analysis. If bands suffer from unknown interferences this can be at once seen as has been found recently with batches where the absorption at 1060 cm-1 shows significant deviations which are absent from other bands.

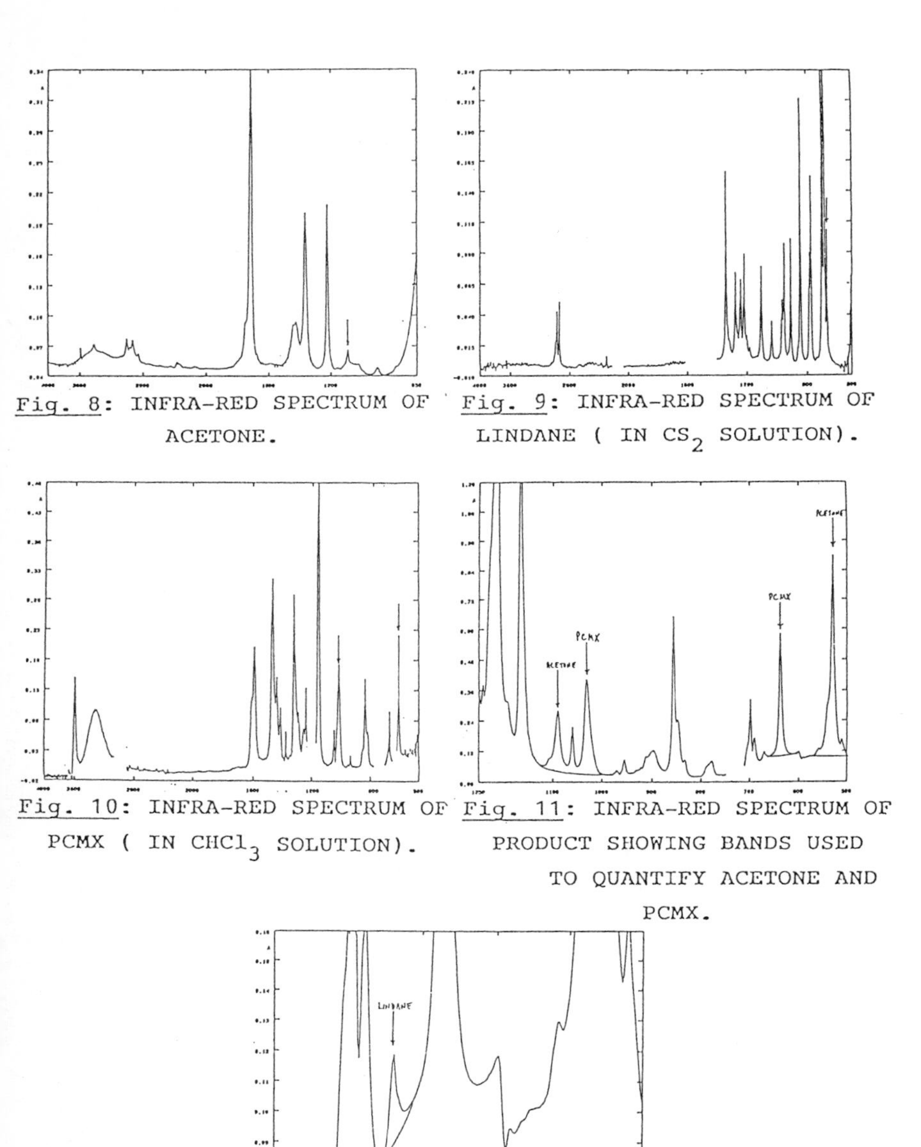

Fig. 8: INFRA-RED SPECTRUM OF ACETONE.

Fig. 9: INFRA-RED SPECTRUM OF LINDANE (IN CS_2 SOLUTION).

Fig. 10: INFRA-RED SPECTRUM OF PCMX (IN $CHCl_3$ SOLUTION).

Fig. 11: INFRA-RED SPECTRUM OF PRODUCT SHOWING BANDS USED TO QUANTIFY ACETONE AND PCMX.

Fig. 12: INFRA-RED SPECTRUM OF PRODUCT SHOWING BAND USED TO QUANTIFY LINDANE.

frequency	PCMX factor	Acetone factor	Lindane factor
529	0.000	0.606	0.000
637	0.718	0.000	0.007
669	0.020	0.000	0.215
691	0.000	0.003	0.708
698	0.270	0.002	0.435
847	0.266	0.000	0.355
856	0.868	0.002	0.069
894	0.054	0.049	0.003
911	0.041	0.031	0.154
1031	0.544	0.000	0.020
1060	0.255	0.010	0.002
1089	0.047	0.172	0.012

Sample 14 Aug 90	Calculated	% deviation	Std 1	Calculated	% deviation
0.7094	0.6969	-1.76	0.6810	0.6908	1.44
0.5174	0.5178	0.07	0.6000	0.6038	0.63
0.0388	0.0389	0.28	0.0374	0.0374	0.11
0.0842	0.0842	-0.05	0.0716	0.0714	-0.30
0.2397	0.2463	2.75	0.2735	0.2708	-0.97
0.2332	0.2320	-0.52	0.2559	0.2575	0.63
0.6339	0.6351	0.19	0.7410	0.7380	-0.40
0.0938	0.0956	1.89	0.1015	0.1015	0.01
0.0818	0.0827	1.13	0.0853	0.0846	-0.86
0.3874	0.3940	1.69	0.4588	0.4589	0.02
0.1694	0.1953	15.31	0.2265	0.2258	-0.31
0.2294	0.2330	1.57	0.2368	0.2367	-0.04

6. SUMMARY:

Quantitative infra-red analysis has been shown to be a convenient technique, even in this case where there are four components in a mixture with the least occurring at below the 0.1% level. Once the analysis is set up the only calibration needed is for the cell path-length. The relative intensities of the infra-red absorbances do not vary from day to day as they are fundamental molecular properties not instrument dependent factors. This makes the technique especially useful when only occasional samples are to be tested: there is none of the setting up and calibration required as there is with chromatographic methods. A single analysis can be readily done in less than half an hour from receipt of sample even if several weeks have elapsed since the method was last used.

7. ACKNOWLEDGEMENTS:

I would like to record my thanks to Dr E.Chipperfield and Ms J.Pollard for provision of samples and wish to dedicate this paper to the memory of the Department of Chemistry of the University of Lancaster (1964-1990), killed in its youth by the University Funding Coucil and a supine Senate of the University.

Section 6

CHEMOMETRICS AND DATA HANDLING

WHAT OTHER SPECTROSCOPIC TECHNIQUES COULD LEARN FROM NIR

Harald Martens and Bjørn Alsberg
CONSENSUS ANALYSIS AS SKI BUSINESS PARK, P.O. BOX 1384, N-1401, SKI, NORWAY
Susan Foulk
GUIDED WAVE, INC., 5200 GOLDEN FOOTHILL PKWY, EL DORADO HILLS, CA 95630, USA
Edward Stark
KES ANALYSIS, INC., 3 M 160 WESTEND AVE, NEW YORK, NY 10023, USA

1 Abstract

Multivariate calibration concerns how to combine many different instrument channels in order to reduce selectivity problems. The foremost application of multivariate calibration today is within Near Infrared (NIR) spectroscopy. The present paper illustrates the quantitative chemometrics involved: Preprocessing of input data, (in this case new methods for multiplicative signal correction and spectral filtering) followed by 'soft' multivariate calibration (in this case PLS regression). Near Infrared spectroscopy relies on multi-channel calibration. Thereby a wavelength region scorned by academic spectroscopists now out-competes traditional UV and IR in many practical applications, since interference effects are removed from the data, not from the samples. The high selectivity problems in NIR raw data are solved by specialized multivariate statistical regression methods. These chemometric tools provide the selectivity enhancement needed for quantitative spectroscopy of e.g. intact biological samples or turbid process mixtures. Their NIR raw data are dominated by several more or less unidentified interferences of chemical, physical or instrumental nature. Still, the multivariate calibration makes the instrument output selective, rapid, relevant and reliable. However, these chemometric calibration techniques, like Extended Multiplicative Signal Correction (EMSC) preprocessing and PLS Regression for calibration modelling, are useful also for standard IR, VIS., UV, and X-ray transmission or fluorescence spectroscopy and other multichannel instruments. General benefits: Less sample preparation, higher reliability and wider range of instrument application.

2 Introduction

2.1 Selectivity enhancement by multivariate calibration

Improved selectivity can be obtained by multivariate calibration in many types of analytical instruments. Multivariate calibration means actively utilizing systematic inter-correlations between signals from different instrument channels. Multivariate calibration [4] is a very general technique applicable to many types of instruments (chromatography, electrophoresis, polarography, spectroscopy, MS, NMR etc). For a given instrument type it can be applied at different levels, ranging from removal of systematic instrument noise in detector signals, via removal of signal interferences due to impurities in samples, to estimation of aggregated sample qualities such as octane number, toxicity or flavour.

Multivariate calibration consists in 'training the computer to find how to convert several non-selective input variables X into selective output variables Y. This process in general requires data preprocessing and multivariate regression modelling.

Thus, multivariate calibration employs advanced statistical procedures. However, commercial software packages are available today to make it sufficiently simple and robust for non-statisticians to employ it successfully.

Near Infrared (NIR) spectroscopy, operating within the wavelength range 900-2600 nm, is the field where multivariate calibration has been put to use first. The reason for this is that NIR data are more or less useless without multivariate calibration of some sort. NIR instrumentation has a high potential in a number of fields, e.g. for applied process instrumentation. But the academic acceptance of NIR instrument has been slow.

The indirect calibration of the multichannel NIR instruments is a challenge to traditional thinking in much of spectroscopy and analytical chemistry. The development of the highly successful NIR technology has been application driven[8], rather than theory driven. Its prime field of application till now, foods & feeds analysis, has rather low standing on the academic status ladder. NIR instruments rely on empirical statistical estimation of transfer functions. For fundamentally inclined scientists this may seem like unreliable 'black magic',- almost like cheating. But it works well in practice, where other analytical methods like IR or UV spectroscopy fail. Consequently, it is very popular in industry, and mostly ignored in universities. The purpose of the present paper is to illustrate that there is solid spectroscopic rationale behind the multivariate calibration techniques in NIR spectroscopy.

Previous NIR experience in complicated systems such as high- speed analysis of intact, 'dirty' material, like wheat flour, has shown that mathematical models based on classical 'first principles' as yet give inferior performance compared to regression models based on purely empirical (X, y) data[1, 4]. The reason is lack of sufficient 'first principles' knowledge,- NIR spectra of complex chemical mixtures cannot just be unmixed as a linear combination of various components' known pure-state NIR spectra. One does not know a priori how to describe unidentified chemical and physical interference sources, complex chemical and physical interactions in the samples, natural instrument variabilities etc.

Thus multivariate NIR calibration represents some cognitive changes compared to classical calibration. This is the reason for the present use of statistical X-Y notation ('Y being predicted from measured X') rather than the causally based notation from classical calibration and control theory; ('Y is caused by X'). The calibration theory for NIR instruments and other multichannel non-selective chemometric sensors is described e.g. in [4]. In the next two paragraphs a summary of the mathematical principles behind multivariate calibration is given.

Those readers who want, may skip those theoretical paragraphs and go directly to *Experimental* and *Results*.

2.2 Multivariate calibration by chemometric regression methods

The goal of the calibration of NIR instruments is to find the transfer function f() that allows us to convert a multichannel input $x_i = x_{ik}, k = 1, 2, \cdots, K$ (say, $K = 500$ NIR wavelengths of a petrochemical sample), into sample quality y_i (say, benzene percentage or octane number):

$$y_i = f(x_i) \tag{1}$$

If general knowledge exists about the mathematical shape of $f()$, one may apply this knowledge in a preprocessing stage in order to simplify the final modelling. An example of this is response linearization and multiplicative corrections. But the final mathematical form and the parameters of the empirically based transfer function $f()$ then need to be estimated statistically, using empirical data.

These calibration data consist of reasonably precisely known data for both chemometric sensor x_i and reference method y_i from a representative set of training samples (objects in calbration set) $i = 1, 2, \cdots, N$. Thus the input data for calibration is an $N \times K$ matrix $X = (x_i, i = 1, 2, \cdots, N)$ and an $N \times 1$ vector $y = (y_i, i = 1, 2, \cdots, N)$. (In this paper, mathematical symbols are designated by italics letters; matrices are always given as capital italics. All vectors are assumed to be column vectors unless transposed (by symbol ').

The principal of least squares is for simplicity used in most applied statistical model estimation. Therefore, such multivariate data analysis is sensitive to outliers, population heterogeneity and other phenomena that require domain specific background knowledge in order to be dealt with properly. Thus it is important to calibrate with interactive and graphically oriented statistical regression techniques that allow domain experts like spectroscopists and process engineers to apply their rich contextual knowledge. In addition, the regressions must use the intercorrelations between the wavelength channels as a stabilizing advantage, not as a 'collinearity problem'. Finally, they must give automatic outlier warnings, and must use mathematical models that are understandable for the domain expert that lacks theoretical statistical background.

Ordinary least squares Multiple Linear Regression, the standard traditional regression technique, fails in several respects as a chemometric calibration method. It does not give insight into the nature of the various interference problems, and it does not allow collinear X-variables.

There are a number of NIR calibration methods available. Bilinear calibration methods have in chemometrics been found to fulfill both the statistical and intepretation needs. The bilinear PLSR (Partial Least Squares Regression), which is used in this paper, was developed[10] specifically to meet the needs in chemometrics. Numerically it can be seen as a statistically truncated version of a conjugated gradient method for matrix inversion [3]. Statistically, PLSR is a flexible intermediate between Multiple Linear Regression (MLR) (y is regressed on every individual X-variable $x_k, k = 1, 2, \cdots, K$) and Principal Component Regression (PCR) (y is regressed on a few 'artificial' variables $ta, a = 1, 2, \cdots,$ which are the major eigenvectors of (centered) XX'): In PLSR, covariances within the block of X-variables stabilize the parameter estimation compared to MLR[2], and X-Y covariances enhances Y-relevant X-structures compared to PCR. The PLSR can have one or more reference variables $Y = y_j, j = 1, 2, \cdots, J$. In its simplest form the PLSR is analogue to linear version of a back-propagation neural net, with two input layers X and Y and one hidden layer T. In such linear modelling, X-Y nonlinearities not corrected during data preprocessing are usually sought modelled implicitly by combining various X-variables with different X-Y curvatures[4]. When implemented with sufficient outlier detection, statistical validation and interpretation graphics, it has proven successful for removing selectivity problems from a number of sensor types,- in particular multi-wavelength NIR data.

2.3 The bilinear regression model

The essential thought behind the bilinear regression model can be summarized by the expression 'The world is under indirect observation'[9]. More detail on the concepts and methods is given in the chemometric literature, e.g. in [4]. In the bilinear regression methods the many individual variables $x_k, k = 1, 2, 3, \cdots, K$ are summarized by a few compact variables t_a, $a = 1, 2, \cdots, A$. Each such latent variable is a linear combination of the X-variables. The block of the few estimated latent variables, matrix $T = t_a, a = 1, 2, \cdots, A$, summarizes the many X-variables in a compact way. A musical analogy would be to regard T as the intensity of the main harmonies (static data) or the main rhythms (dynamic data) in the complicated sounds from an orchestra of multi-string (multi-channel) X-instruments. These few compact bilinear factors, represented by T, are then in turn used as regressors for both the Y-variables and the X-variables themselves. The basic bilinear data model for a single Y-variable is:

$$X = 1\bar{x}' + TP' + E \quad (2)$$

$$Y = 1\bar{y}' + Tq' + f \quad (3)$$

with bilinear products (scores x loadings):

$$TP' = t_a p_a', a = 1, 2, \cdots, A \quad (4)$$

$$Tq' = t_a q_a', a = 1, 2, \cdots, A \quad (5)$$

where:

X-loadings: $p_a = p_{ka}, k = 1, 2, \cdots, K$ and Y-loadings: q_a, X-scores: $t_a = t_{ia}, i = 1, 2, \cdots, N$. Index $i = 1, 2, \cdots, N$ is here the number of calibration samples (objects), $k = 1, 2, \cdots, K$ is the number of X-variables (e.g. wavelength channels) and $j = 1, 2, \cdots, J$ is the number of Y-variables (e.g. the number of analytes calibrated for). Index $a = 1, 2, \cdots, A$ represents the number of bilinear factors. T is called the score matrix and depicts the main between-objects variations in X in a compact form. P and q are the loading matrices for X and y, respectively. These depict the main between-variables relationships. The compact bilinear products TP' and Tq' are intended to span the variation in the X- and Y-spaces that are statistically relevant, usually represented by a low number of phenomena $a = 1, 2, \cdots, A$. They are estimated by least squares regression of each individual variable in X and y on $t_a, a = 1, 2, \cdots$ The residuals E and f are intended to pick up the rest of the variation in the variables X and y, due to noise, model errors etc.

There are many ways of using this bilinear model. For computational and cognitive reasons it may be advantageous to have certain parameters orthogonal to each other: Thereby one can think of - and estimate -each factor separately without loss of generality. In the traditional PCR, and in the present PLSR version developed mainly by the chemometrician Svante Wold, both the scores and the loading weights are orthogonal: The score vector t_a for factor a is estimated by

$$t_a = (X - 1\bar{x}' - \cdots - t_{a-1}p_{a-1}')w_a \quad (6)$$

such that the score vectors are orthogonal ($T'T$ is diag$(t_a't_a)$, $a = 1, 2, \cdots, A$). For similar practical reasons the loading weights $W = w_a, a = 1, 2, \cdots$ in equation 6 are defined to be orthogonal. To avoid scaling problems, problems, either T or W

is usually scaled to fixed length. In the present paper we have chosen to scale W to orthonormality ($W'W = I$). The bilinear loading weights $w_a, a = 1, 2, \cdots$ can be estimated in a variety of ways, depending on what 'interestingness criterium' is used. In traditional PCR, the criterium is maximum described X-X covariance: w_a for each subsequent factor is chosen so as to maximize the X-X covariance described, i.e. $max(t'_a t_a, a = 1, 2, \cdots)$, which are related to the singular values of (centered) X. In the PLSR the concept of 'Y-relevant X-X covariance' is introduced, in order to enhance Y-relevant X-structures and reduce the importance of X-structures with no correlation to Y: Each vector w_a is chosen to maximize the described remaining X- Y covariance,- this means that $(y - 1\bar{y})'t_a$ is maximized. If there is only one Y-variable, this algorithm, which is termed 'PLS1', is non-iterative and fast. If there are more than one Y- variable ($J > 1$), the PLSR is then termed 'PLS2', and requires a certain amount of iterations.

For linear instruments one expects to find a number of factors A that corresponds to Aexpected, the number of independent chemical or physical phenomena affecting the X-data. However, the optimal number of factors may be lower than A expected if the initial calibration data set is small and noisy, and it may be higher than A expected if there are unexpected interferents, curvatures etc. Explicit validation methods are therefore used to decide the optimal of factors, A. Full cross validation is a conservative statistical validation method; graphical inspection for e.g. recognizable spectroscopic structure in loadings $p_a, a = 1, 2, \cdots$ forms a mental validation method; ideally the two should indicate more or less the same number of factors. The final predictor function $f()$ in equation 1, using factors $a = 1, 2, \cdots, A$, can be summarized by the formulation

$$y_i = b_0 + x'_i b \tag{7}$$

where

$$b = W(P'W)^{-1}q' \tag{8}$$

and

$$b'_0 = \bar{y} - \bar{x}'b \tag{9}$$

This can be used for predicting the Y-variables from X- measurements in new objects $i = 1, 2, \cdots$ However, in practice, the same predictions of y_i are obtained whether this simplified formula or a full prediction (via t_i) is used; the latter gives better outlier warnings.

2.4 Preprocessing

2.4.1 Extended Multiplicative Signal Correction (EMSC)

Linear and bilinear calibration models like PLS regression are not particularly suited for approximating multiplicative effects, - i.e. uncontrollable variations that scale all the X-variables by the same factor ([4], chapter 7). One such effect is varying unknown optical path length in spectroscopy, due to varying sample thickness or varying light scattering level. Similarly, in chromatography and electrophoresis, unknown variations in total amount of sample applied to the instrument represent a multiplicative interference. In order to get good additive calibration modelling, such multiplicative effects should be removed prior to e.g. PLS regression.

However, most analytical signals can only be approximately modelled, and that holds true also for multiplicative effects. Unidentified nonlinearities, interactions and unexpected phenomena make purely multiplicative models imperfect. For improved fit to data, multiplicative models may therefore be given an additive term as well. But if there are additive effects (e.g. seen as baseline shifts) on top of the multiplicative effect, the multiplicative effect cannot be removed from the data by simple normalization, like dividing each spectrum by its sum or by a single internal standard. The Multiplicative Signal Correction (MSC) was developed to accomodate such situations under certain conditions. In MSC modelling, the input data Z_{ik} for object i at wavelength k are transformed into MSC treated data (X_{ik}) by

$$X_{ik} = \frac{Z_{ik} - a_i}{b_i} \quad (10)$$

were the scalar a_i is an estimate of the general additive (baseline) offset in object i, and the scalar b_i is an estimate of the general multiplicative (path length) factor in object i. Parameters a_i and b_i are estimated statistically for each object, by linear regression of input spectrum $z_{ik}, k = 1, 2, \cdots, K$ on some reference spectrum, e.g. the average spectrum in matrix $Z = (z_{ik}, i = 1, 2, \cdots, N; k = 1, 2, \cdots, K)$.

In the standard MSC, large spectral effects due to changing chemical composition etc. can erroneously affect the estimation of a_i and b_i. As outlined in [4] p. 350, Stark and Martens in 1989 developed an extension of MSC in order to overcome this problem. In this Extended Multiplicative Signal Correction (EMSC)[7] prior knowledge about the approximate spectral signatures of the main analytes and interferents are included in the estimation of a_i and b_i.

EMSC is used in the present work in order to divide away simulated multiplicative path length variations in the presence of simulated baseline variations.

2.4.2 Signal Interference Subtraction (SIS)

As a further extension of the EMSC it is possible to simplify the subsequent PLS regression. In the Spectral Interference Subtraction (SIS)[6] the above mentioned approximate knowledge about the analytes' and known interferents' spectral signatures is employed to subtract much of the known additive interferences, so that their spectral signatures do not have to be estimated statistically from the calibration data. At the end of this paper, SIS is applied in order to eliminate the effects of known solvent interferents.

3 Experimental

3.1 Input data

Fifty mixtures of organic solvents (benzene, toluene and xylene) in various known ratios were prepared. This sample set is intended to simulate samples from an industrial process stream of an impure solvent. Benzene is the solvent,- the analyte to be calibrated for, while toluene and xylene are to be treated as 'unidentified interferents'. The NIR data of these mixtures were measured in a Guided Wave Model 200-45 process spectrophotometer, using a 2m single-strand optical fiber and a transmission probe configuration. The transmission (T) spectra were obtained at every 2nm between 1600 and 1800nm. The data were linearized by the conventional transform into optical density.

$$OD = \log(\frac{1}{T}) \qquad (11)$$

These original spectral measurements were then modified mathematically in order to simulate grave, unknown pathlength and offset variations (let 'inputdata' be Z_{ik}):

$$Z_{ik} = OD_{ik}b_i + a_i \qquad (12)$$

where a_i and b_i were random uniformly distributed numbers.

Percentage of benzene is employed as the only Y-variable to be calibrated for. This means that the concentration of the other two solvents, toluene and xylene, is considered *unknown* in the data analysis; their interference effects are to be picked up and corrected for automatically by the multivariate PLSR calibration regression. The 50 available samples were split into two representative parts. A calibration set of 27 samples was used for model estimation in the preprocessing and calibration modelling. An independent test set of 23 samples was then used to test the predictive performance of the various models obtained. Within each of these two data sets the samples were sorted according to increasing analyte concentration, to give better graphical representation.

3.2 Data analysis

Mathematical modification of measured O.D. data to generate the 'input data' Z was done in MATLAB[1]. Spectral preprocessing (EMSC, SIS) was performed in the Quantitative Inference Engine Toolbox (QUIET) version 0.9 [2]. PLS regression and plots were performed in the UNSCRAMBLER program, version 2.3[3]. All data analysis was executed on a PC under MS-DOS.

4 Results

4.1 Input data: Fiber-optic process NIR spectra

Since this paper is intended to show a general approach to selectivity enhancement in analytical chemistry, the NIR specific aspects of the present data will not be focused upon. Thus, the actual wavelength identification (every 2 nm from 1600 to 1800 nm) is ignored, and the 101 spectral variables are simply regarded as 101 'instrument channels'. Figure 1 shows the O.D. spectrum of the analyte, benzene in pure state. This is the analyte to be quantified fiber-optically in mixtures containing other 'unidentified' solvents with similar spectra, and with possible other interference phenomena such as chemical interactions, temperature effects, path length variations, baseline shifts etc.. Due to possible spectral overlap from the interferents, only parts of this benzene spectrum can probably be used for quantitative determination. It may even be that the benzene spectrum changes with its chemical environment. Hence empirical multivariate calibration is necessary. Figure

[1]MATLAB is a softwareproduct from MathWorks Inc., 20 North Main St., Suite 250 Sherborn, MA 01770 USA
[2]QUIET is a software package from Consensus Analysis AS, N-1401 SKI, Norway
[3]UNSCRAMBLER is a program package from CAMO AS, N-7041 Trondheim, Norway

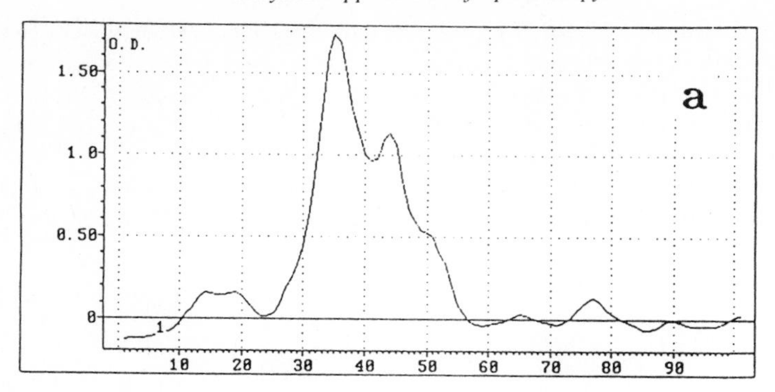

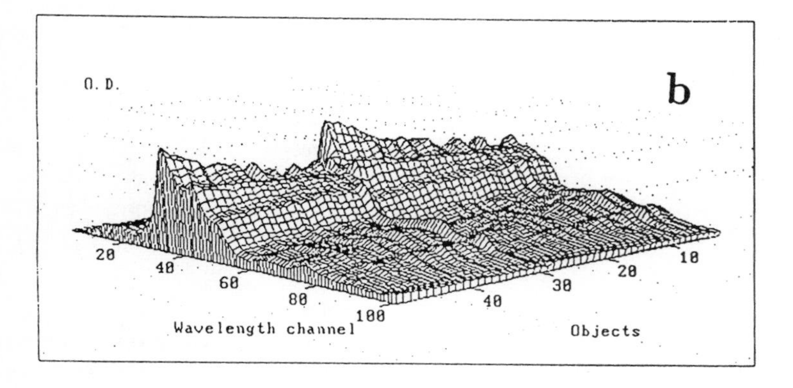

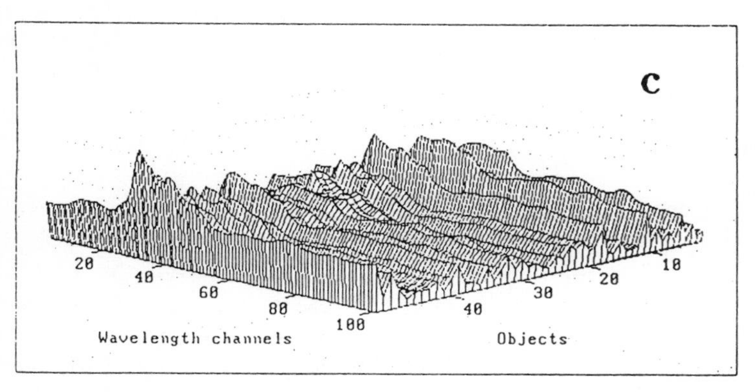

Figure 1: a: Fiber optic NIR O.D. spectrum of the analyte Benzene, b:Originally measured spectra prior to modification, c: Input spectra Z, which have been modified to simulate uncontrolled random path length and baseline variations

1b shows the NIR O.D. spectra of the originally measured spectra of the mixtures in the calibration set and test set, prior to artificial modification. These data are not used in the actual data analysis. Figure 1c shows the modified spectra of the same samples in the same order, with artificially modified variations in 'path length' and 'base line'. These are from now on considered as as 'input spectra' Z. The purpose of the data analysis is now to determine how to convert these spectra into selective and precise determinations of the concentration of benzene.

4.2 Extended Multiplicative Signal Correction (EMSC)

Figure 2 gives the 'input spectra' z_i of two of the calibration samples in Figure 1c, object $i = 1$ and object $i = 23$, together with the average spectrum of the calibration set in Figure 1c. How much of the individual objects' deviations from the average spectrum is due to changing analyte concentration, how much is due to changing levels of unknown interferents (toluene, xylene, plus real chemical interactions etc), how much is due to multiplicative path length effects and how much is due to trivial baseline shifts? In Figure 3 all three spectra in Figure 2 are plotted against the average spectrum in Figure 2. Obviously, the average plotted against itself generates a straight diagonal line. The other two samples $i = 1$, $i = 23$ generate curves with different general linear slope and offset trends plus some individual spectral patterns due to differences in chemical composition. In conventional MSC one would consider these 'chemical' features to be small enough to be ignored, while the linear trends would represent the physical information to correct for. One would therefore have estimated the slope b_i and offset a_i by averaging over the 'non-linear' features. In the new EMSC one attempts to compensate for the 'non-linear' features in the estimation of a_i and b_i in a statistically stable way. Figure 4 shows the spectra of the two chemical interferents toluene and xylene. Information about the spectra of benzene, toluene and xylene was used in the EMSC estimation of a_i and b_i. By this preprocessing the two curves in Figure 2 were transformed as shown in Figure 5. The application of this EMSC preprocessing converted the 'input spectra' in Figure 1c to those in Figure 6.The beneficial consequence of the preprocessing can be seen in Figure 7a and b.

Here, the 'best' individual wavelength channel for benzene, $k = 35$ (1668 nm), is plotted against the benzene concentration, for the test samples ($i = 28, 29, \cdots, 50$). The univariate performance of the 'input spectra' is given in Fig.7a and that of the EMSC-treated spectra in Fig.7b. However, even after EMSC the spectra data still display undesirable variability making them unsuitable for such single-channel univariate calibration.

4.3 PLSR calibration

Figure 7c gives the corresponding results, using all 101 wavelength channels in multivariate PLSR calibration on the EMSC-treated spectra. The EMSC treated spectra in Figures 6 were here used as X-variables in multivariate calibration for $y =$ benzene concentration. Full cross validation within the calibrations sample set was used in order to find how many of the PLSR factors had predictive ability. Three factors were found to have predictive ability: one major factor, one intermediate and one very small factor.

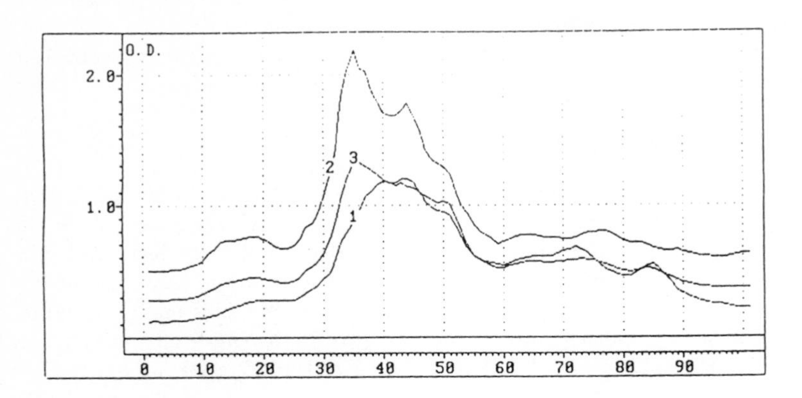

Figure 2: Before EMSC: 'Input' spectra Z; curve 1 =sample #1, curve 2=#23, and curve 3=mean spectrum of calibration set

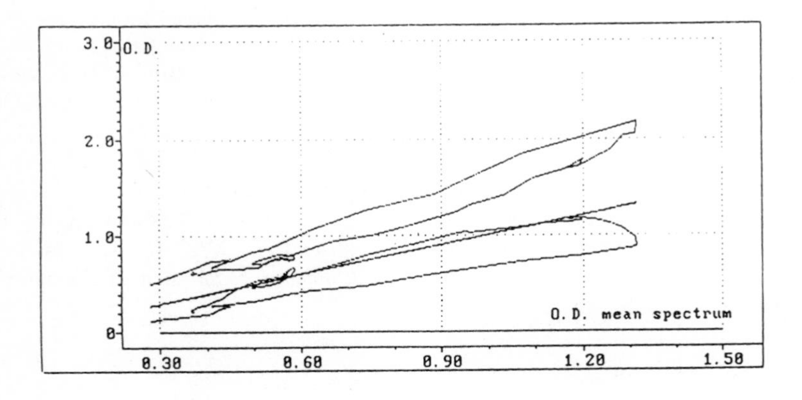

Figure 3: Spectra Z of samples # 1, 23 and mean spectrum vs. mean spectrum

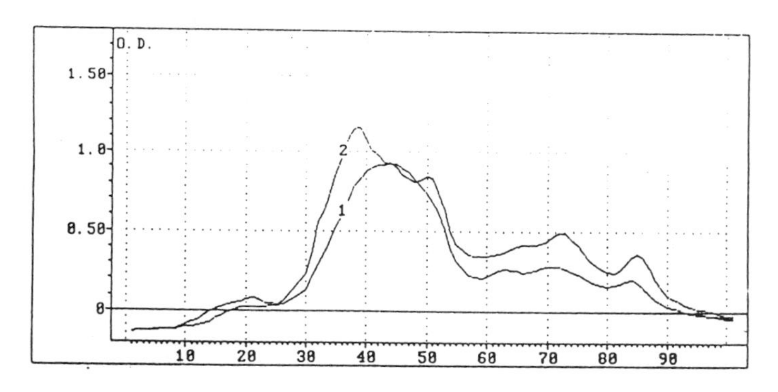

Figure 4: Spectra of the interferents, toluene and xylene

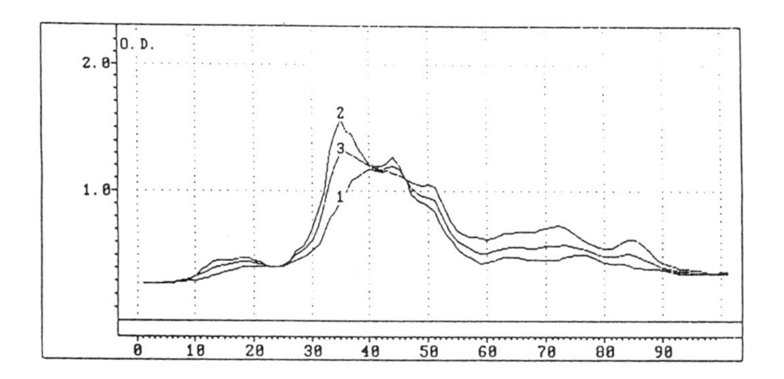

Figure 5: After EMSC: Spectra X; curve 1=sample #1, curve 2=#23, and curve 3 = mean spectrum of calibration set

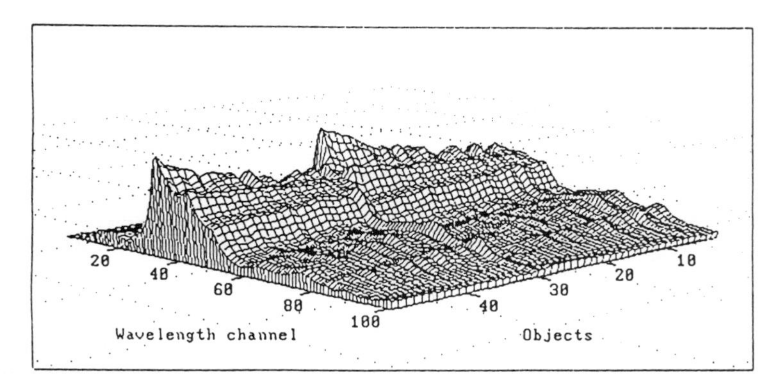

Figure 6: Effect of spectral preprocessing Spectra after EMSC: Spectral data X

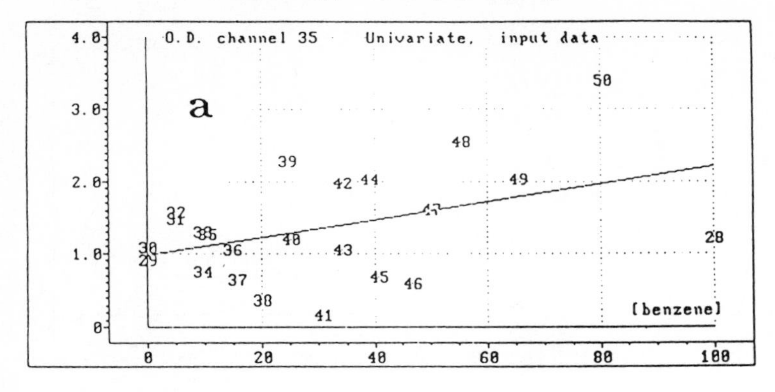

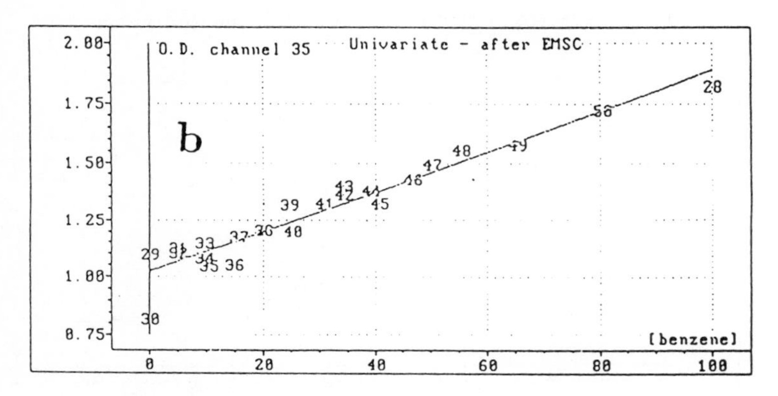

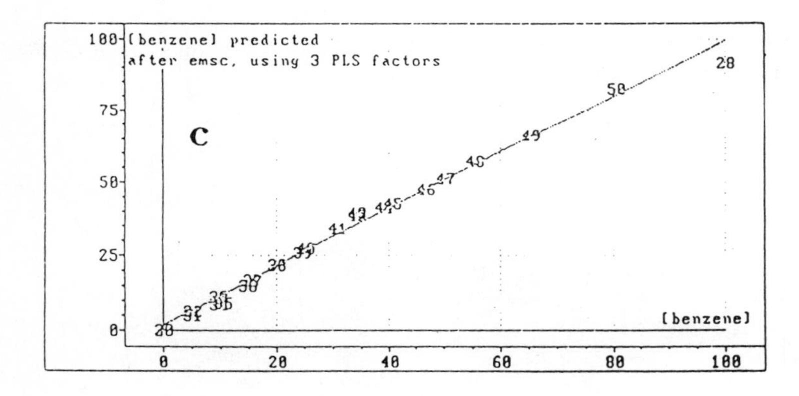

Figure 7: Predictive performances in test set. Univariate relationship between analyte concentration and wavelength channel # 35 (1668 nm): a:Before EMSC, b: After EMSC c:Multivariate modelling after EMSC, using 3 PLSR factors.

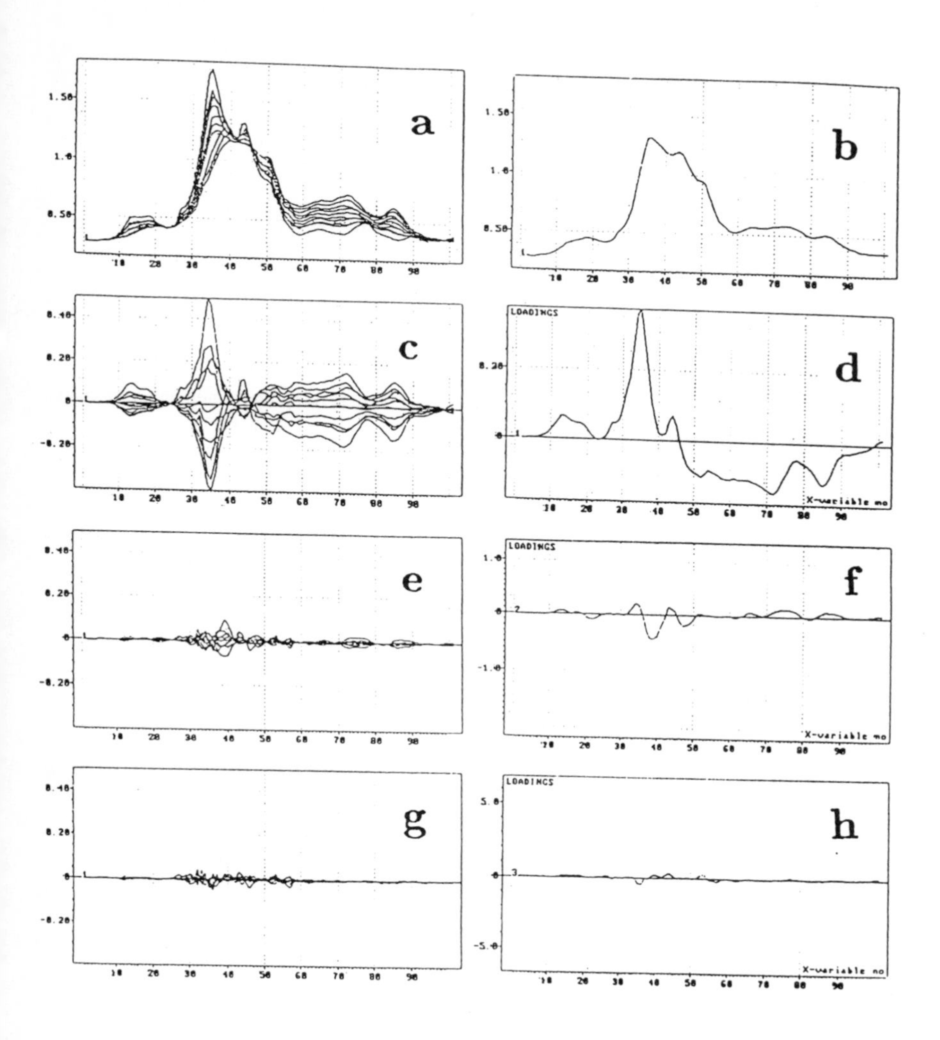

Figure 8: Multivariate calibration modelling. Left side: Spectra for the 10 first calibration samples. Right side: Model parameters estimated by PLSR

8a	Input data X after EMSC and SIS	8b	$\bar{x}$
8c	$E_0 = X - 1\bar{x}'$	8d	Loading p_1'
8e	$E_1 = E_0 - t_1 p_1'$	8f	Loading p_2'
8g	$E_2 = E_1 - t_2 p_2'$	8g	Loading p_3'

Analyte y = benzene percentage was calibrated for by PLSR in the twentyseven calibration samples. The left side of Figure 8 illustrates how the bilinear modelling proceeded for ten of the samples, and the right side of the figure shows model parameters estimated. The PLSR loading parameters are scaled according to the relative importance of each factor to the modelling of the NIR spectra in the calibration set.

Figure 8a shows the NIR spectra of calibration samples, and Figure 8b shows their mean spectrum $\bar{x}$. After subtraction of $\bar{x}$ the variation in the NIR data X looks like Figure 8c. The first bilinear factor's loading p_1 (Figure 8d) summarizes most of this variability. After subtraction of this first factor, Figure 8e shows the residual NIR variation left in X. The second factor's loading p_2 (Figure 8f) summarizes much of this remaining NIR variation. Figure 8g shows the residual NIR variation after subtraction of the second factor as well. Figure 8h gives the third factors' loading, p_3. The fourth and the subsequent factors appeared to contain only random noise and are not shown. Thus, Figure 8 shows how various types of systematic variation in the X-data are picked up and subtracted by the self-modeling PLSR method. No spectral interpretation was required at this stage, except a visual check for apparent random noise structures. Both the graphical inspection (Figure 8) and the statistical cross validation prediction testing indicated one major and two minor factors to be validly estimated from the present calibration dataset. This calibration model was used in order to predict y from X in test set, as shown in Figure 7c.

Figure 9 explains how the multivariate selectivity enhancement works geometrically. Each point now represents a test mixture.

Figure 9a shows corresponding benzene percentage y of both the calibration set (objects 1- 27) and the the test set (objects 28-50) plotted against their scores for the first two PLSR factors from the NIR data, t_1 and t_2. Seen from a certain angle (corresponding to the ratio between the Y-loadings q_1 and q_2) a highly selective relationship is seen between y and these two linear combinations of NIR data X.

Thus, the multivariate PLSR calibration has converted non-selective spectra X (Figures 6, 7c) into very selective analyte predictions, as also seen in Figure 7 , without using any information about the interferences!

For an analytical chemist it may be strange to accept that good predictive ability can be attained in spite of serious non- selectivity in the raw data - particularly when the causes of interference are unknown. The reason why it works is that there is sufficient systematic information in the calibration data themselves to model the unidentified interference effects automatically. However, multivariate calibration - initial calibration as well as updating- should be a causal learning experience as well as a statistical process. Bilinear calibration methods such as the PLSR give graphical access into the covariance matrices involved in multivariate analysis.

Figure 9b shows the same as Figure 9a, but seen from another angle, and with vertical lines indicating the three-dimensional coordinates of the various samples. In the test set samples of pure benzene, toluene and xylene were included for illustration, and the position of these are marked explicitly, and a triangle is drawn between them. The figure shows that the samples fall inside the triangle that has the analyte benzene and the two 'unknown' interferents, at its corners. So, the self-modeling PLS regression has been able to pick up and describe both the analyte variation and the variation within the unknown interferents, even though neither pure spectra nor concentration data were given for the interferents. The two first

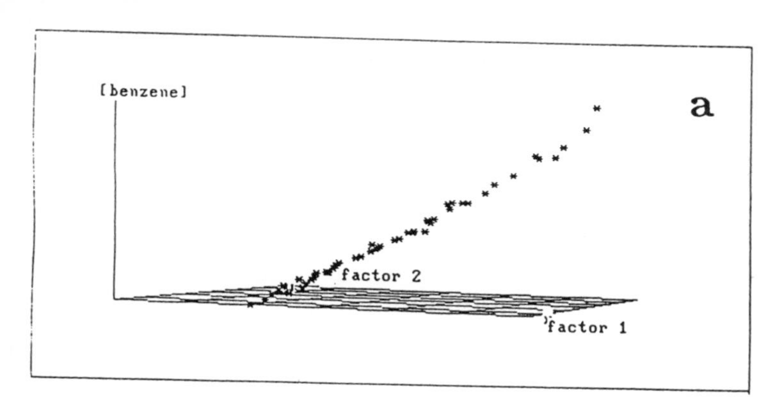

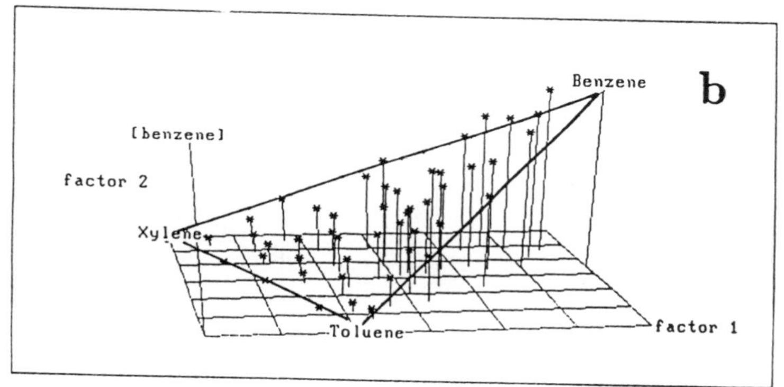

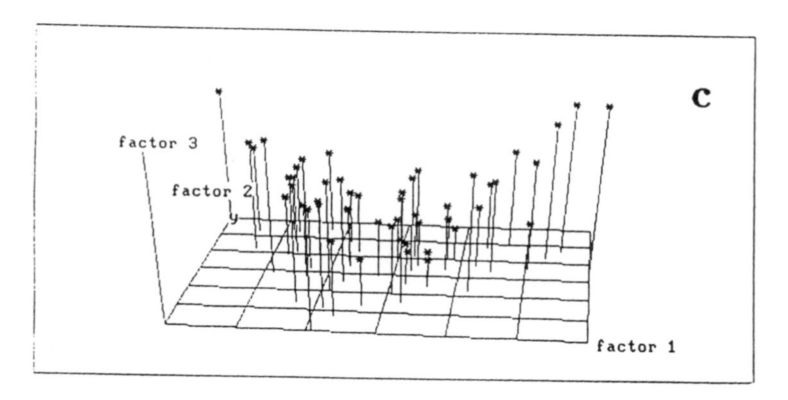

Figure 9:a: Multivariate PLSR modelling: y vs NIR scores t_1 and t_2. b: Same as a:, but seen from another angle. c: A look into X-space:t_3 vs. t_1 and t_2.

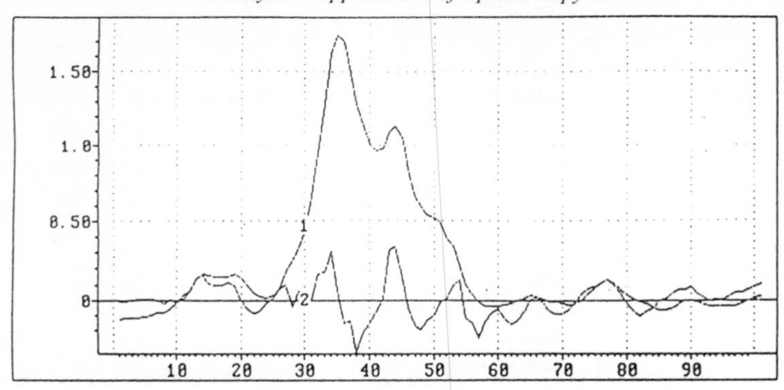

Figure 10: Interpretation of the interference problem. Curve 1: O.D. spectrum of the analyte benzene. Curve 2: Regression coefficient spectrum for benzene determination in the solvent mixtures, b, using 3 PLSR factors.

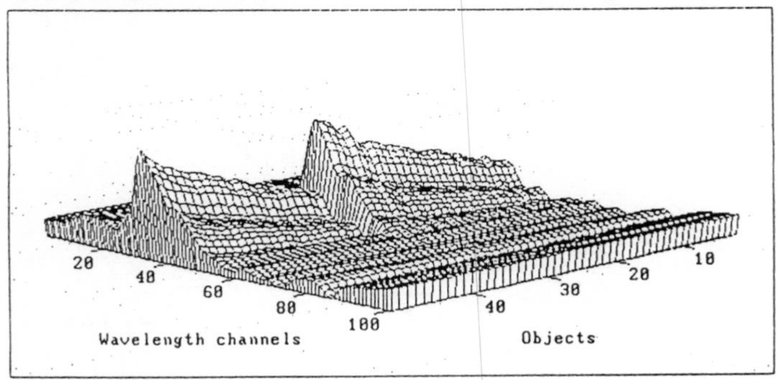

Figure 11: Spectral Interference Subtraction (SIS) preprocessing of spectra after EMSC

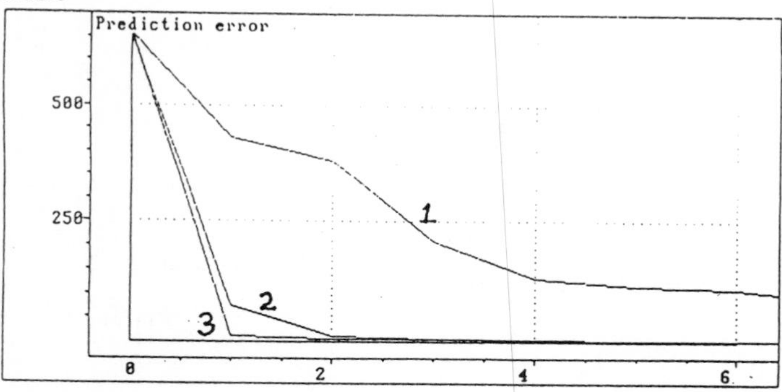

Figure 12: Model comparison : Percent correctly explained variance in y = benzene concentration, estimated in test set (objects 28- 50) as a function of number of PLS regression factors. 1: Raw simulated 'input NIR data' Z. 2: EMSC treated NIR data X 3: EMSC and SIS treated NIR data

factors thus spanned the two-dimensional variation of a three-component mixture system (3 components summing to 100%). Seeing this plane from the side gave the selectivity enhancement required (Figure 9a) and seeing it from the top gave chemical interpretability (Figure 9b).

What was the third, minor but valid factor? In Figure 9c this factor score, t_3, is plotted against the two first factors' scores, t_1 and t_2. The figure shows two things: The ternary mixtures generally lie down in a 'pocket' or valley in the middle, with higher scores t_{i3} for the pure solvents and for binary mixtures. This nonlinear effect may be due to solvent interactions of some sort. However, scores for the third factor seem generally more irregular and noisy than those for the first two factors. This is not unexpected, considering how small variations were available for its estimation (Figures 8g and 8h). But by the third factor the self-modeling bilinear PLSR has been able to account for some of the nonlinear structure in the data. But this effect is small, so it was difficult to model precisely. It would in this case have no grave consequence to leave it out of the calibration model.

4.4 Information content in the analyte spectrum

Figure 10 shows the resulting PLSR calibration coefficient spectrum, vector b, for predicting the fraction of benzene from the 3-factor PLS regression model.

On top of b the figure repeats the benzene spectrum from Figure 1a, for comparison. The figure shows that the regression b vector for benzene relies primarily on the main peak benzene peaks (near channels 35 and 45) but has strong negative corrections at various wavelengths to account for the interference due to toluene and xylene variations. Thus, the b vector can be regarded as the 'net analyte signal',- the aspects of the analyte spectrum that is unique in the presence of the given interferents. Thus, multivariate calibration concerns how to find this net analyte signal, and at the same time to interpret the underlying patterns found in the data.

4.5 Spectral Interference Subtraction (SIS)

The EMSC preprocessing assumed that we knew the approximate spectra of at least some of the interferents present in the model. However, in the subsequent PLSR calibration regression we made no such assumptions; the estimation of the calibration model was purely data driven, based on the intercorrelation structure in the spectral data. This was done in order to illustrate how multivariate calibration works very well in 'dirty' systems. However, the cost of such purely data driven regression modelling is that the calibration data set must be large and precise. Sometimes the calibration sample set is limited, and it is then advantageous to reduce the need for statistical parameter estimation by the use of a priori knowledge. The EMSC treated data (Figure 6) were submitted to further pre-processing by Spectral Interference Subtraction (SIS)[6], where each spectrum was modelled in terms of the spectra of the three constituent spectra;benzene, toluene and xylene. The spectra were then reconstructed without the interferents toluene and xylene. The results are shown in Figure 11.

Comparison with Figure 6 shows a certain simplification, as expected. These SIS treated spectra were submitted to PLS regression, again using samples 1-27 for PLSR calibration. Crossvalidation yielded one very major and one very minor

PLS factor with predictive ability. Hence the knowledge driven SIS pretreatment simplified the data driven statistical calibration model estimation.

4.6 Summary

Figure 12 summarizes the predictive performance of the three ways in which the simulated 'input spectra' were pre-treated.

The figure gives the mean squared error of prediction (average squared deviation between actual and NIR- predicted benzene concentration in the test set) for the different pre- treatments, as functions of number of PLS factors used. The corrupted 'input spectra' Z (curve 1) needed four to seven factors to give minimum prediction error, and this prediction error was very high; rms average error of prediction (rmsep)

$$(\text{ rmsep}) = \sqrt{130} = \pm 11.4\% \tag{13}$$

using 4 PLS factors. Purely multiplicative interferences (same scaling factors at every wavelength) thus generate problems in additive regression models. As discussed by [4](p. 349), this problem is particularly important in practice when very correct instrument response linearization has been found: Ironically, if a less perfect linear instrument linearization had been used, e.g. using T instead of $\log(\frac{1}{T})$, the additive modelling of this multiplicative effect might have been more sucessfull! Multiplicative correction EMSC (curve 2) reduced the rmsep to 2.6% using 3 factors. Good estimation and subsequent division compensation of the multiplicative effect was clearly benefitial in this case. This EMSC is substantially better than conventional MSC treatment ([7]; not shown here).

Thus, using correct a priori knowledge in estimating the multiplicative effects is advantageous. Subtraction of a priori known interferents SIS (curve 3) further reduced rmsep to $\pm 2.5\%$ after 2 factors. Thus, applying correct a priori knowledge to subtract known interferences led to simpler PLS models. This is important for interpretation purposes. The predictive ability was slightly improved as well,- this is probably due to the fact that fewer statistical parameters had to be estimated from the given calibration data.

A typical NIR and IR application of the EMSC technique would probably be spectroscopy of water containing samples with varying turbidity. The water spectrum, plus its temperature variation, should then be included among the a priori spectra used in the EMSC.

Hence the use of 'self-modelling' PLS regression allowed 'automatic' selectivity enhancement for NIR spectroscopy data. Application of a priori knowledge improved the performance of this PLS calibration modelling.

Experience, as well as theory, has demonstrated that these conclusions are relevant for many other types of analytical data- UV/vis/IR /microwave spectroscopy, (reflectance, transmittance, fluorescence), chromatography, electrophoresis, etc. It is important for a chemical laboratory to deliver rapid, reliable and relevant results. However, it is a mistake to spend a lot of time on sample preparation and method optimization in order to make the raw input measurements selective. Quantitative chemometrics can convert non- selective input data into selective output data.

References

[1] Hruschka, W. R. & Norris, K.H. (1982) Least squares curve fitting of near infrared spectra predicts protein and moisture content of ground wheat. Applied Spectroscopy 36, 261-165.

[2] Höskuldsson, A. (1988) PLS regression methods, J. of Chemometrics 2, 211-228.

[3] Lanczos, C. (1950) An iteration method for the solution of the eigenvalue problem of linear differential and integral operations. J. of Research of the National Bureau of Standards, 45, 255-282.

[4] Martens,H. and Naes, T. (1989) Multivariate Calibration. J.Wiley & Sons Ltd. 1989 ISBN 0 471 90979 3.

[5] Martens, H., Westad, F., Foulk, S. and Berntsen, H. (1990). Updating multivariate calibrations of process NIR instruments. Submitted to ISA/90, New Orleans.

[6] Martens, H. and Stark, E. (1990) Spectral Interference Subtraction. In prep.

[7] Stark, E. and Martens, H. (1990) Extended Multiplicative Signal Correction. In prep.

[8] Williams, P.C. and Norris, K.H., eds. (1987). Near-infrared technology in agricultural and food industries. Am.Assoc.Cereal Chem. St.Paul Minnesota.

[9] Wold, H.(1981) Soft modelling: The basic design and some extensions. in: Systems under indirect observation. Causality- structure-prediction. (K.G.Jöreskog and H.Wold, eds.) North Holland, Amsterdam.

[10] Wold, S., Martens, H. and Wold, H. (1983) The multivariate calibration problem in chemistry solved by the PLS method. Proc.. Conf. Matrix Pencils, (A.Ruhe, B Kågström, eds) Lecture notes in Mathematics, Springer Verlag, Heidelberg 286-293.

NEAR INFRARED REFLECTANCE SPECTROSCOPY AND OTHER SPECTRAL ANALYSES

F. E. Barton, II and D. S. Himmelsbach

USDA-AGRICULTURAL RESEARCH SERVICE, RICHARD B. RUSSELL AGRICULTURAL RESEARCH CENTER, ATHENS, GA 30613, USA

1 INTRODUCTION

Near Infrared Reflectance Spectroscopy (NIRS) has changed the way we view spectral analyses. NIRS is a rapid spectral analysis technique that requires minimal sample preparation and yields precise accurate results. It incorporates the same components as any instrumental method; an instrument (spectrometer), a reference to relate a spectral characteristic to an analytical value, and the algorithm that describes the relationship. These components are the same all across the spectrum, from the ultra-violet, (UV) through the radio frequencies (Nuclear Magnetic Resonance, NMR). It is the same for the use of a simple standard to establish an extinction coefficient for a Beer's Law relationship to the use of Partial Least Squares (PLS) model development. Yet as we become more sophisticated and require more information from our analytical system we tend to lose sight of the reference component and its importance to the accuracy of the method. We have used the mid-IR and C-13 NMR to help interpret the NIR region of the spectrum and improve the accuracy of the models developed for the various empirical procedures used for the compositional analysis of agricultural commodities.

These type of studies enabled us to obtain certification of NIRS for the analysis of Crude Protein (CP), acid detergent fiber (ADF) and moisture in feed materials. In this paper we will describe the problems associated with the certification process and NIRS in general, discuss improved reference methods, improved interpretations, and the future of chemometrics and spectroscopy as an analytical system.

Methods and Certification of NIRS

The use of any quality assessment method entails two essential items; the quantitative value and the physical or biological interpretation of that value. Early work to compare the tissues that correspond to fiber analyses values by Akin et al.,[2] Barton and Akin,[3] and Barton et al.,[5] with warm- and cool-season grasses showed that the ADF residues, when compared by Scanning Electron Microscopy (SEM), were not the same across the species. Basically, the findings were: (1) warm-season grass ADF residue still contained tissues which could be degraded by rumen microorganisms and the cool-season grass did not. (2) When lignin was removed as a barrier, the rate and extent of digestion were greatly increased for all samples. Finally (3), that ADF "apparent digestibility coefficients" were not very meaningful as they were over-estimated by the procedure because of the double treatment of digestion "in vitro" with rumen microbial innoculum followed by the standard ADF (2% hexadecyl-trimethyl-ammonium bromide) treatment. As an alternative method and a means of comparison to existing fiber analyses, Windham et al.[10] used extractive hydrolysis with tri-fluoro acetic acid (TFA) in a study of the digestible hydrolyzable polysaccharides from forages. These authors found that the sum of materials extracted by TFA was very close to the In Vitro Dry Matter Digestibility (IVDMD) of the sample and correlated better with IVDMD than any other measure of digestibility currently in use. A more viable alternative to the above problem is to use NIRS. For forage analysis, it is essential that the reference data for all samples correlate to the spectrum in the same manner. Barton and Burdick,[4] showed that a different calibration was needed for the analysis of warm-season grasses, particularly bermudagrasses. The wavelengths selected for ADF, NDF, lignin, and protein (CP) were different for warm-season and cool-season grasses and with the limitations in calibration sample set size (45 samples) separate calibrations were required. Later work by Abrams et al.[1] and Barton et al.[6] showed that broad based calibrations could be developed when a sufficiently large and diverse data set was assembled. The proper selection of a data set requires the use of an algorithm which selects samples based on maximizing the spectral diversity with the minimum number of samples.

In order for the NIRS technique to become a certified procedure, collaborative studies must be conducted to demonstrate the accuracy of the method and the among laboratory errors. The first study was conducted in the mid-1980's and an "Official Method" status was awarded by the Association of Official Analytical Chemists (AOAC) in 1988.[7] The two analyses certified were ADF and CP. This is the first NIRS/Chemometric method to receive AOAC "Official Method" status. The data in Table 1 summarize the among laboratory errors for the chemical and NIRS determinations for ADF and CP. In addition, the data from the original chemical studies are included for comparison. The among laboratory errors of 0.42 for CP and 1.14 for ADF compare favorably with the chemical data from these samples of 0.50 for CP and 1.39 for ADF.

TABLE 1. Standard error of a difference (%) associated with acid detergent fiber and crude protein analyses between collaborative laboratories[h]

Method of analysis and effects	Quality parameters					
	Crude protein			Acid detergent fiber		
	RSD_R[a]	S_R[b]	CV[c]	RSD_R	S_R	CV
Chemical						
Among labs[d]	0.50	0.19	2.58	1.39	0.25	1.76
Among labs[e]	0.75	---	5.60	2.24	---	5.94
Among labs[g]	0.21	---	10.0	0.40	---	1.02
NIRS						
Among labs[f]	0.42	0.15	2.89	1.14	0.35	2.54
Among labs[d]	0.49	0.22	3.03	1.08	0.44	2.07
Among labs[e]	0.83	---	7.26	1.81	---	4.01

[a] RSD_R = Reproducibility standard deviation.
[b] S_R = Standard deviation of overall mean.
[c] Coefficient of variation.
[d] Chemical and NIRS analyses at the following laboratories: A, B, D.
[e] Data adapted from Templeton et al., Proc. XIV Int. Grassl. Cong. 1981. 528-531.
[f] NIRS analyses at the following laboratories A, B, C, D, E, F.
[g] Bates et al., (1955) for CP, Van Soest (1973) for ADF.
[h] Reprinted from J. Assoc. Off. Anal. Chem. 71:1162-1167, 1988.

Perhaps most important is that not only did these same laboratories improve their precision of NIRS analysis over a previous collaborative study (0.83, CP; 1.81, ADF), the errors in the chemical data were improved significantly (0.75, CP; 2.24, ADF). The value of having a second method to verify results is that you can really obtain a measure of accuracy and not be totally dependent on precision. Thus, quality assessment processes can be more closely controlled when the magnitude of all errors are known.

The determination of moisture/dry matter (DM) by the traditional air oven methods is an example of an analysis where the precision is quite good, but the level of accuracy is subject to sampling, presence of other volatiles, and atmospheric errors which are not obvious because there is no independent verification of the results. With the NIRS technique a calibration equation has been developed with moisture data from Karl Fischer analysis that works well with warm- and cool-season grasses, legumes, silage, and mixed feeds (10). The among laboratory errors for the preliminary moisture collaborative study are in Table 2. Since moisture determinations

TABLE 2. Standard error of a difference associated with moisture analysis between collaborating laboratories[e]

Methods of analysis and effects	SED,[a] %	SED(C),[b] %	Coefficient of variance, %
NIRS KF			
Between collaborative sample sets[c]	0.16	0.15	0.17
Between collaborating labs	0.27	0.23	0.24
Between collaborative sample sets[d]	0.23	0.23	0.25
NIRS AO			
Between collaborating labs	0.39	0.35	0.39
Air Oven			
Between collaborating labs	0.63	0.57	0.62
NIRS KF vs NIRS AO			
Between collaborating labs	0.42	0.38	0.32

[a] Standard error of a difference due to random and systematic errors.
[b] Standard error of a difference corrected for systematic error.
[c] SED between collaborative sample sets at referee's laboratory prior to collaborative study.
[d] SED between collaborative sample sets at referee's laboratory after collaborative study.
[e] Reprinted from J. Assoc. Off. Anal. Chem. 71:256-262, 1988.

are very susceptible to environmental changes and time of storage, it was necessary to determine the variation of the sample set before and after they were sent to collaborators. The standard error of difference, corrected for bias (SED(C)), of 0.16 before is quite close to the value of 0.23 after. In fact it is almost identical to the SED(C) among laboratories, 0.27. The SED(C) for NIRS with air oven data was 0.35 (about 50% greater) and the SED(C) for air oven methods was more than twice as large 0.57. This study shows that NIRS is more precise (NIRS air oven vs. air oven, 0.35 vs. 0.57 respectively) and with accurate Karl Fischer data more accurate than laboratory air oven methods (0.23 vs. 0.57, respectively).

Improved Reference Methods

The most important criteria for future development of NIRS or any chemometric procedure is to devise reference methods which correlate to the information in the spec-

UNIVERSITY OF GREENWICH LIBRARY

trum. New methods are required which more closely relate the desired quality parameter to the information in the spectrum or chromatogram. For agricultural commodities such as forages that are utilized by ruminant animals this means the analysis of plant cell wall for protein, fiber, and lignin. Fiber has been assayed by gravimetric extraction techniques. Work by Akin and Barton[2] has shown that what the analysis yields as a quantitative number is not botanically the same for all forages. Thus what we find in the spectrum may not be what the spectrometer sees. For this reason studies were begun to examine an enzymatic method to assay fiber that would better describe fiber physiologically and provide a better quantitative number for the calibration of an NIRS instrument. Preliminary results indicated that a better definition of fiber based on tissue types and amounts removed could be made.[11] The quantitative work is still in its initial stages.

Another way to obtain analytical information is to assay for the monomeric compounds within the cell wall. Microspectrometry offers the advantage of mapping the presence of a compound across a 10-100 micron area of a cross section. This technique has been employed by Akin, et. al.[13] and Hartley, et. al.[12] to study the monomeric phenolic compounds in the individual tissue types of Bermudagrass cell walls. If one has the appropriate location in the plant cell wall and can identify the specific compound, then an analytical as opposed to an empirical assay could be developed. The use of these analytical values could simplify model development and produce a more robust model.

Correlative Spectroscopy and the Future

Chemometrics has been defined as the application and use of mathematical and statistical methods to extract useful chemical information from spectral and chromatographic data. This means it is an approach which will try to find a model which establishes a relationship between what you can measure with assured certainty and want you want to measure. The "working curve" from a "Beer's Law" Intensity vs. Concentration plot is an example of a one term, two dimensional chemometric model. The heart of any chemometric system is the model development module. Current systems used in the analysis of agricultural commodities uses one of three statistical treatments, or variations of them. The best known treatment is "Multiple Linear Regression" which the ARS National NIR Research Project used and explained in Handbook #643. The newer techniques of "Principal Component Analysis/Principal Component Regression" (PCA/PCR) and "Partial Least Squares" (PLS) offer the advantages of more robust model without the problems of collinearity and overfitting. The PLS methods also offer the advantage of plotting the "loadings" or principal components (PC) which can be interpreted as the spectrum of the entity being modeled. When techniques such as "Cluster Analysis" (CA) and "pattern recognition" are mixed with a chemometric model, an expert system can be developed to allow the instruments to function unattended and make decisions and control processes. The future of agricultural analysis is with these techniques and the gains in time and product quality and uniformity they can provide.

Current research indicates that it is possible to correlate one region of the spectrum against another. For example, the mid-IR against the NIR to provide an interpretation of the NIR spectrum (Fig. 1). A map of the squared correlation coefficient can be plotted to show frequencies of highest correlation. Since the mid-IR is more easily interpreted because you can see the fundamental vibrations, this assists the spectroscopist with the difficult task of assigning the observed absorbances in the NIR where the combination and weaker overtones overlap and possess considerable intercorrelation. In Figure 1 the 2800 and 1800 cm^{-1} fundamental vibrations from the NIR for C-H structure in lipids and waxes show sharp high correlations above 0.90 (R^2) at 1386.7, 1730.7, and 2307.1 nm in the NIR, places where statistical correlation to lipid content usually occurs. Note also the high level of general intercorrelation across the entire NIR region.

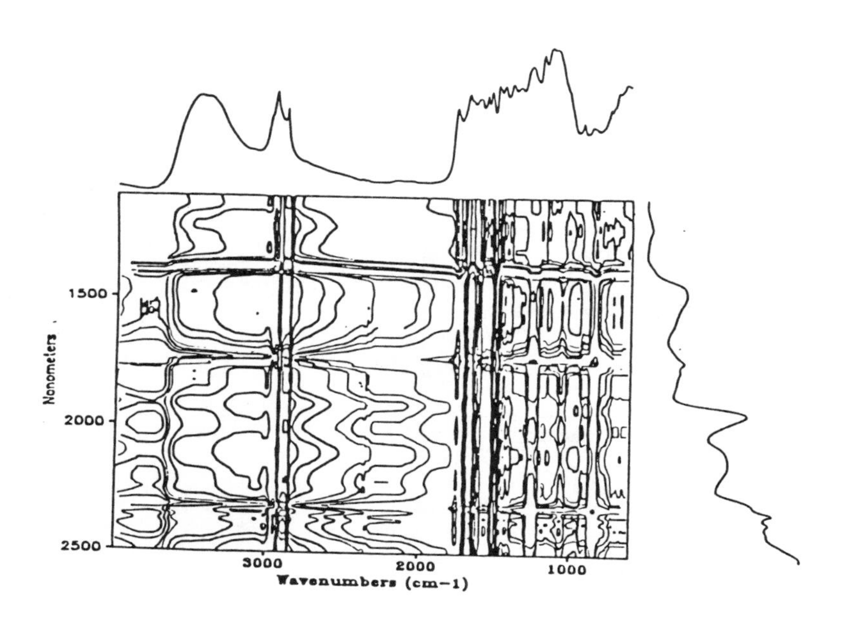

Figure 1 Two-dimensional correlation of the NIR and MIR

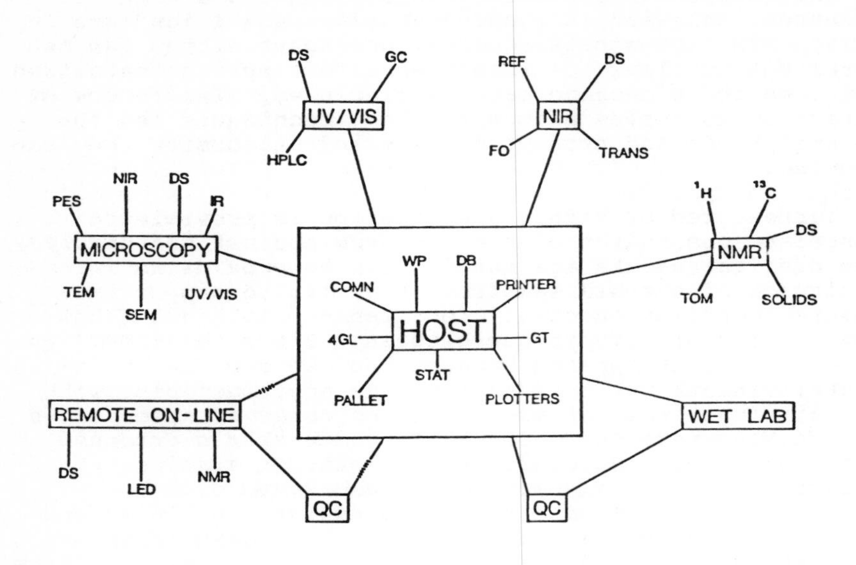

Figure 2 21st century agricultural laboratory

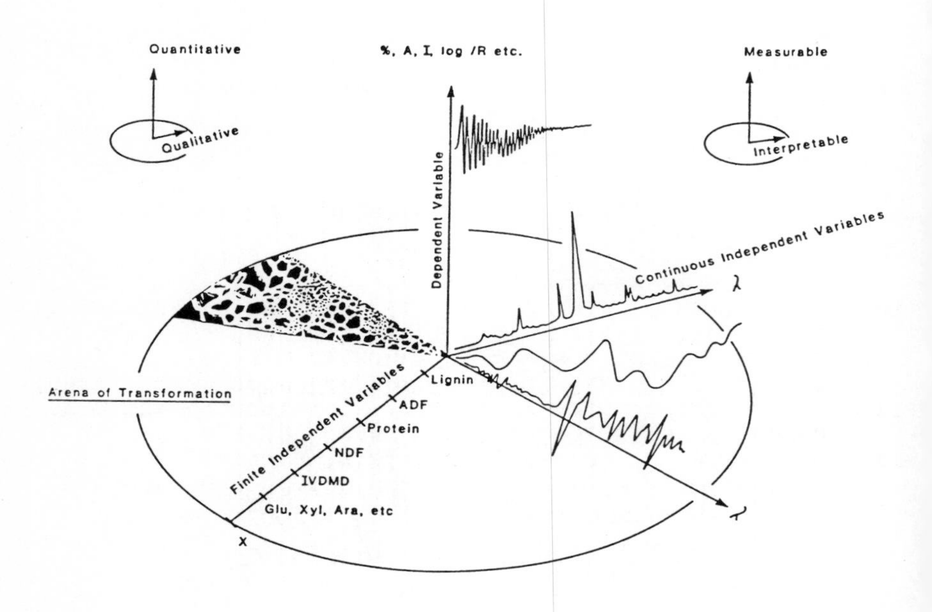

Figure 3 The dimensions of chemical information

Progress in this area has been aided by the appearance of instruments which cover multiple regions of the spectrum. The sample remains in place while the visible (VIS), NIR, and mid-IR spectrum are taken. Thus sampling error and multiple instrument error are removed. Additionally, software packages are available which will combine data from multiple regions and instruments. In the future an analytical laboratory will resemble that shown in Figure 2. Most of the technology is in place to set-up such a laboratory. The concept of what we will be doing is illustrated by Figure 3. Here the analyst is challenged to consider the interpretable and the measurable; the qualitative and the quanitative aspects of an analysis. The possibilities for the extraction of new information from spectral and reference data are limited only by our ability to conceptualize the models involved.

REFERENCES

1. S.M. Abrams, J.S. Shenk, M.O. Westerhaus, and F.E. Barton, II, J. Dairy Sci., 1987, 70, 806.
2. D.E. Akin, F.E. Barton, II, and D. Burdick, J. Agric. Food Chem., 1975, 23, 294.
3. F.E. Barton, II and D.E. Akin., J. Agric. Food Chem., 1977, 25, 1299.
4. F.E. Barton, II and D. Burdick, J. Agric. Food Chem., 1979, 27, 1248.
5. F.E. Barton, II, D.E. Akin, and W.R. Windham, J. Agric. Food Chem., 1981, 29, 899.
6. F.E. Barton, II and G.W. Burton, J. Assoc. Off. Anal. Chem., 1990, 73, (In Press).
7. F.E. Barton, II and W.R. Windham, J. Assoc. Off. Anal. Chem., 1988, 71, 1162.
8. G.C. Marten, J.S. Shenk, and F.E. Barton, II (Eds), "Near infrared reflectance spectroscopy (NIRS): Analysis of forage quality." U.S. Dep. Agric. Handb. No. 643, U.S. Government Printing Office, Washington, DC, 1989.
9. W.R. Windham, J.A. Robertson, and R.G. Leffler, Crop Sci., 1987, 27, 777.
10. W.R. Windham, F.E. Barton II, and J.A. Robertson, J. Assoc. Off. Anal. Chem., 1988, 71, 256.
11. J.H. Woodward, D.E. Akin, and F.E. Barton II, Proc. 46th Ann. Mtg. Electron Microscopy Soc. Amer., 1989.
12. R.D. Hartley, D.E. Akin, D.S. Himmelsbach, and D.J. Beach, J. Sci. Food Agric., 1990, 50, 179.
13. D.E. Akin, N. Ames-Gottfred, R.D. Hartley, R.G. Fulcher, and L.L. Rigsby, Crop Sci., 1990, 30, 396.

PRINCIPAL COMPONENTS ANALYSIS FOR FTIR SPATIAL MAPPING AND TIME RESOLVED DATA

R. E. Aries, J. Sellors, and R. A. Spragg

PERKIN-ELMER LIMITED, BEACONSFIELD, BUCKINGHAMSHIRE

INTRODUCTION

Modern FTIR instruments are often used in measurements which produce a series or array of spectra. There are considerable advantages in using principal components analysis (PCA) as a data reduction method for these applications. In PCA the original spectra are represented in terms of a set of independent spectrum characterised by the contributions of these principal components to the spectrum, called the principal component scores. The relationships between the spectra are found by examining plots of these scores. The spectral features corresponding to any trends of differences in scores are identified by forming the corresponding combinations of the principal components.

We have applied PCA to a number of areas including kinetics studies, microscopy, and evolved gas analysis. The procedures and the potential are illustrated here by an example involving curing of an epoxy resin.

EPOXY RESIN CURING

PCA can provide more information about a reaction such as curing of an epoxy resin than the traditional IR method of monitoring absorbance at a single frequency or thermal analysis methods. The reaction of a bisphenol A epoxy resin with excess of an aliphatic diamine at 80°C was followed by recording FTIR spectra at 45 second intervals.

The changes in the spectra are relatively small but are evident in the regions above 300cm^{-1} and 900-1200cm^{-1}. Three principal components account for 99.97% of the variance in the data. The scores for the first principal component form a monotonic curve which is similar to that observed by

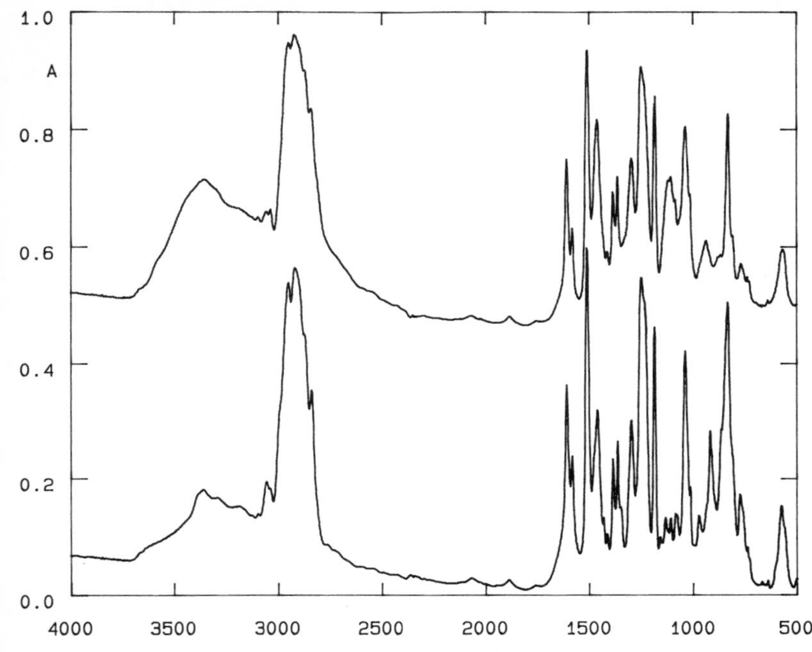

1. FTIR spectra of a thin film of an epoxy resin during curing.
Lower : initial, upper : final.

monitoring the 916cm^{-1} absorption band from epoxy groups. The first principal component shows major features associated with the epoxy groups of the starting material and the hydroxyl groups of the end product.

Plots of the scores for the second and third principal components both show turning points. This indicates that qualitatively different changes are occurring at different times. A plot of the scores for these two principal components against each other reveals three distinct stages which appear as linear regions in the plot. During the first two minutes (spectra 1-4) the score on PC2 decreases rapidly and that on PC3 increases. For the next four minutes the score on PC2 is almost constant while that on PC3 decreases (spectra 5-10). Thereafter both scores increase slowly to their final values (spectra 11-40).

The changes involving PC2 and PC3 show how the reaction at any point differs from the overall reaction associated with PC1. These differences can be examined by constructing the combinations of PC2 and PC3 corresponding to the directions observed in the scores plot. During the first stage there is an overall decrease in amplitude and

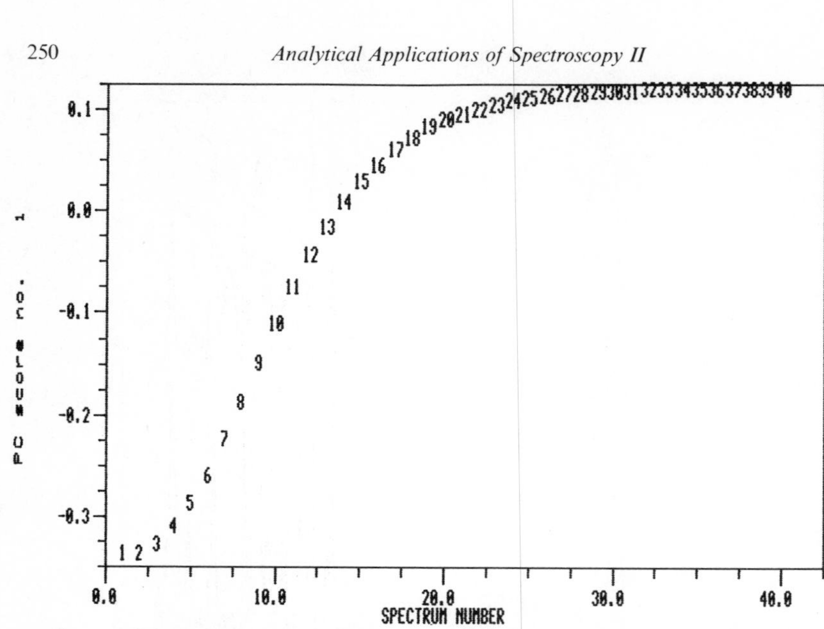

2. Scores on PC1 for individual spectra recorded during curing.

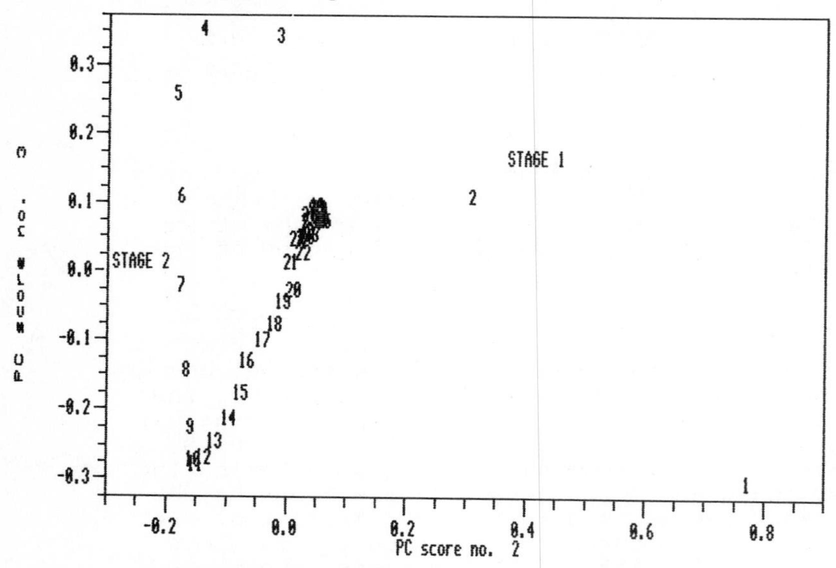

3. Scores on PC2 versus scores on PC3 for individual spectra.

slight broadening of all the spectral features of the resin component. This probably reflects a physical change rather than chemical reaction. In the second stage the main changes are seen in the $3100\text{-}3500 cm^{-1}$ region.

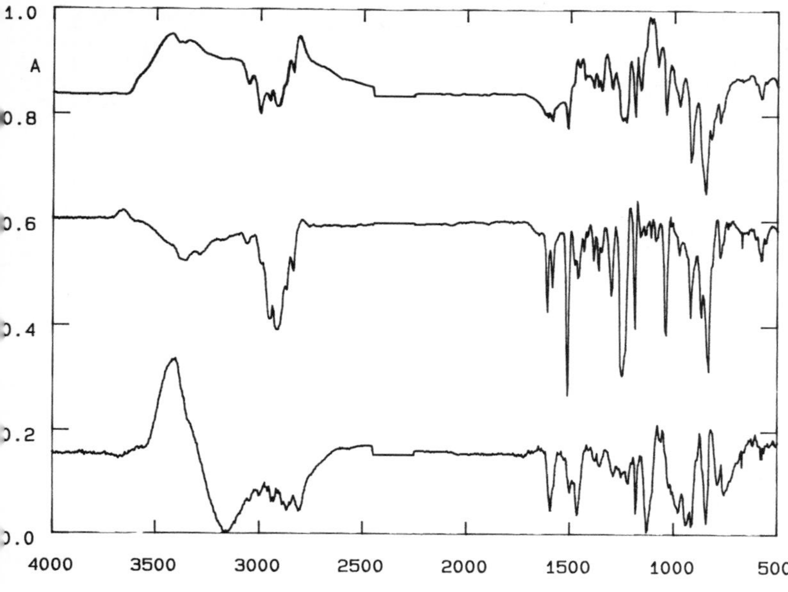

4. Top : PC1. Middle : combination of PC2 and PC3 corresponding to stage 1. Bottom : combination of PC2 and PC3 corresponding to stage 2.

PCA clearly reveals more detailed information than other ways of examining these data. Similar results have been achieved with series of spectra obtained during solid-liquid phase changes and from mapping across inhomogenous samples with an FTIR microscope. Even when PCA reveals no more other information than specific frequency plots it is much more sensitive and does not rely on the user selecting appropriate frequencies. It is also very effective at revealing unexpected features.

The software used for these analyses was part of the Perkin-Elmer CIRCOM package.

QUANTITATIVE NIR FT-RAMAN SPECTROSCOPY – A CORRELATION BETWEEN DIESEL FUEL QUALITY AND FT-RAMAN SPECTRA USING MULTIVARIATE ANALYSIS

Kenneth P. J. Williams

BP RESEARCH, SUNBURY RESEARCH CENTRE, CHERTSEY ROAD, SUNBURY-ON-THAMES, MIDDLESEX, TW16 7LN

Rupert E. Aries, David J. Cutler, and David P. Lidiard.

PERKIN-ELMER LIMITED, POST OFFICE LANE, BEACONSFIELD, BUCKS, HP9 1QA

1. INTRODUCTION

Cetane number and cetane index are the two main parameters used in the petroleum industry to measure diesel fuel quality. The cetane number of a diesel fuel is a measure of the ignition delay characteristics of the fuel; the higher the cetane number, the shorter the ignition delay. By definition the cetane number of a fuel is the percentage of cetane in a blend of cetane and α-methyl naphthalene which has the same ignition qualities as the test fuel; when tested in the same engine under the same conditions. Cetane (n-hexadecane) and α-methyl naphthalene are arbitrarily assigned values of 100 and 0, respectively, i.e. cetane has very good ignition delay qualities and α-methyl naphthalene very poor.

Experimental measurements of the cetane number require the use of a "standard" test engine and are difficult to make. The current ASTM standard quotes a repeatability of 0.6 - 0.9 and a reproducibility of 2.5 - 3.3 (both depend on the cetane number measured). It can be seen from these figures that repeatability and reproducibility are poor. Furthermore, there are only a few places where engine testing for cetane number is carried out, however, this is still the referee method for the determination of diesel fuel quality.

The cetane index of a fuel is a calculated parameter. This measure has a complex relationship which depends on the density of a fuel and its boiling range. The calculated cetane index can be used to estimate the cetane number where a cetane engine is not available, or the quantity of sample is too small for engine testing. Once the cetane number of a fuel has been established, the cetane index provides a useful cetane number check on subsequent samples, providing the source and method of manufacture are unchanged. The cetane index equation is not, however, applicable to fuels containing additives for raising the cetane number, or to pure hydrocarbons, synthetic fuels or coal tar products.

The Institute of Petroleum has recently proposed an equation for the calculation of cetane index, IP method 380, which requires the density together with the 10%, 50% and 90% recovery temperatures. This equation has been shown to produce a better correlation with the cetane number, than previous methods (1). The quoted correlation coefficient of 0.94 and standard error of 2.09 cetane units is substantially better than the previous measurements.

Clearly, the level of accuracy and repeatability of diesel fuel quality is far from ideal and scope exists within the oil industry to improve this situation. Molecular spectroscopy in general terms offers many advantages over the current technology in that it is often non invasive and non destructive. In addition it provides data which is highly specific and has good reproducibility and repeatability.

Indeed, Raman spectroscopy has been widely used in this way over recent years in particular, both for in-situ reaction monitoring and kinetic studies (2,3). However, the two major drawbacks of the method for achieving a universal range of applications have been problems of laser induced sample fluorescence and sample decomposition. The intensity of the laser induced fluorescence is often orders of magnitude more intense than the Raman scatter and often prevents any meaningful analysis. Many groups over the past decade have tried various methods to overcome these problems which have included, discrimination between the fluorescence and the Raman photons on a temporal basis (4), the use of far red and ultraviolet laser excitation (5,6) and surface enhanced Raman spectroscopy (7). Problems of sample decomposition have been overcome in a variety of ways including sample spinning and cooling.

In 1986 the demonstration of Fourier transform (FT)-Raman spectroscopy, which made use of near infrared (NIR) Nd:YAG laser source at 1.064 um, provided a method with the potential for overcoming many of the fluorescence and sample decomposition problems (8). Since the original demonstration of the method many new applications of Raman spectroscopy, from samples previously not amenable to analysis, have been published (9-11).

One publication, to date, has detailed the opportunities for obtaining quantitative NIR FT-Raman data from a fuel mixture (12). This study used a multivariate analysis method to model and predict the composition of synthetic fuel mixtures generated from the individual pure components. The work clearly demonstrated that an individual component concentration could be predicted, however, no underlying chemistry was deduced from the analysis. Further, the samples analysed were sythesised mixtures rather than real world samples in which the number of sources of variation is likely to be very much higher.

This paper presents a quantitative study from a series of diesel fuels which have been analysed using a proprietary multivariate software package called CIRCOM. It is shown that a good correlation between the NIR FT-Raman data and the fuel quality can be obtained. We also demonstrate the value of our approach for understanding the basic chemistry of complex systems.

2. EXPERIMENTAL

NIR FT-Raman spectra were recorded using an experimental system comprising a modified Perkin-Elmer 1700 X near infrared Fourier transform spectrometer fitted with a germanium detector operating at 77 K. The instrument characteristics have been described in full elsewhere (13). Laser excitation at 1.064 um was provided from a Nd:YAG laser supplied by Spectron Lasers Limited.

Raman data from a series of 18 diesel fuels were recorded from a 10 mm pathlength cuvette. This had been silvered on three sides to increase the amount of Raman light collected by the instrument. Data were obtained between 4000 and 400 cm^{-1}, 4 cm^{-1} resolution for 32 accumulations.

Laser powers and sampling conditions were kept constant throughout the course of the experiment. Repeat runs of the 18 samples were made on a daily basis for three days to model any of these variations, this resulted in 54 spectra used for our multivariate calibration set.

The statistical package used in this study was proprietary software called CIRCOM from Perkin-Elmer Limited. The method is a modified form of a principal component regression (PCR) and is described in detail elsewhere (14,15). The NIR FT-Raman data used in the CIRCOM method covered the spectral range 3200 to 600 cm^{-1} using an interval of 4 cm^{-1}. The data were mean centred. Values of cetane number and cetane index for the diesel fuels studied were obtained by conventional methods.

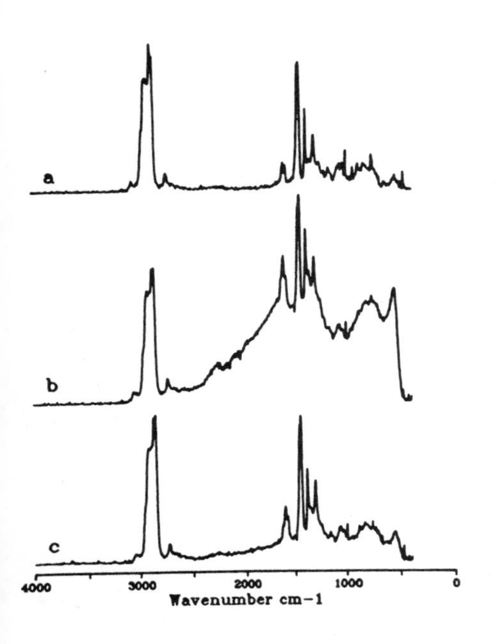

Figure 1

NIR FT-Raman spectra of diesel fuels in arbitary intensity units. (a) Sample with low cetane character, cetane number 42, cetane index 39.1. (b) Sample with high laser induced fluorescence, cetane number 45.4, cetane index 47.2. (c) Sample with high cetane character, cetane number 50.6, cetane index 51.1.

3. RESULTS AND DISCUSSION

The main goals for this research were two fold, initially and foremost; to assess the scope for the FT-Raman method for quantitative analysis using multivariate methods. Secondly, it was hoped that the underlying chemistry could be extracted from the statistical analysis.

Figure 1 illustrates three NIR FT-Raman spectra recorded from diesel fuel samples. The data represent the extremes of the spectral information recorded from the series of samples investigated. The most striking differences in the data are in the intensity of the laser induced sample fluorescence (cf. Fig 1(a) and 1(b)) and variations in the chemical compositions between samples (cf. Fig 1(a) and 1(c)). These chemical variations relate to fuel performance and reflect the relative proportions of the aliphatic and aromatic (naphthalenic) components in the diesel fuel. The spectra clearly contain information which relate to fuel performance as measured by cetane index and cetane number. If this information can be extracted and isolated from the other sources of variation within the data set, using multivariate calibration techniques, then a quantification should be possible.

In total 54 spectra were used in our data set. The CIRCOM analysis revealed five significant principal components (PC's) which expressed 99.4% of the total variance (see Figure 2). PC 1 is identified as being principally due to variations in sample fluorescence (94% variance) which as stated previously can be very marked. The second PC (4.3% variance) has the appearance of a Raman spectrum with an intense peak at 1378 cm^{-1} together with other peaks of medium intensity at 1662, 1602, 1578 and 1432. These peak positions were found to fit extremely well with those recorded from a sample of pure α-methyl naphthalene. PC's 3 and 4 were very similar and express 0.7 and 0.3% of the total variance, respectively, a 50° rotation of these factors revealed intense Raman peaks at 2,852, 1436 and 1304 cm^{-1}. These have been shown to correlate well with the Raman spectrum of n-hexadecane. The final PC used PC 5 (0.1% variance) is statistically significant but contains no chemical information.

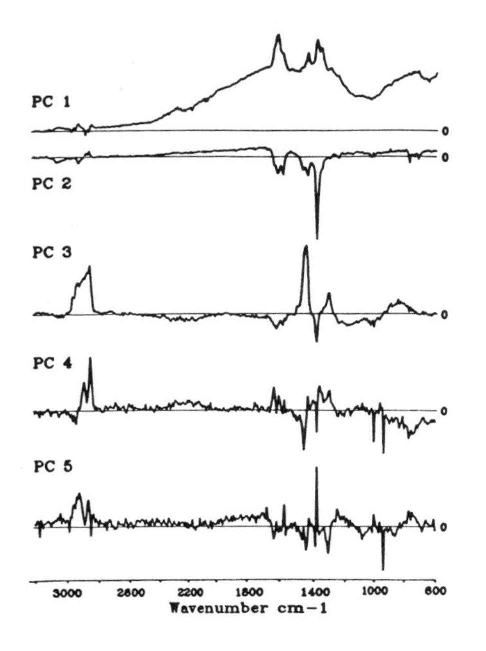

Figure 2

Significant principal components (PCs) identified by CIRCOM.

The application of multiple linear regression to the PC scores for each sample against the known fuel quality parameters of cetane index and cetane number produced significant correlations. The statistical values obtained are given in Table 1. It is apparent that the correlation coefficient obtained against cetane index (r^2 - 0.93) is significantly better than for cetane number (r^2 - 0.77). However, this is not totally unexpected since the errors associated with cetane number measurements using the method discussed previously, are much greater than calculated values for cetane index. The values for the standard error of estimates and standard error of prediction for cetane index of 1.33 and 1.22 cetane units, respectively, are very encouraging and well within the repeatability and reproducibility values quoted using traditional methods. The same measures for cetane number are higher (see Table 1) which again reflect the quality of the primary data.

<u>TABLE 1</u> Summary of CIRCOM Analysis of Fuel Quality Parameters

	Cetane Number	Cetane Index
Maximum Value	50.6	52.7
Minimum Value	39.1	39.7
Average Value	45.1	46.1
Correlation Coefficient (r^2)	0.77	0.93
Standard Error of Estimate (SEE)	2.03	1.13
Standard Error of Prediction (SEP)	2.19	1.22

Property coefficient weighting spectra were generated from CIRCOM for both cetane index and cetane number. These were largely indistinguishable which indicates that the underlying chemistry of the two properties is identical. These data could be interpreted further by multiphoton by the average NIR FT-Raman spectra, calculated by CIRCOM. The resulting "spectra" indicated positive and negative features associated with aliphatic and naphthalene species. The agreement between this and the previously known chemistry which influences fuel performance is excellent.

4. CONCLUSIONS

The NIR FT-Raman data obtained from a series on diesel fuel samples have been shown to correlate well with fuel quality parameters using multivariate analysis methods. The correlation was better for the laboratory derived parameter cetane index. The less rigorous correlation achieved between the FT-Raman data and cetane number is attributed to the errors in measuring fuel quality from ignition delays in an engine test.

The chemistry extracted from our analysis has provided us with knowledge in respect of the components which influence fuel quality and performance. This has correlated well with previous information. However, the method offers scope for use in a predictive manner which is chemically less well defined.

5. ACKNOWLEDGEMENT

Permission to publish this manuscript has been granted by British Petroleum plc.

6. REFERENCES

1. C.J.S. Bartlett, Petroleum Review, June 1987, 48.

2. D.L. Gerrard and H.J. Bowley, Anal Chem 1988, 60, 365R.

3. D.L. Gerrard, Anal Chem 1986, 58, 6R.

4. N. Everall, R.W. Jackson, J. Howard and K. Hutchinson, J. Raman Spectrosc., 1986, 17, 415.

5. K.P.W. Williams and D.L. Gerrard, Optics and Laser Tech, 1985, 245.

6. W.D. Boreman and T.G. Spiro, of Raman Spectrosc., 1980, 9, 369.

7. P. Hildebrandt and M. Stockburger, J Phys. Chem, 1984, 88, 5944.

8. T. Hirshfeld and B. Chase, Appl Spectrosc., 1986, 40, 133.

9. K.P.W. Williams, S.F. Parker, P.J. Hendra and A.J. Turner, Microchim Acta (Wien) II, 1988, 231.

10. P.J. Hendra, P-Ce Barenzer and A. Crookell, J Raman Spectrosc., 1989, 20, 35.

11. K.P.J. Williams and S.M. Mason, Trends in Anal Chem, In Press (1980).

12. M.B. Seaholtz, D.D. Archibald, A. Lorber, and B.R. Kowalski, Appl Spectrosc., 1989, 43, 1067.

13. A. Crookell, P.J. Hendra, H.M. Mould and A.J. Turner, J Raman Spectrosc., 1990, In Press.

14. P.M. Fredricks, J.B. Lee, P.R. Osborn and D.A.J. Swinkels, Appl Spectrosc., 1985, 39, 303.

15. P.M Fredricks, P.R. Osborn and D.A.J. Swinkels, Anal Chem 1985, 57, 1947.

WHAT'S ν IN NEAR-INFRARED HADAMARD TRANSFORM SPECTROSCOPY

R. A. Hammaker, A. P. Bohlke, J. M. Jarvis, J. D. Tate, J. S. White, J. V. Paukstelis, and W. G. Fateley

CHEMISTRY DEPARTMENT, KANSAS STATE UNIVERSITY, MANHATTAN, KANSAS 65506, USA

1 INTRODUCTION

Two popular methods for performing optical spectrometry are dispersive spectrometry (DS) and Fourier transform spectrometry (FTS). Now there is third alternative, Hadamard transform spectrometry (HTS), which is a merging of dispersive and multiplexing techniques yielding some of the most desirable features of both DS and FTS.[1] A description of the Hadamard transform (HT) spectrometer is described in the literature,[1-3] as are the specifics of our HT spectrometer.[3-5] Also, the theory and practical examples of the most efficient way to carry out a multiplexing technique such as HTS are available elsewhere.[1,2]

In a dispersive HT spectrometer, the radiation from a source is dispersed and presented spatially onto a multi-slit mask mounted in a focal plane. An encoding scheme is based on Hadamard mathematics and allows the permitted of particular combinations of spectral resolution elements that are allowed to pass through the mask. The radiation exciting the mask is recombined (i.e. dedispersed) into pseudo-white light and passed to the single detector. The number of Hadamard encodements needed is the same as the number of spectral resolution elements, usually the number is 2^{n-1}. The data collected from the HT spectrometer is a plot of detector response versus encodement number, called an encodegram by analogy to the interferogram as a plot of detector response versus optical retardation, or equivalent time, for the Michelson interferometer in the Fourier transform (FT) spectrometer.

Our development has been the introduction of the stationary Hadamard encoding mask.[2-5] The use of electro-optical materials to generate the mask elements provides a stationary Hadamard encoding mask and eliminates the mechanical problems associated with continuously moving mask.

The advantages of FTS compared to DS are: (1) the multiplex or Fellgett advantage,[6] (2) the frequency precision or Connes' advantage and (3) the throughput or Jacquinot advantage.[6] The HT spectrometer possesses a multiplex advantage. When a stationary Hadamard encoding mask is used in the HT spectrometer, the whole instrument has no continuously moving parts. Consequently, the spectral character of a spectral resolution element passing through a given mask element remains unchanged during the experiment. This feature provides an advantage analogous to the frequency precision or Connes' advantage in FTS. Our HT spectrometer exhibits good co-addition and spectral subtraction in practice.[3-5,7,8] Since dispersive spectrometers generally have f-numbers considerably larger (>f/4) than those of interferometers (f/1-f/4), no throughput or Jacquinot advantage is expected for HT spectrometer. A recently designed concave grating with f/2 collection has changed this. A good test of any dispersive instrument is the ability to collect Raman species. A general comparison of HT and FT Raman spectrometers in Table 1.

Table 1. Instrumentation: A Comparison

	HT-Raman	FT-Raman
Spectrometer	dispersive	interferometric
Spectral collection	optical encodement via grating and mask	moving mirror (after filtering)
Rayleigh line rejection	field stop mask (spatial filtering)	optical filtering
Transformation	fast HT	fast FT
Multiplexing (Fellgett advantage)	yes	yes
Jacquinot advantage	no	yes
Co-addition /subtraction (Connes' advantage)	yes	yes
Resolution	low	high
Spectral range	narrow	wide
Dynamic range	small	large
Simple electronics and computing	yes	no
Continuously moving parts	no	yes

Inspection of Table 1 suggests that for simplicity and cost the clear choice is the HT-Raman spectrometer but for overall capability the clear choice is the FT-Raman spectrometer. The most important advantage of the HT-Raman spectrometer is the use of spatial filtering for Rayleigh line rejection. The major potential limitations of HT spectrometers are: (1) low resolution and (2) limited spectral range for a given grating position. In addition to its two advantages, (1) the Fellgett or multiplex advantage and (2) the Connes' or frequency precision advantage, the HT-Raman spectrometer has the three simplicity factors: (1) Rayleigh line rejection by spatial filtering, (2) no continuously moving parts, (3) the fast Hadamard transform, (4) the selective multiplex advantage. That is the elimination of regions of the spectra. In many cases, the two limitations are counter-balanced by the two advantages and the four simplicity factors.

2 EXPERIMENTAL

The Hadamard transform spectrometer used to collect the spectra has been previously described in detail.[3-5,7-9] Briefly it consists of a modified Jarrell-Ash Model 25-100 double Czerny-Turner spectrograph. The spectrograph was converted to a polychromator by replacing the intermediate slit with an encoding mask. The second half of the instrument was reconfigured to dedisperse the light that passed through the encoding mask and onto the detector. The detector used was a liquid-nitrogen-cooled indium-gallium-arsenide (InGaAs) photodiode. A continuous wave Nd:YAG laser (CVI Laser Corp. model C-95), operated in a multimode at 1.064 μm, was used as the Raman excitation source. The first generation of stationary Hadamard encoding masks used a polarizer-analyzer pair surrounding a nematic liquid crystal (LC) to generate the opaque and transmissive state by absorption and transmission, respectively. The second generation of stationary Hadamard encoding masks utilizes polymer dispersed liquid crystal (PDLC) materials and operates on the principle of a refractive index mismatch causing scattering and a refractive index match causing transmission to generate the opaque and transmissive states, respectively. Table 2 compares the parameters used with the two generations of masks.

Table 2. A comparison of data acquisition parameters required for the 127 element nematic LC Hadamard encoding mask and the 255 element PDLC Hadamard encoding mask.

Parameter	127 element nematic LC Hadamard mask	255 element PDLC Hadamard mask
Laser power at sample (mW)	600 - 900	300 - 400
Data acquisition time (min.)	6	3.2
Entrance slit width (μm)	800	270
Resolution (cm^{-1})	8 - 12	4 - 6
Points coadded/encodement	5000	400 - 600
Delay time (ms)	500	400

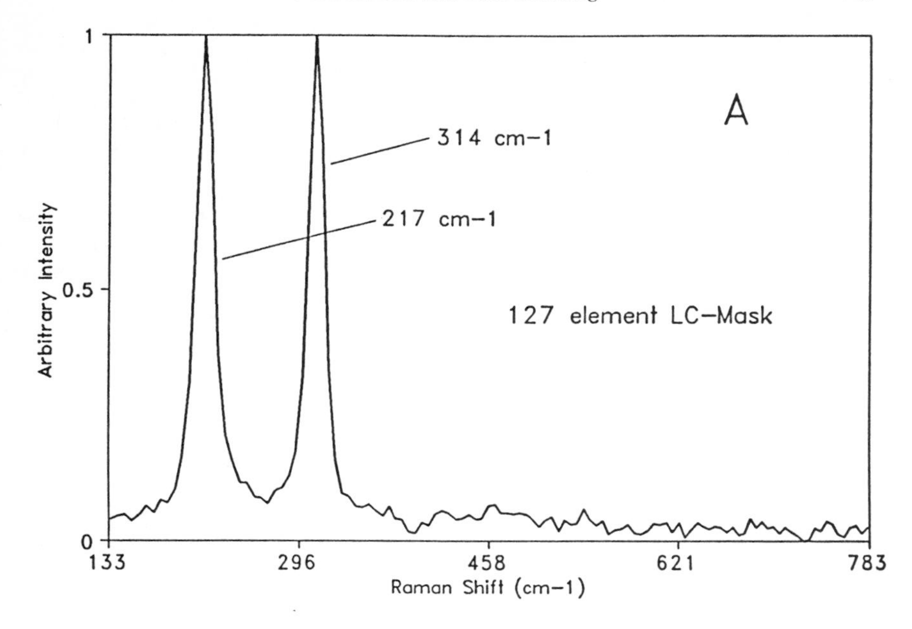

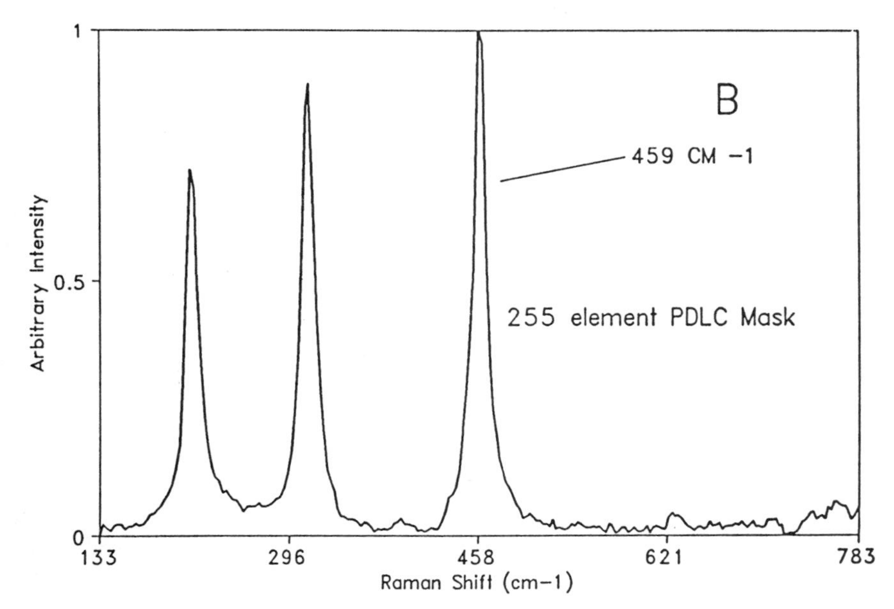

Figure 1 The NIR Hadamad transform Raman spectra of carbon tetrachloride. (A) The HT-Raman spectrum obtained with the 127 element nematic LC Hadamard encoding mask. (B) The HT-Raman spectrum obtained with the 255 element PDLC Hadamard encoding mask. From ref. 10.

3 RESULTS AND DISCUSSION

The ability to collect near-infrared (NIR) Raman spectra was demonstrated using the 127 element nematic LC mask; however, intense strongly polarized Raman bands were either non-existent or strongly attenuated due to absorption by the NIR polarizers. The 255 element PDLC mask does not employ a polarizer-analyzer pair; therefore, the problem is avoided. For CCl_4, the 459 cm^{-1} band, normally the most intense in the spectrum, is completely absent from the spectrum from the LC mask and most intense band in the spectrum from the PDLC mask

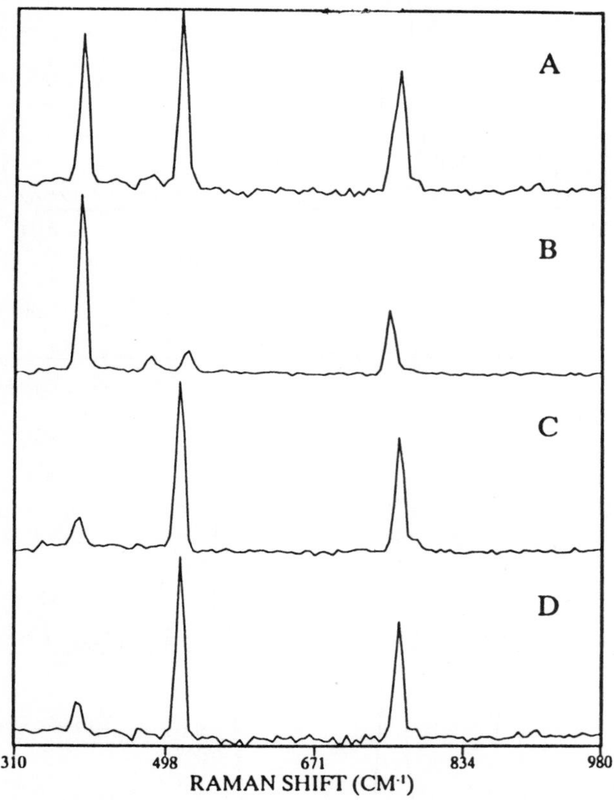

Figure 2 An example of spectral subtraction. (A) An NIR Hadamard transform Raman spectrum of a mixture of two solids, anthracene and naphthalene. (B) An NIR Hadamard transform Raman spectrum of pure anthracene. (C) An NIR Hadamard transform Raman spectrum of naphthalene. (D) The NIR Hadamard transform Raman difference spectrum of naphthalene obtained by subtracting the spectrum of pure anthracene (B) from the mixture spectrum (A). From ref. 8.

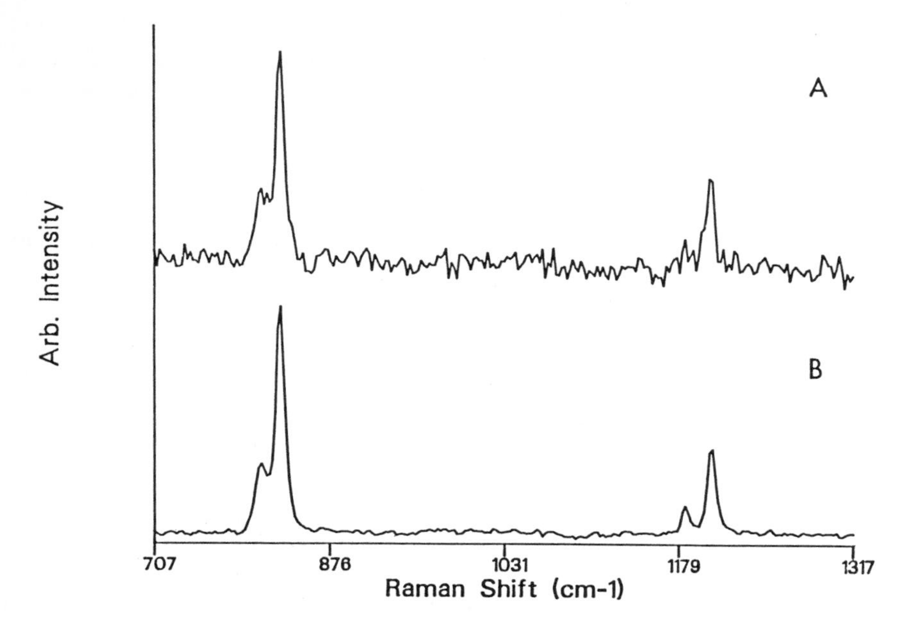

Figure 3 A demonstration of the multiplex advantage in HT-Raman spectrometry. (A) The pseudo grating scanned Raman spectrum of p-xylene. (B) The HT-Raman spectrum of p-xylene. From ref. 10.

(Figure 1)[10]. For bis-methylstyrlbenzene, the spectrum from the LC mask shows the bands near 1200 cm^{-1} to be the most intense while the spectrum from the PDLC mask correctly shows that the bands near 1600 cm^{-1} are three times as intense as those near 1200 cm^{-1}.[10] In spite of its problems with the polarizer-analyzer pair, even the LC mask works well in spectral subtraction mode. For example, the spectrum of pure naphthalene matches nicely with the spectrum of naphthalene obtained by the scaled subtraction of the spectrum of pure anthracene from the spectrum of a mixture of anthracene and naphthalene (Figure 2)[8]. The theoretical multiplex advantage is demonstrated with p-xylene by comparing a simulated grating scan spectrum to a Hadamard transform spectrum. The grating scan is simulated by keeping all mask elements opaque except for a single transmissive element sequentially scanned across the mask (Figure 3)[10]. For the same data acquisition time, the signal-to-noise ratio (SNR) is improved by a factor of 3.95 for the Hadamard transform spectrum (theoretical improvement is 4.01[11]). An additional important feature of HTS, the stationary Hadamard encoding mask provides a selective multiplex advantage since any mask element may be selectively turned off during data acquisition to imitate

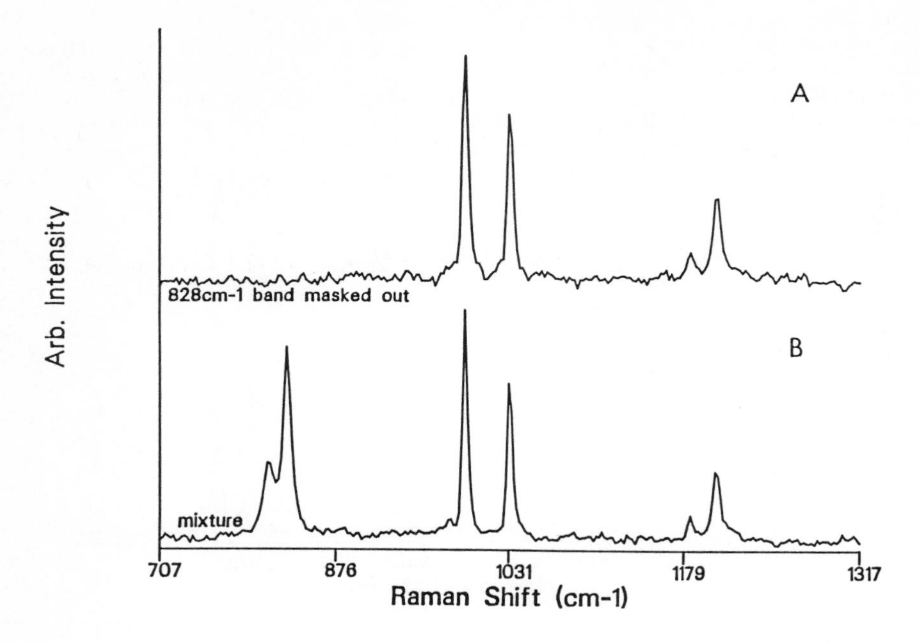

Figure 4 A demonstration of the selective multiplex technique. (A) The HT-Raman spectrum of a mixture of pyridine and p-xylene with the 828 cm^{-1} band of p-xylene masked out. (B) The HT-Raman spectrum of a mixture of pyridine and p-xylene. The 990 and 1030 cm^{-1} bands are due to pyridine and the 828 and the 1205 cm^{-1} bands are due to p-xylene. From ref. 10.

a tunable notch filter for unwanted spectral features. The application of this feature eliminates solvent bands as we try to develop surface enhanced HT-Raman spectrometry as a molecular detector for high performance liquid chromatography (HPLC). Demonstrations of this capability are the elimination of the strong 828 cm^{-1} band of p-xylene from the spectrum of a mixture of p-xylene and pyridine (Figure 4)[10] and the elimination of the major band of anthracene near 1400 cm^{-1} to make the weaker anthracene bands more prominent (Figure 5)[10].

A most important benefit of HTS for quantitative chemical analysis is the ability to utilize the encodegram directly.[12] This procedure circumvents the requirement to Hadamard transform the encodegram data first into the spectral domain before quantitative information about the chemical system can be extracted. Full spectrum, factor based techniques, such as principal components regression or partial least squares regression, are used to develop the calibration equation and perform predictions on unknowns. Resulting calibrations using encodement data are equivalent to, or in some cases better than, those performed in the spectral domain. Utilization of the selective multiplex advantage, a capability unique to HTS, permits masking of portions of the spectrum that have low information content thus increasing the precision of the calibrations.

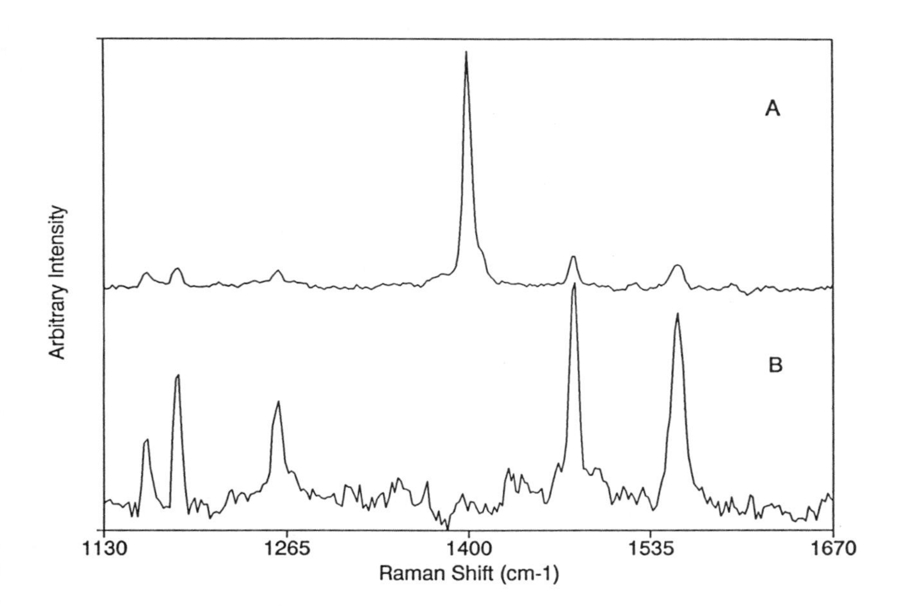

Figure 5 Improving the dynamic range using the selective multiplex technique. (A) The HT-Raman spectrum of anthracene. (B) The HT-Raman spectrum of anthracene with the 1405 cm-1 band masked out. From ref. 10.

4. ACKNOWLEDGEMENT

Work supported by the U.S. Department of Energy, Chemical Sciences Division, under Contract Number DE-FG02-95ER 13347.

REFERENCES

1. M. Harwit and N.J.A. Sloane, Hadamard Transform Optics (Academic Press, New York, 1979).

2. R.M. Hammaker, J.A. Graham, D.C. Tilotta and W.G. Fateley, "What is Hadamard Spectroscopy?" in Vibrational Spectra and Structure, 15, edited by J.R. Durig (Elsevier, Amsterdam, 1986), pp. 401 - 485.

3. A.P. Bohlke, D. Lin-Vien, R.M. Hammaker and W.G. Fateley, "Hadamard Transform Spectrometry: Application to Biological Systems, a Review", in Spectroscopy of Inorganic Bioactivators, Theory and Applications-Chemistry, Physics, Biology, and Medicine NATO ASI Series C, 280, edited by T. Theophanides (Kluwer, Dordrecht, 1989), pp. 159 - 189.

4. D.C. Tilotta, R.M. Hammaker and W.G. Fateley, Appl. Spectrosc. 41, 727 (1987).

5. D.C. Tilotta and W.G. Fateley, Spectroscopy 3(1) 14 (1988); R.M. Hammaker, W.G. Fateley and D.C. Tilotta, Spectrosc. Int. 1(2), 10 (1989).

6. P.R. Griffiths and J.A. de Haseth, Fourier Transform Infrared Spectrometry (Wiley, New York, 1986) pp. v, 78, 220, 248-249, 269, 274-276, 281-282, 387, 521.

7. D.C. Tilotta, R.D. Freeman and W.G. Fateley, Appl. Spectrosc. 41, 1280 (1987).

8. A.P. Bohlke, J.D. Tate, J.S. White, J.V. Paukstelis, R.M. Hammaker and W.G. Fateley, J. Molec. Struct. (Theochem.), 200, 471 (1989).

9. D.C. Tilotta, R.C. Fry and W.G. Fateley, Talanta, 37, 53 (1990).

10. A.P. Bohlke, J.M. Jarvis, J.S. White, J.D. Tate, J.V. Paukstelis, R.M. Hammaker and W.G. Fateley, J. Molec. Struct. (Theochem.), accepted June 15, 1990.

11. D.C. Tilotta, R.M. Hammaker and W.G. Fateley, Appl. Opt. 26, 4285 (1987).

12. John Jarvis, Ph.D.Thesis, Kansas State University, unpublished 1990.

ACCURACY OF E.S.R.-SPECTROMETRIC CHEMICAL ANALYSIS: INFLUENCE OF REFERENCE COMPOUND

V. Yu. Nagy and T. A. Orlova

INSTITUTE OF MICROELECTRONICS TECHNOLOGY PROBLEMS, USSR ACADEMY OF SCIENCES, CHERNOGOLOVKA, MOSCOW REGION, 142432, USSR

1 INTRODUCTION

Unlike most analytical methods, electron spin resonance (ESR) spectrometry is based on measuring signals produced only by ions with specific charges rather than all the ions and atoms of the element present in the sample. On the other hand, this spectrometric method does not require dissolution of the sample, i.e. the process which may bring about changes in the original ratio of concentrations of different ions of an element. This rather rare combination of features makes ESR spectrometry useful for reliable determination of concentrations of specific chemical forms of elements. The need for such determinations increases with the advancement of science and technology.

As with other spectrometric analytical methods, ESR analysis can be considerably simplified if performed in its relative variant, namely, by comparing the signal from the sample to be analyzed to that from the reference sample with known content of the paramagnetic ion to be determined. This approach, however, requires standard samples with kindred composition and reliably determined concentrations of the element in specific oxidation states. Very few such standard samples are now commercially available. Because of this, ESR analysis of solid samples is mostly carried out with use of reference samples which are rather different in nature from the sample to be analyzed.

Such determinations are based on the implicit assumption that all paramagnetic particles with the same spin quantum number produce signals of equal specific intensity (the area under the absorption curve per particle) and that specific signal intensities of paramagnetic particles with different spins are related to each other by a simple ratio:

$$I_a = \frac{S_a*(S_a+1)}{S_b*(S_b+1)}, * I_b$$

where S_a and S_b are the spin quantum numbers of each particle. The validity of this assumption is not obvious a priori since the transition probabilities, i.e. the coefficients in the equation relating signal intensity to the number of absorbing particles, may, in principle, be unequal both for different paramagnetic ions and for one and the same paramagnetic ion in different coordination spheres. The effect of transition probability on ESR signal intensity was pointed out many years ago;[1] theoretical evaluation of its magnitude was then performed for some simple cases. Nevertheless, this influence seems to be entirely ignored by analysts nowadays.

The main aim of the present work was to check the significance of the above factor on analytical results by a careful experimental comparison of specific intensities of ESR signals produced by paramagnetic ions of different elements in various compounds.

2 EXPERIMENT

Paramagnetic Compounds

To achieve maximum accuracy when determining the number of paramagnetic particles in the samples prepared for ESR experiment, we used mainly individual high purity compounds of transition metal paramagnetic ions. We intended to make up a representative series comprising compounds of paramagnetic ions with various spin quantum numbers as well as various compounds of one and the same ion; a typical stable organic nitroxide radical was also included. The following compounds were studied:

S=1/2	Cu(2+):	sulfate, chloride, perchloride, di-(2-ethyl)hexylphosphate, acetylacetonate, benzoylacetonate, thenoyltrifluoro-acetonate;
	VO(2+):	sulfate, acetylacetonate;
	>N-O$^{\cdot}$:	tempol (stable nitroxide radical).
S=3/2	Cr(3+):	sulfate.
S=5/2	Fe(3+):	sulfate, mixed ammonium sulfate.
	Mn(2+):	zinc sulfate doped with Mn(2+).

ESR Measurements

Equipment. ESR spectra were recorded with an X-band SE/X-2543 spectrometer ("Radiopan", Poznan, Poland) using 100-kHz magnetic field modulation. An H_{102}-type resonant cavity was used. During spectrum recording both the magnetic field strength and microwave frequency were continuously measured. The sensitivity of the spectrometer was $5x10^{10}$ spin/Oe.

The analog signal was transformed into digital form by means of a 12-bit converter and spectra were stored in a computer memory (4096 points per spectrum). The visualization system made possible observation of the stored spectrum with a resolution up to a single point.

The spectrum processing was performed using the "EPRMON" software package (CAMI, Kishinev, USSR) as well as the programs drawn up in this laboratory.

Procedure. In the course of the ESR-spectrometric experiment every effort was made to achieve maximum accuracy and precision. Special attention was paid, in particular, to the following factors.

Signals were recorded at microwave power levels well below saturation.

The possible variations of resonant cavity Q-factor were steadily taken into account by normalizing the signal of the sample to that of the adjacent standard fixed in the cavity (synthetic ruby).

The nonuniformity of microwave and modulation fields in the cavity were quantitatively studied before measurements. Upon spectrum recording a sample was always kept within the region of approximate field uniformity where the signal intensity variations due to sample displacement did not exceed 0.3%. The reproducibility of the sample positioning in the cavity was ensured by using one and the same sample tube with a rigidly fixed holder.

Cavity microwave field perturbation by the material of the sample tube and the sample itself was unified by the use of the only sample tube and very small samples (about 1 mm^3 in volume).

Measurements within a single comparative series were taken at constant values of microwave field power and modulation amplitude, this enabled us to avoid possible errors of normalization.

The speed of magnetic field sweeping was sufficiently low to ensure the time of the narrowest line peak-to-peak recording to be at least 10 times greater than time constant used.

The powdered samples used were 0.001-0.01 g in weight, i.e. they were sufficiently large to afford precise weighing, yet, did not induce line broadening as a result of the Q-factor decrease upon resonance passage.[2]

The experiments were carried out at room temperature, thus the errors resulting from the difference in Curie temperatures were negligible.

Particular care was taken in the determination of the area below the absorption curve. High signal intensity allowed for avoiding baseline drift correction and eliminated noise-produced errors. Double integration was performed by the Simpson method over all the 4096 points. All spectra were recorded and processed over the range exceeding 10 halfwidths of the widest line. The method of linear anamorphism was used to analyze the lineshapes prior to integration. In the case of spectrum lines retaining Lorentzian lineshape over the whole range indicated, theoretical correction was made for the neglected area below the "cut-off" wings.[3,4]

The spectrometry procedure described enabled us to achieve a sufficiently good reproducibility of specific signals from different samples of the same material with an anisotropic ESR spectrum, the relative standard deviation being about 2%.

Chemical analysis

Though most of the reagents used were high purity inorganic compounds with impurity contents below 10^{-5}%, the real amount of paramagnetic ions in a weighted portion could be different from the theoretical one because of the losses of crystallization water and partial oxidation or reduction of not very stable ions. Therefore, just before measuring ESR signals, the composition of the reagents was checked up using the most precise gravimetric and titrimetric methods of chemical analysis, providing RSD less than 0.5%.

3 RESULTS

The measurements yielded the ratios of specific intensities of various paramagnetic compounds presented in Table 1. Note, that specific intensity of Cu(II) signal in $CuSO_4*5H_2O$ is used as a unit. All the specific intensities were normalized on spin value according to the expression presented in Introduction.

4 DISCUSSION

The results indicate that the widespread belief in the possibility of direct using of practically any paramagnetic compound as an ESR spectrometry standard is true in the first, very rough, approximation only. Indeed, when an error up to several tens per cent is tolerable, simple comparison of the signal from the sample to be analyzed to that from any paramagnetic substance available is sufficient.

Table 1. Specific intensities of ESR signals produced by diverse paramagnetic centres in various compounds

(Specific intensities are presented in arbitrary units resulted from normalization on spin value and the specific intensity of Cu(2+) in $CuSO_4 \cdot 5H_2O$)

Paramagnetic compound	Specific intensity
$CuSO_4 \cdot 5H_2O$	1.00
$CuCl_2 \cdot 2H_2O$	1.00*
$Cu(ClO_4)_2 \cdot 6H_2O$	1.14±0.06
Copper(2+) di-(2-ethyl)hexylphosphate	1.00*
Copper(2+) acetylacetonate	0.95±0.02
Copper(2+) benzoylacetonate	0.89±0.02
Copper(2+) thenoyltrifluoroacetonate	0.85±0.02
$VOSO_4 \cdot 2H_2O$	0.76±0.03
Vanadyl acetylacetonate	0.79±0.03
Tempol (nitroxide radical)	0.85±0.03
$Cr_2(SO_4)_3 \cdot 6H_2O$	0.87±0.06
$Fe_2(SO_4)_3 \cdot 9H_2O$	0.80±0.05
Zinc sulfide doped with Mn(2+)	0.95±0.03

* Discrepancy between the specific intensity of the compound and that of $CuSO_4 \cdot 5H_2O$ is not statistically significant (P=95%).

However, if RSD of less than 5% is to be attained, careful account of difference in transition probabilities is necessary. Really, even kindred compounds of one and the same paramagnetic ion show significant differences in specific signal intensity (see the series of copper beta-diketonates). It therefore follows, that when quite identical reference materials are not commonly available, a working procedure should comprise instructions on the use of alternative standard substances, providing an analyst with the corresponding correction factors, carefully determined during development of the procedure.

With such an approach, it is possible to propose almost any easily available paramagnetic compound as a reference material.

From our viewpoint, copper sulfate pentahydrate is quite suitable for this purpose, yet, it should be recrystallized and kept in a damp atmosphere. Its storage under conventional laboratory conditions may lead to significant water losses and a corresponding increase in the number of copper ions per unit sample mass. Studies of 10 samples of this compound received from different laboratories (different lots from different suppliers) revealed a 10% excess of the "specific" signal over the signal of the recrystallized reagent. Another important point, the substance should be properly grinded to achieve effective averaging in particle orientations with respect to magnetic field.

However, nitroxide radical tempol appears to be a more convenient universal reference material. It is easy to purify by the chromatographic methods, fairly stable and has stable composition, the spectrum of its powder contains a much narrower line.

REFERENCES

1. R. Aasa and T. Vanngard, J. Magn. Reson., 1975, 19, 308.
2. B. Vigouroux, et al., J. Phys. E., 1973, 6, 557.
3. I.B. Goldberg, J. Magn. Reson., 1978, 32, 2334.
4. H.S. Judeikis, J. Appl. Phys., 1964, 35, 2615.

THE RESOLUTION OF BANDS IN SPECTROSCOPY

A. S. Gilbert

DEPARTMENT OF PHYSICAL SCIENCES, WELLCOME RESEARCH LABORATORIES, BECKENHAM, KENT

1. INTRODUCTION

Because bands have breadth and are also tailed they commonly overlap with others in a spectrum. The breadth is a consequence of one or more factors which are either intrinsically derived, for example from molecular motions, quantum effects, *etc.* or alternatively result from extrinsic causes such as slits or applied magnetic fields. Control of these factors may allow separation or resolution of overlapped bands but if this is not possible then resort must be made to mathematical manipulation of the spectral data.

A number of different mathematical techniques are available, the technique of choice being dependent on the prior knowledge of the system under consideration and any assumptions that can be made about the final results. Table 1 below summarises the situation for the four techniques that will be considered here.

TABLE 1

Prior knowledge and assumptions required for various methods of band resolution.

Method	Prior Knowledge	Assumptions about final results
(a) Differentiation	None	None
(b) Linear Deconvolution	Lineshape	None
(c) Iterative Deconvolution utilising the Principle of Maximum Entropy	Lineshape, noise on data	All results (data) must be non-negative
(d) Curve Fitting	Lineshape, noise on data, number of bands	None

As a general rule the more that is known and the greater the number of assumptions that can be made, then the 'better' the final result that is possible. Therefore it is desirable to utilise a technique that appears as low down in Table 1 as possible [*e.g.*(b) rather than (a)].

Note that there is no reason why more than one technique should not be used in stepwise fashion; thus a deconvolution method might be employed at first to find the total number of bands in a heavily overlapped spectrum followed by curve fitting.

2. DIFFERENTIATION AND LINEAR DECONVOLUTION

Differentiation is the traditional method for sharpening bands chiefly because in the days before computer interfacing to spectrometers it could be implemented electronically. Even where data is available in digital form it is the simplest of the mathematical methods. Even order differentiation produces band-like profiles but sidelobes and negative values are generated and relative area relationships between bands are lost. In addition the signal to noise ratio SNR across the spectrum is reduced. Nevertheless, despite these drawbacks, differentiation has been and is still extensively applied to UV-visible spectra in analytical situations, one reason being the typically high SNR's encountered in such spectra.

A more elegant technique for resolving bands by narrowing them is to deconvolute the line or bandshape. The process is best explained by considering the Fourier domain and Figure 1 shows a band with Lorentzian lineshape alongside its Fourier transform (*NB* only the cosine coefficients are shown). The cosine waves of the transform can be seen to decay as exp(-ky) going left to right from the origin of the domain; k is a constant that is proportional to the band-width in the spectral domain. Point by point multiplication with exp(+ky) of the Fourier domain as shown in the picture therefore gives a set of pure cosine waves which when transformed back to the spectral domain yields an infinitely sharp line. Thus the Lorentzian lineshape has been completely removed (*i.e.* deconvoluted) and as the multiplication involves just one step the process is termed Linear Deconvolution.

Differentiation is equivalent to multiplication of the Fourier domain by non-exponential functions, for instance for second order differentiation the function is $-y^2$. The result is to remove a fixed and arbitrary lineshape from the band to leave a somewhat unreal though narrower profile.

Unfortunately the infinite degree of narrowing shown in Figure 1 cannot be realised in practice. This is because the data is always accompanied by noise which comes both from the detector and also the computational process due to the finite word length of the computer. The noise is in general just about equally distributed across the Fourier domain. The signal (the data representing the spectral bands) is mostly concentrated near the origin. But the 'rising' functions necessary for deconvolution magnify the other end of the domain where noise usually predominates and as noise in any part of the Fourier domain appears in all parts of the spectral domain the net result is swamping of the deconvoluted peak(s). However, this problem can be partially alleviated by apodization which involves using a second function to reverse the rise of the first in regions away from the origin of the Fourier domain. There is a penalty unfortunately in that the apodization function imposes some breadth on the resultant band(s) but nevertheless significant degrees of narrowing can be achieved. In addition there is almost always a decrease in SNR *i.e.* apodization causes SNR to be traded off against resolution.

The practical combination of Fourier deconvolution with apodization is known as either Fourier Resolution Enhancement[1] or Fourier Self Deconvolution[2] (FRE or FSD) and with the widespread availability of Fast Fourier Transform routines has become increasingly popular in recent years. Figure 2(a)-(c) compares the results of applying FRE and second order differentiation to a synthetic noisy Lorentzian doublet. The apodization function employed in the FRE is a second power exponential which reverses the rise of the first power exponential that deconvolutes the Lorentzian lineshape. The combined function for multiplication of the Fourier domain thus takes the form $\exp(ky\text{-}jy^2)$ and the effect of this apodisation is to impose Gaussian lineshape onto the resultant bands. It is obvious which method has given the better results. Differentiation has introduced side-lobes and has generated a higher overall noise level than FRE which has both reduced the band widths (widths at half height) and also derived benefit from the Gaussian lineshape which is less tailed than Lorentzian.

Both the degree of deconvolution and apodization can be varied separately in FRE and Figure 3(a) demonstrates how stronger apodization trades off some resolution with a gain in SNR while Figure 3(b) illustrates the dangers of over-deconvolution. The latter case has important implications where deconvolution is carried out on overlapped bands whose widths (and lineshapes) are not well known and indicates the possibility of generating negative sidelobes that get hidden under neighbouring positive going bands with consequent reduction in band strength and distortion of shape. In circumstances where bands of differing widths are present complete deconvolution of the broader bands leads to over deconvolution of the narrow ones but this can sometimes be avoided by using a lower degree of deconvolution to first separate the sharp peaks which can then be removed before proceeding to tackle the rest of the spectrum.

Finally, the trade-off of SNR against resolution (or noise against final bandwidth) is not linear but biased to the detriment of the SNR in exponential fashion. Thus Lorentzian deconvolution with Gaussian apodization might increase the noise by about three times for a bandwidth reduction of a factor of two while increasing it by more than three orders of magnitude for a four times width reduction.

3. ITERATIVE DECONVOLUTION, THE MAXIMUM ENTROPY PRINCIPLE AND CURVE RESOLUTION

Deconvolution and differentiation suffer from two basic defects. Firstly the noise is not separated from the deconvolution process whereas in many types of spectroscopy the noise (or most of it) only appears during measurement *i.e.* after the bands have acquired their breadth. The detrimental effects on SNR are a consequence of this. Secondly [as demonstrated in Figures 2(b), 3(b)] negative values can be produced and this is plainly not sensible for instance in the case of an emission spectroscopy such as the Raman effect.

These deficiencies can be addressed by deconvoluting in an iterative fashion with utilisation of the Principle of Maximum Entropy. For the purpose of discussion it is helpful to introduce the parlance of the astronomers who were the first to use this method. Thus the aim of the deconvolution process is to obtain the 'object' which is the true or real

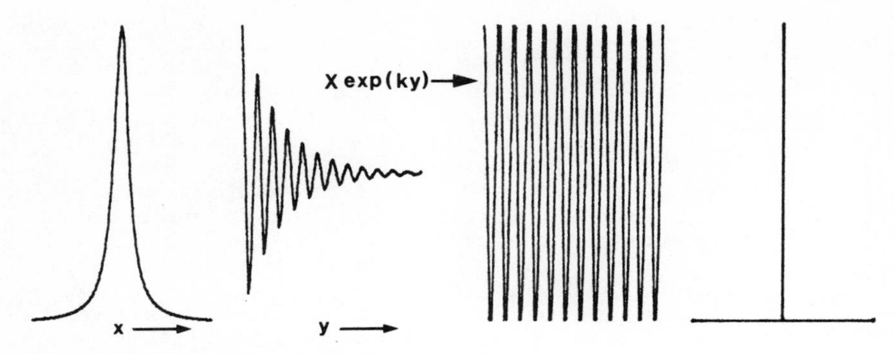

Figure 1 - Deconvolution of Lorentzian lineshape

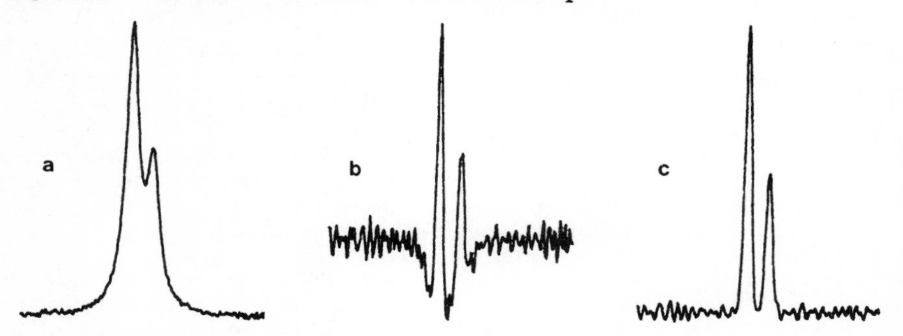

Figure 2 - The comparison of differentiation and linear deconvolution for resolution of a Lorentzian doublet (a) in the presence of noise. The components have equal bandwidths and one is half the height of the other. The second differential (b) has been inverted. The apodization factor j (see text) is such that a bandwidth reduction factor of 2 has been achieved for linear deconvolution (c).

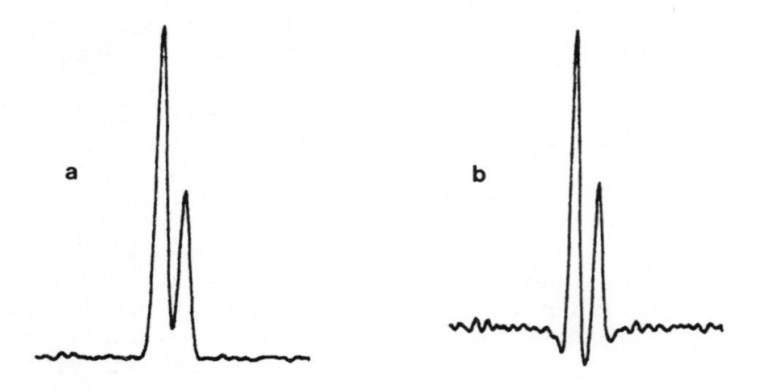

Figure 3 - Varying the degree of apodization and deconvolution in the FRE of the Lorentzian doublet (Fig. 2(a)). In (a) the apodization factor j has been chosen so that a bandwidth reduction factor of only 1.4 has been achieved. For (b) the deconvolution factor k is larger than that employed to give Fig. 2(c) while j is the same.

spectrum. This is 'blurred' by one or many of the intrinsic or extrinsic factors mentioned in the Introduction and then degraded by noise to give an 'image' (*i.e.* the measured spectrum).

If a guess is made at the object and this trial object is blurred and the result subtracted from the image the residuals will, unless one is very lucky, be plainly distinguishable from the noise itself assuming of course that the overall noise characteristics are known. The residuals, however, can be utilised to modify the trial object with the proviso that no values are allowed to be negative and this new trial can then be blurred and new residuals computed and so on until a set of residuals is obtained which cannot be distinguished from the noise. This iterative procedure is illustrated in Figure 4.

The trial object at this point is however not necessarily the object itself. This is because there are very many sets of residuals that are indistinguishable from the noise (which by definition is only known by its overall characteristics *e.g.* root mean square value, frequency distribution, *etc.*) and therefore very many compatible trial objects, are selected by the iterative cycle being an arbitrary outcome of the initial guess and the mechanics of the procedure.

Fortunately it is possible to make a logical choice[3] as to which candidate is most likely to be the real object. If a spectrum is regarded as a finite number of photons divided among a number of channels then a figure for the likelihood of that spectrum can be given by the ways W that the photons can be arranged among the channels.

$$W = \frac{N!}{(N-n_1)!n_1!} \times \frac{(N-n_1)!}{(N-n_1-n_2)!n_2!} \times \dots\dots\dots$$

$$= \frac{N!}{\prod_{x=1}^{x=M} n_x!}$$

where n_x = number of photons in channel x
M = number of channels
N = total number of photons

The likelihood is largest where all n_x are equal (*i.e.* the spectrum is a featureless straight line) and smallest where all the photons are in just one channel.

A more useful form of the equation is obtained by using Stirling's approximation (where there are large numbers) to get the configurational entropy

$$S = \text{Log}_n W = K - \sum_{x=1}^{x=M} P_x \log_n P_x$$

where P_x = n_x/N
K = a constant

The principle of Maximum Entropy states in the present context that

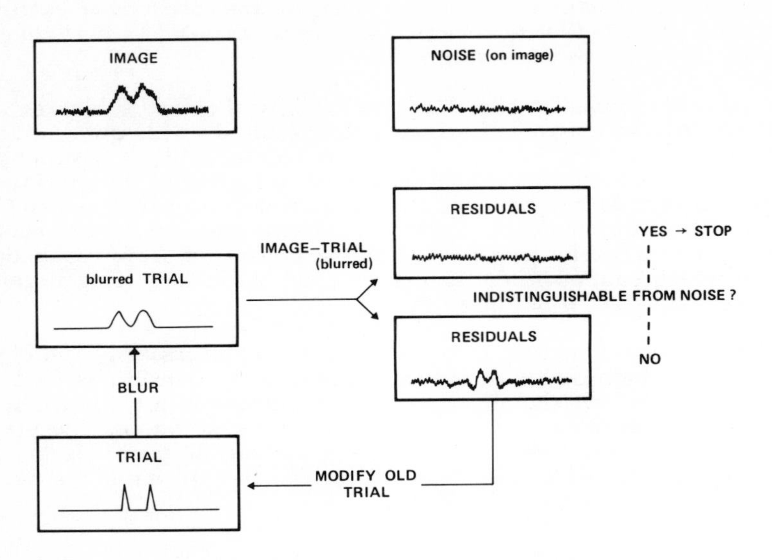

Figure 4 - Iterative deconvolution

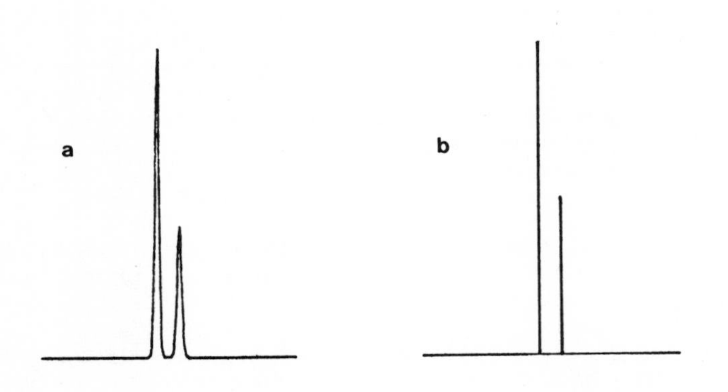

Figure 5 - MEM deconvolution (a) of the Lorentzian doublet of Fig. 2(a). Though the height of the smaller peak is less than half of the larger, the ratio of the band areas is very close to 2. The blurring factor utilised was the known Lorentzian lineshape of the synthetic doublet. The true object is shown as (b).

the candidate possessing the greatest value of S is most likely to be the object. The rationale for utilising the Principle is discussed in a recent review by Jaynes[4]; the associated methodology is termed the Maximum Entropy Method, MEM. Because of the form of the entropy the chosen object will be flatter or smoother than other candidates which is consistent with the true object being by definition noiseless.

Of course the number of candidates in any real situation is very large thus it is by no means a simple task to find the object, nevertheless iterative cycles can be designed for this purpose. Essentially the task is to modify the trial object so as to maximise the function

$$Q = S - \lambda C$$

on each step of the iteration, where C is the chi-squared statistic and λ is an adjustable multiplier. The iteration is stopped when C is equal to unity. One way of performing the maximization for example is to utilise the Newton-Raphson method thus

$$O_{new} = O_{old} - Q'/Q''$$

where O is a trial object.

A particularly efficient and robust algorithm has been developed by Skilling and co-workers[5] who from general considerations of probability have derived a slightly different form of the entropy[6]. Their MEMSYS software package has been used for all examples of MEM deconvolution shown here.

Figure 5(a) illustrates the application of MEM to the Lorentzian doublet shown in Figure 2(a). The result is considerably better than FRE both in the degree of deconvolution achieved and the preservation of SNR. It is not however perfect for Figure 5(b) shows the true object which is a pair of infinitely sharp lines. This object could be inferred from curve fitting assuming full knowledge of the blurring function; therefore in real situations of this kind where it is the precise peak positions that are sought, curve fitting is most appropriate. But where the blurring factors (lineshapes) are not known accurately then curve fitting loses its advantage and if in addition the number of bands is also unknown its performance will be severely degraded.

4. APPLICATIONS TO REAL SPECTRA

Figure 6 illustrates the differing abilities of the methods discussed above in deconvoluting the slit function from the Raman spectrum of the C-Cl stretching mode region (400-500cm^{-1}) of carbon tetrachloride. Note that curve fitting cannot be used in this situation.

The slit width used to record the original spectrum (Fig. 6(a)) is sufficiently broad (the band-pass was of the order of 3cm^{-1}) to blur out the isotope pattern which can be seen clearly in a spectrum recorded with much narrower slits (about 1cm^{-1} for Fig. 6(e)). The natural linewidth of these individual peaks is about 2cm^{-1}.

Differentiation (Fig. 6(b)) of the spectrum in 6(a) gives a very poor result while linear deconvolution (Fig. 6(c)) gives some semblance of the pattern but possesses numerous interfering side lobes. These are the

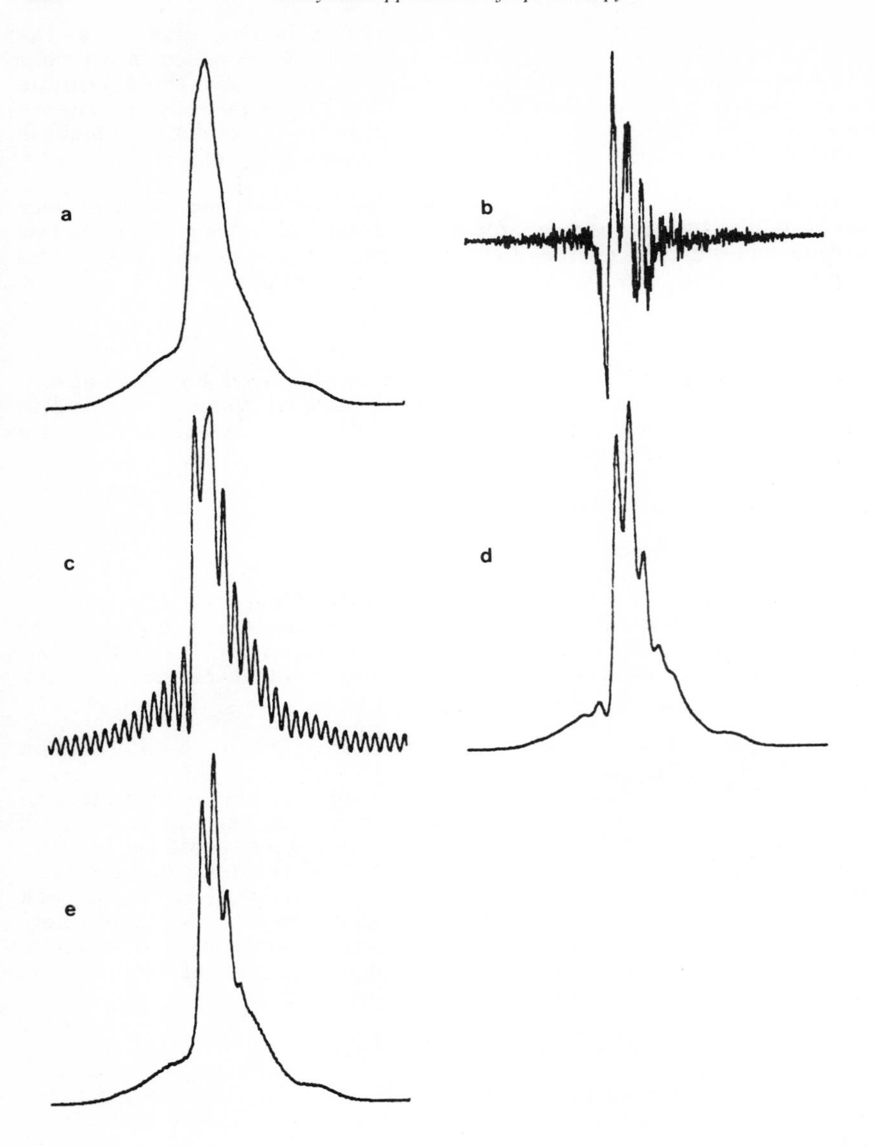

Figure 6 - The removal of a slit function from a Raman spectrum of CCl_4. Original spectrum (a) of the C-Cl stretching region recorded with 'wide' slits. The inverted second differential is (b) while the linear deconvolution of the slit function is (c) (to control the noise the Fourier domain was truncated followed by Norton-Beer apodization). The MEM result is (d) while a spectrum recorded with much narrower slits is shown as (e). Note that a Gaussian lineshape was used for (c) and (d) as the slit function.

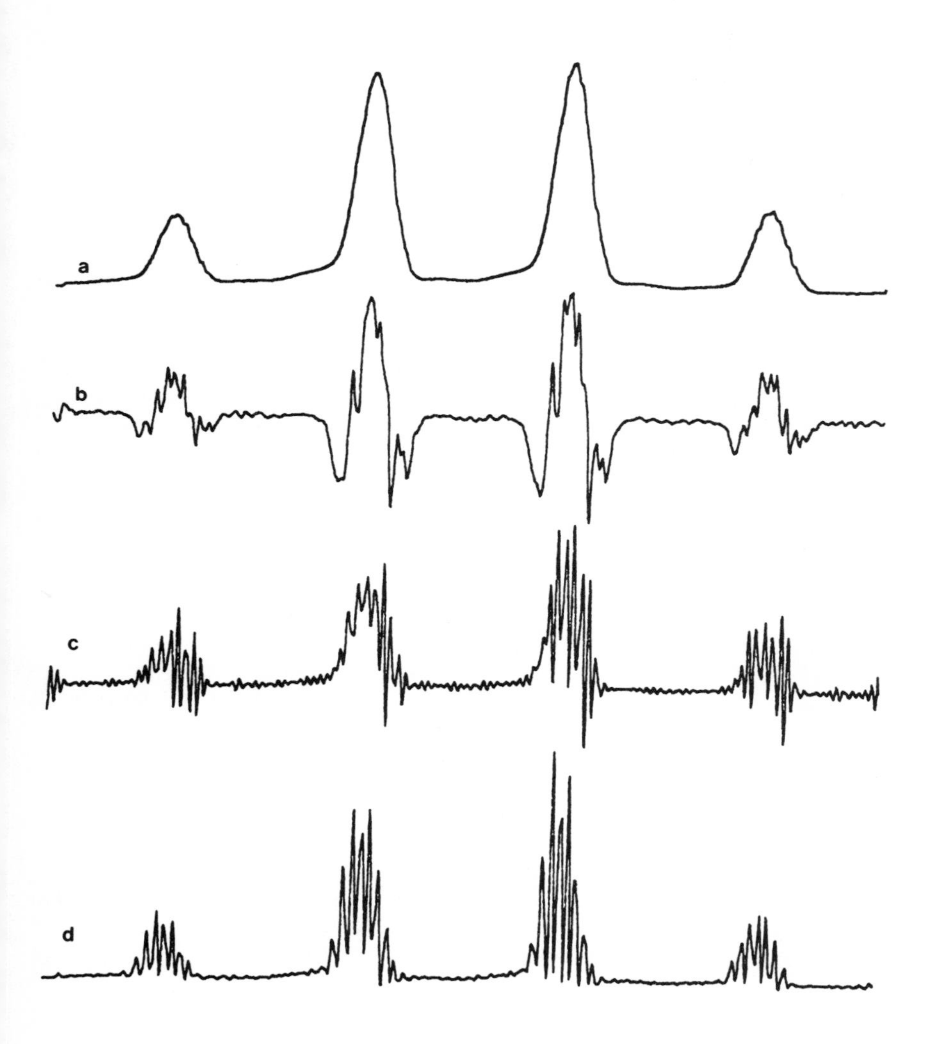

Figure 7 - The resolution of the long-range splitting pattern on the methylene protons of the ^{1}H nmr spectrum of ethyl-benzene. Original spectrum recorded at 200 MHz is (a). The inverted second differential is (b) while linear deconvolution using a modified Lorentzian lineshape with Gaussian apodization gives (c). The MEM result using the same modified Lorentzian lineshape is (d). Original spectrum courtesy of Dr. R.D. Farrant and Mr. M.J. Seddon (Wellcome Research Laboratories).

consequences of the need to severely truncate the Fourier domain to reduce the very high noise generated by the deconvolution which would otherwise swamp the bands. Apodization was also employed.

The result of MEM is shown in Figure 6 (d). Though not perfect the method has effectively removed a large part of the blurring induced by the wider slits. It should be noted that the slit function used for linear and MEM deconvolution was a Gaussian approximation to the true function for the type of spectrometer utilised (Czerny-Turner mounting) which is a slightly asymmetric Gaussian-triangular mixture.

Finally Figure 7 shows the various outcomes from tackling an nmr spectrum. The original (Fig. 7(a)) shows the signals from the methylene protons of ethyl benzene, the four peaks are the result of coupling with the adjacent methyl group. Each of these peaks are made up of eighteen much narrower signals blurred together by the relatively low external magnetic field (200 MHz) and resulting from long range coupling with the aromatic protons. The pattern is essentially one of triplets of triplets of doublets in each case.

As in Figure 6 MEM has been most successful. Linear deconvolution has generated negative lobes, distorted relative peak heights and yielded higher effective noise. A modified Lorentzian lineshape was used in both cases as a model for the blurring function. Even MEM however does not reveal the full pattern for various reasons, one of which was insufficient data points across the spectrum and another is the lack of knowledge of the actual lineshape of the individual peaks which would probably not make it worthwhile to employ curve fitting.

5. SUMMARY

Curve fitting has only been briefly considered here because its successful application requires a considerable degree of prior knowledge, not normally available, of the system under study. In certain circumstances, for instance in the removal of slit functions in order to reveal an unknown natural spectrum with bands that still possess breadth, it cannot be used at all.

Of the deconvolution methodologies, MEM is undoubtedly the most powerful but does require rather more computer time than linear deconvolution, though this is not excessive for the examples shown. Differentiation as demonstrated yields very poor results.

REFERENCES

1. J.C. Lindon and A.G. Ferrige, J.Mag.Res., 1981, 44, 566.
2. J. Kauppinen, D.J. Moffat, H.H. Mantsch and D.G. Cameron, Appl.Spectrosc., 1981, 35, 271.
3. B.R. Frieden, J.Opt.Soc.Am., 1972, 62, 511.
4. E.T. Jaynes, Proc.IEEE, 1982, 70, 939.
5. J. Skilling and R.K. Bryan, Mon.Not.R.Astron.Soc., 1984, 211, 111.
6. S.F. Gull and J. Skilling, IEE Proc.(F), 1984, 131, 646.

STRUCTURE DETERMINATION USING N.M.R. SPECTROSCOPY

Andrew E. Derome*

THE DYSON PERRINS LABORATORY, UNIVERSITY OF OXFORD, OXFORD OX1 3QY

1 INTRODUCTION

In contrast with most of the other contributions to this meeting, we shall be concerned here with the spectroscopy of pure substances and, since we wish to use nuclear magnetic resonance, with relatively large amounts of material (of the order of several µmoles). The goal is to see how structure determination should now be approached in the light of the developments in NMR spectroscopy over the last decade or so, with the choice of experimental techniques being guided by the desire for a balance between information content and practicality. By this I mean to exclude those methods that, though apparently yielding irresistible quantities of *information*, require for their success outlandish quantities of *material* (for instance, identification of ^{13}C–^{13}C coupling networks at natural abundance). Hopefully we will thereby establish the set of experiments that should be available to all chemists engaged in structure determination, whether of synthetic or natural products, and achieve some perspective on when and how they should be employed.

Traditional Structure Determination

The availability of various new correlation techniques to be described later demands a significantly different approach to structure determination than has been usual in the past, and to emphasise the contrast it is useful to consider this traditional method. Faced with a variety of spectral data and, often, prior chemical information, we proceed by a cycle of three stages. The first step is to collect inferences from all the available data. These might take the form of hints about functional groups from the IR and UV spectra, while the mass spectrum yields the molecular formula and perhaps a collection of likely segments of the structure *via* the fragmentation. From the ^{1}H NMR we often infer much about small regions of the molecule; for instance, the presence of an ethyl group or a phenyl ring may be obvious. At a more detailed level we can often identify signals in a high–field ^{1}H spectrum which must correspond with only one or two protons, and for reasonably well–functionalised molecules we may be able to find enough of these signals to account for most of the protons in the structure. These then form the building blocks we would like to assemble to identify the backbone of the molecule. ^{13}C NMR may provide some information of a similar kind, but a particularly useful

* Associate Member of the Oxford Centre for Molecular Sciences

feature of this spectrum is its ability to identify symmetry in the structure. Since, for functionalised compounds, it is almost always the case that every distinct carbon site will give a separate ^{13}C signal provided the experimental conditions are selected properly, comparison of the number of signals with the molecular formula provides evidence for symmetry, or lack of it.

A final very important piece of prior inference is, unfortunately, usually applied almost sublimally, and that is our expectation of what the structure should be. In synthetic work, in fact, this one factor may override the most compelling experimental evidence, because we *know* what we wanted to make! Even in the natural product area one may unconsciously expect a terpene or alkaloid and bias the interpretation accordingly. A purist would argue that one should try to rigorously exclude such dangerous thoughts and proceed only from the experimental data, but realistically there can be few structures which have been determined in that frame of mind. The true position is that the problem is too difficult: the evidence available from spectra is rather ill defined, and when we come to the second stage of structure determination – proposing candidate structures – an element of educated (i.e. unconsciously biased) guesswork must enter in. So we take what information we have, and try to think of compatible structures. It will be readily apparent that this is the nub of the problem of structure determination; it is easy to tabulate spectral features and the corresponding inferences, and relatively easy to work back from a proposed structure to check whether it is compatible with the available data (the third step), but the leap forwards from a rather nebulous set of inferences to a set of concrete proposed structures can be a formidable obstacle. This is clear when trying to teach structure determination, because at this stage all that can be said is something along the lines of "well, you look at the data and try and think of some structures". With long experience one finds that it does indeed become possible to "think of some structures", but it is very hard to convey to the beginner just what mode of thought that experience has developed.

The exciting thing about the range of more modern NMR experiments now available is that they bear directly on this central problem in structure determination – building up the molecule from various subunits – and they do so in a way which comes closer to the purists' ideal. By combining the information from three principal experiments, it is possible in favourable cases to build up large portions of the structure in a truly well defined way. Furthermore, this can be done with little recourse to the notoriously unreliable correlations of chemical shifts with structural features that play a prominent role in the traditional method. In the remainder of this talk, the background to these experiments and a protocol for their application will be outlined.

2 Modern NMR

Everyone is aware by now that hundreds of new NMR experiments exist; indeed it can seem rather overwhelming at times. To keep things in perspective, it is useful to reflect that only three fundamental concepts underly the whole operation. I list them here in what we may well find to be increasing order of difficulty:

Time domain measurement and the use of Fourier analysis
Dipolar coupling and the Overhauser effect
Scalar coupling, coherence and coherence transfer

I will begin by briefly introducing each of these ideas, before showing you how they can be combined to make useful experiments.

Time Domain Measurement

No doubt everybody here is well familiar with the merits of signal measurement in the time domain. The efficiency gained in this mode is especially welcome in NMR, because the low intrinsic sensitivity mandates signal averaging. However, there proves to be a second more subtle advantage, in that certain states into which we can put a spin system, even if they do not give rise to NMR signals in themselves, turn out to be able to modulate some other signal which we measure as a function of a time interval defined in our experiment. Application of the FT technique to this time interval may thus let us obtain a "spectrum" of otherwise inaccessible modes, or it can simply allow us to construct experiments that measure more than one span of NMR frequencies at a time – for instance, ^{1}H combined with ^{13}C. This is part of the basis of two–dimensional NMR, of which more later.

The Overhauser Effect

The dipolar coupling often strikes chemists as an esoteric concept, but it is nothing more than the direct magnetic interaction between spins (just like the forces experienced on bringing two bar magnets into proximity, but on a quantum mechanical scale). In an immobile system of two spins–1/2, the possibility for each spin to be in one of two states means that the dipolar coupling leads to a different local field at one spin for each state of its partner. The spectrum of each spin therefore consists of two lines, separated by the coupling constant, which is orders of magnitude bigger than the more familiar scalar coupling J. Such a spectrum could certainly be observed, for a single crystal standing in a fixed relation to our static field, but more usually we measure spectra in solution where the molecules are tumbling at random. Thus it is necessary to consider the effects of motion on the coupling, and for sufficiently fast motion it proves to average to zero. The splittings we do see in proton NMR are also due to the relative spin states of neighbours, but transmitted in a more subtle way which normally involves the bonding electrons of the molecule.

Even though it does not appear directly in spectra, the dipolar coupling is of considerable significance, because it is the interaction underlying the nuclear Overhauser effect. A proper outline of the origin of the Overhauser effect would require a digression into the mechanism of relaxation of a system of interacting spins, for which there is unfortunately insufficient time, but the very essence of the effect can perhaps be appreciated along the following lines. The Overhauser effect is a change in intensity of lines in the spectrum, so we must consider the population differences across the transitions that give rise to the lines of interest. Any particular spin system has an equilibrium state, which it gets to after standing in the magnet for a while, and the population differences across its transitions in this state are determined by their energies and by the energy available from the environment (i.e. the temperature) according to the Boltzmann distribution. In the Overhauser experiment we perturb the population differences of selected transitions; any perturbation will do, but a common one is *saturation* of all the transitions belonging to a single spin (so that the associated energy levels cease to have different populations). Then we wait for the system to get back to a new equilibrium position and measure the spectrum again. Now, if we take it that there is a tendency for the *total* population difference across all the transitions of the system to have some value, and consider that we have *removed* some of this difference by our perturbation, perhaps we might expect differences elsewhere in the system to *increase* to compensate. I add hastily that this is a gross simplification, but actually it *is* what happens for molecules which are tumbling rapidly in solution. Since various population differences elsewhere in the system increase, we see a corresponding increase in the intensity of some of the signals when we measure the spectrum. The extra twist that makes the experiment useful is that only spins that are interacting, *via* dipolar coupling, with the one we have saturated are able to take notice of its change in populations, and since the dipolar coupling depends on the internuclear distance, so does the Overhauser effect. This makes it a useful probe of three–dimensional structure.

Coupling and Coherence

Scalar coupling is the familiar interaction that causes multiplet structure in ^1H spectroscopy. It has a very close link with molecular structure, because the principal mechanism by which it arises involves transmission of the information about nuclear spin states through the bonds of the molecule. For ^1H NMR, in fact, it is roughly the case that scalar coupling is only of significant magnitude for protons within two or three bonds of each other (common exceptions to this, such as allylic coupling, will no doubt be familiar to you already). Detection of scalar coupling is therefore of prime importance in structure determination, and its existence is what makes NMR the most effective technique in this area. Traditionally, couplings have been identified either by simply examining the appearance of multiplets, or by the familiar homonuclear decoupling experiment. Homonuclear decoupling is a potentially unambiguous way of locating coupling connections, but it suffers from at least two deficiencies. First of all, in order to perform a decoupling it is necessary to selectively irradiate one multiplet whilst observing the multiplicity change in another. Either aspect of this exercise may be thwarted by overlap in a complex spectrum, and in practice since the dispersion of ^1H NMR is not so great, this problem can arise even in relatively simple molecules. A second criticism of this experiment relates to measurement efficiency, and in fact is another example of the frequency domain *versus* time domain question. Suppose we need to perform a lot of decouplings in order to determine a structure (and let us set aside the question of whether they can be done selectively for the time being). The experimental protocol would consist of measuring a series of spectra, each obtained presumably in the FT mode, with the decoupling frequency set to a different value each time. It is interesting to note that this is actually a *two–dimensional* NMR experiment: the results are a function both of proton chemical shifts in the "natural" measured dimension, and of the decoupling frequency, which has introduced a new "artificial" dimension to the experiment. However, although we have begun doing 2D NMR, we have made a mistake! The new dimension has been measured in the frequency domain, so the usual inefficiency of this mode of operation will apply. What is really needed is a time–domain equivalent of this experiment, and we shall see that this can be obtained, and that in doing so we also solve the selectivity problem inherent in homonuclear decoupling.

The existence of coupling means, amongst other things, that spins possess more than one non–degenerate transition, and since these transitions differ in energy we have the opportunity to manipulate them separately in various ways. This permits a number of very interesting manouevres that can be classed as the creation and transfer of *coherence*. Coherence is a generalisation of the description of the relationship between energy levels in a spin system, and though it is a strictly quantum–mechanical concept, the character of it can be appreciated as follows. We are quite used to referring to some aspects of the state of a system as a function of the relationship between levels, as for instance when we speak of population differences. The *number* of spins in a particular state, however, is not sufficient to characterise it properly; they also have an associated *phase*. A physical picture of what this means is not always readily available, but for a simple two–level system one can at least get an inkling of what it might mean. Suppose we have some number of nuclei in one of two states, for instance with the *z*–component of their magnetisation aligned either with or against an applied field (the z–direction being defined by that field). If we count up the number of spins in each state and find an excess in one of them, then we can say that there should be a net component of magnetisation directed along the z axis proportional to the z–magnetisation of a single nucleus multiplied by the number in excess in one state; thus population difference is a sufficient parameter to determine the z magnetisation. However, if we consider whether there is any magnetisation perpendicular to the z–axis we encounter a complication. An individual nucleus certainly has an xy component of magnetisation, which precesses about the z–axis at the Larmor frequency, but in evaluating the net result for a set of spins we need to know whether all the xy components precess together in alignment or whether they are randomised. This difference reflects the phase relationship between the states, and it must be specified in order to give a complete description of the system. At thermal equilibrium, since no direction in the xy plane differs in energy from any other direction, nuclei precess *incoherently* or with random phase; when we apply a pulse to the sample it aligns the xy components of spins and creates a *coherence*. More generally, all pairs of levels in a coupled spin system may or may not possess phase relationships; not all such pairwise relationships give rise directly to NMR signals (the selection rules specify that only coherences between levels at either end of a single transition can do that), but they do all have characteristic frequencies with which they evolve in time.

The most fascinating property of coherences is the way in which they can be manipulated by sequences of pulses. A single pulse always creates only coherences across single transitions, but application of further pulses can cause transfer of information from one such coherence to another and also the creation of more exotic relationships between energy levels. I shall mention today only experiments involving the former process of *coherence transfer*; the essence of it is that the phase and amplitude of signals from a spin can be influenced by the prior experiences of its partners in coupling. A characteristic of transfers of this kind is that they require the development of different population differences across various transitions within the same multiplet; and of course this is only possible if the transitions are in fact different – hence the requirement for coupling to enable coherence transfer. This is an interesting contrast with the Overhauser effect, which can arise whether or not the spins involved have scalar coupling.

Two Time Dimensions

Now we have seen some of the underlying ingredients of modern NMR experiments, we need to consider how to make a 2D experiment which is a true time–domain measurement in each dimension. The trick is, in fact, remarkably simple, and can be summarised as in Figure 1. We do something to the sample, wait a while and then do something else; provided either the phase or the amplitude of the resulting NMR signals has a dependence on the duration of the time interval t_1, we can vary this interval in order to sample a second "artificial" time dimension. It is possible to illustrate the idea of modulation resulting from the variation of this time interval with a very simple system: a single NMR line

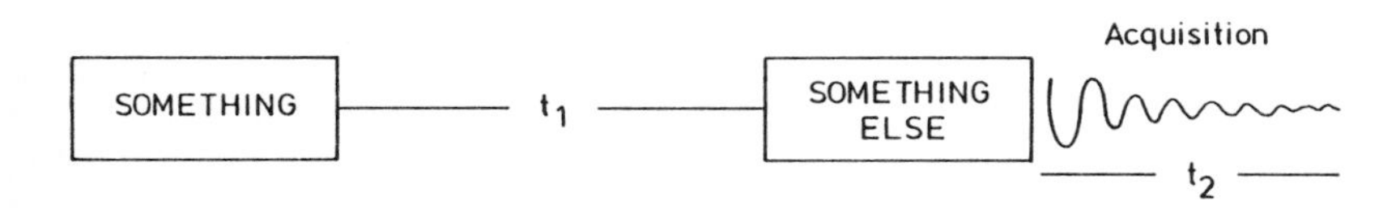

Figure 1: The basic scheme for all two–dimensional NMR experiments

with offset frequency ν, subject to a pair of 90° pulses separated by the time interval t_1 (Figure 2). After the first pulse, we can picture the state of the sample magnetisation as a vector rotating around the z axis with frequency ν, so at the beginning of t_1 it has moved through an angle $2\pi\nu t_1$. The second pulse is a rotation around the x axis in this case, so the magnetisation component proportional to $\cos(2\pi\nu t_1)$ along the y axis is returned to the z axis, while the component proportional to $\sin(2\pi\nu t_1)$ remains along the x axis.

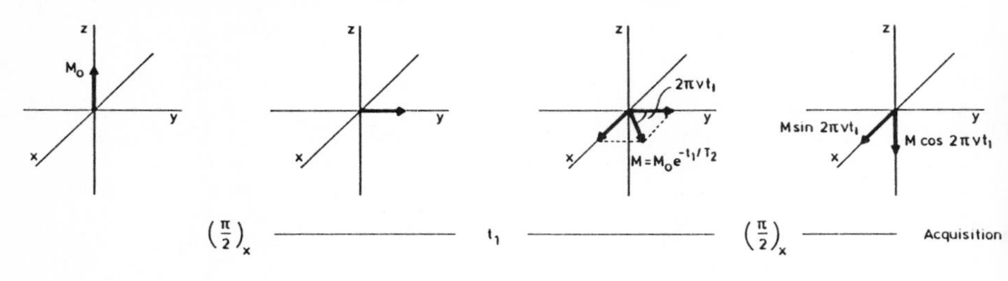

Figure 2: A simple two–pulse sequence causing amplitude modulation of an NMR signal

Thus the NMR signal measured after the second pulse has amplitude proportional to $\sin(2\pi\nu t_1)$, so that varying t_1 gives the result shown in Figure 3 after each spectrum has been transformed relative to its natural time dimension. Further Fourier transformation relative to the artificially sampled dimension therefore gives a 2D spectrum with a single peak on the diagonal (Figure 4), since the magnetisation experienced the same evolution frequency ν in each time dimension. This is a 2D experiment with no added information content, but in the presence of coupling it is possible to arrange that lines evolve with different frequencies during each time interval, leading to experiments with useful applications. In the remainder of the talk I will classify the various 2D experiments in use at the moment, without attempting to describe in detail the mechanism by which evolution during t_1 influences the behaviour of signals during t_2.

3 2D NMR AND STRUCTURE DETERMINATION

Types of 2D Experiment

Following the basic recipe I have described for constructing 2D experiments, we can obtain various types of spectra that share in common the fact that they map out diagrammatically the relationship between one frequency variable and another. The connection between the variables depends on the mechanism we choose to exploit during the transfer step after t_1. 2D experiments can generally be identified as belonging to one of three categories:

Separation of interactions
Shift correlation *via* the Overhauser efect
Shift correlation *via* scalar coupling

The first of these, which includes the various forms of J–spectroscopy, achieved prominence during the early development of 2D NMR, but is now generally agreed to offer no useful advantages in structure determination applications, so we

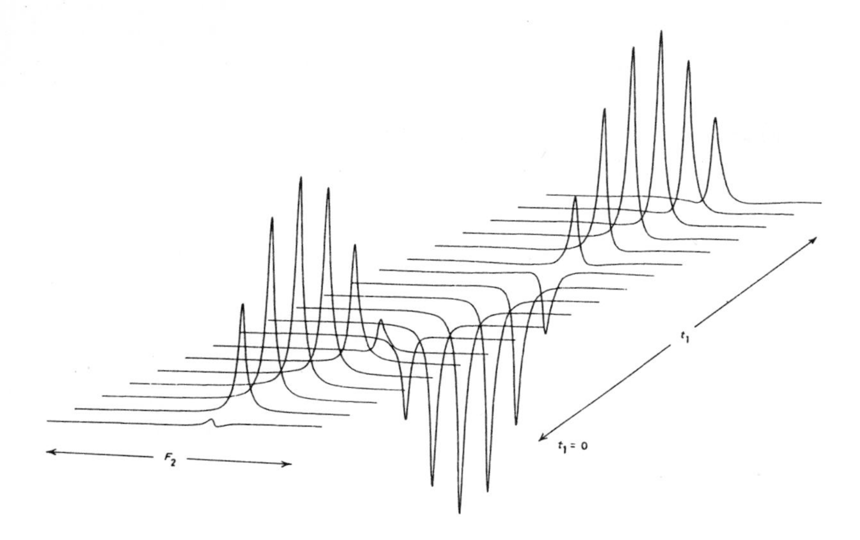

Figure 3: Experimental results obtained with the two–pulse sequence, after transformation relative to the "natural" time dimension

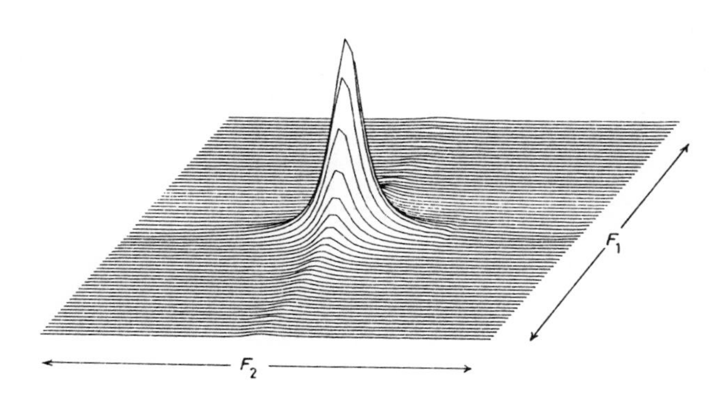

Figure 4: Further transformation relative to the "artificial" time dimension yields the complete 2D frequency spectrum

will not consider it further. Similarly, although the Overhauser effect is certainly very useful in structural studies, and the related 2D experiment is of central importance to the application of NMR to the determination of protein tertiary structure, the magnitude and the rate of growth of NOE's in small molecules makes their determination by 2D measurement problematical. A method known as

ROESY[1] (for NOE spectroscopy in the rotating frame) has considerable promise here, but at present there are a number of difficulties with the application of this technique, and a reasonable recommendation is to stick with the determination of equilibrium NOE's by the traditional 1D difference spectroscopy at present. I am therefore going to restrict the rest of the discussion to the third category, shift correlation through scalar coupling, and in fact to three examples of experiments of this kind: homonuclear shift correlation[2](COSY), heteronuclear shift correlation through 1–bond proton–carbon couplings[3], and heteronuclear shift correlation through 2– and 3–bond proton–carbon couplings[4].

Shift Correlation

In the COSY experiment, we use proton–proton coupling as the link between the two dimensions. The resulting spectrum has peaks along its diagonal corresponding with the normal proton spectrum of the sample, and off–diagonal peaks ("cross–peaks") between pairs of coupled protons (Figure 5). This is the proper 2D equivalent of the series of decoupling experiments I described earlier, and it suffers from none of the disadvantages of the latter approach. The question of selectivity of irradiations does not arise, and the unambiguous identification of couplings requires only that cross–peaks are adequately dispersed. Since cross–peak locations depend on a *pair* of chemical shifts, they are much more likely to be separated one from another than are multiplets in a 1D spectrum. Also, the experiment is a time domain measurement in each dimension, with the increase in efficiency that that implies. The information content here is precisely like that obtained from the familiar decoupling experiment, so that it is easy to apply COSY to routine problems. Really exciting advantages arise, however, only when it is used in conjunction with the next two experiments to be described.

If we use proton–carbon couplings as the connecting interaction in our 2D experiment, then we obtain various forms of heteronuclear shift correlation. Since $^{1}J_{HC}$ is substantially different in magnitude to $^{2}J_{HC}$ and $^{3}J_{HC}$ (125 – 200 Hz *versus* 0 – 15 Hz), it is possible to construct experiments to detect selectively one or the other interaction. The 1–bond experiment simply associates each proton shift with the shift of the carbon to which it is attached, which is not necessarily very useful in itself, but provides an essential link between the proton and carbon spectra. The long–range correlated experiment detects *geminal* and *vicinal* proton–carbon relationships, and in my view this is the most important information available to assist in structure determination, as I shall show shortly. First of all, however, since the detection of these long–range couplings involves some interesting experimental aspects, let us digress for a moment to consider them.

Both the usefulness and the difficulty with proton–carbon long–range couplings is that there are very many of them – sketch any structure you like and count how many protons are within two or three bonds of each carbon and you will immediately appreciate this. This is useful because it provides much information about the structure, but it means that it is difficult to optimise the coherence transfer step so as to ensure adequate intensity in each cross–peak. The usual mode in a heteronuclear correlation experiment is to allow proton frequencies to evolve in the t_1 interval, and then to transfer magnetisation to ^{13}C, which is measured as the normal NMR signal. In the long–range case, one effective scheme for this approach is known as COLOC, and generally this permits the detection of these correlations with reasonable sensitivity. An alternative approach, which has become popular recently, is to turn the experiment round so that carbon frequencies evolve during t_1 while proton signals are detected in t_2 (usually referred to as the HMBC experiment, or "inverse" detection). Now, at first sight there should be a major sensitivity gain here, because the NMR signals from ^{1}H nuclei are intrinsically 64 times bigger than those of ^{13}C, and

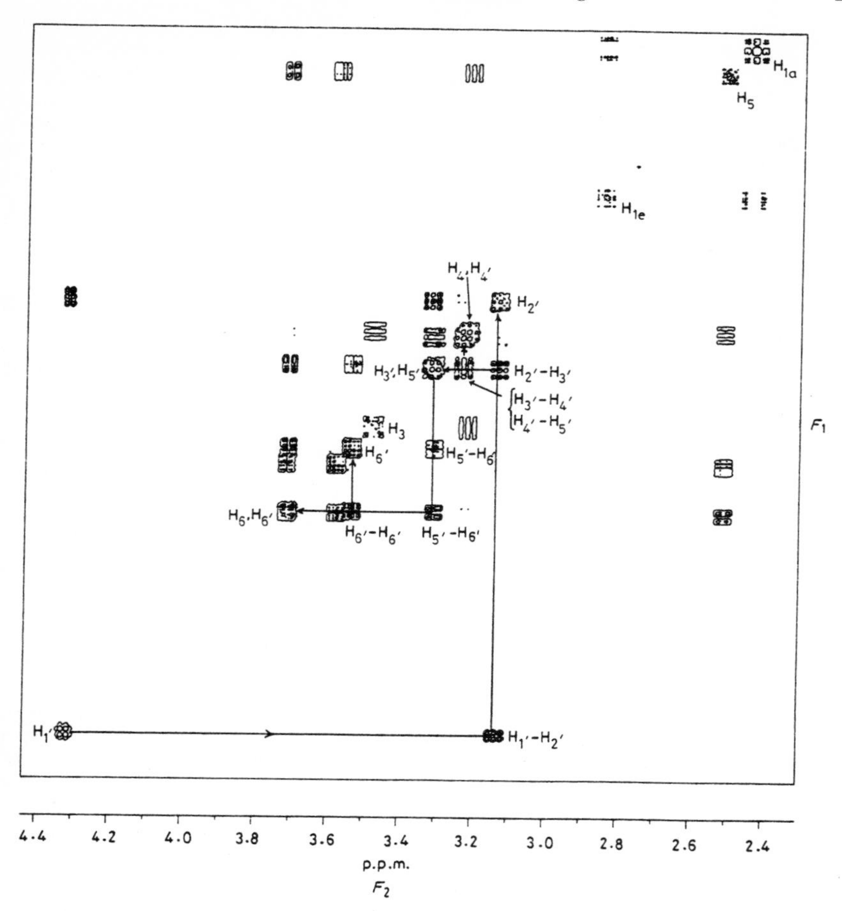

Figure 5: COSY spectrum of an unknown. The lines drawn on the spectrum trace out the route by which the coupling network is followed.

proponents of this experiment cite this as its major advantage. Unfortunately things are not quite so straightforward, since we are concerned not with *signal*, but *signal to noise ratio* in an NMR measurement, and the noise also depends on frequency (as its square root). We thus lose a factor of two immediately in this case. In addition, since the "forwards" correlation experiment involves transfer of proton population differences to carbon transitions, we have up to another factor of four reduction in effective advantage of the "inverse" approach. Further difficulties arise because the proton signals to be detected in "inverse" mode are almost invariably multiplets, perhaps with four or eight lines, whereas in the COLOC–type experiment there is no multiplet structure in either dimension – another loss in sensitivity according to the number of lines over which the signal is spread. It is also necessary to suppress the intense signals arising from protons attached to ^{12}C atoms in the "inverse" detection mode, which is a significant experimental difficulty. Thus, the argument about the relative sensitivity of the two approaches is actually rather finely balanced, and it might be as well to assume that there is little difference in practice. There is, however, a key advantage of the HMBC experiment for certain special applications, and that is that the multiplet structure of each cross–peak only contains a contribution from *one* long–range heteronuclear coupling. So if you want to actually measure the coupling constant, this is definitely the experiment to use, and applications in this area have appeared in the literature. However, we are concerned today only with detection of connections, not measurements of couplings, so we can continue operating "forwards".

Structures

Now let us consider how we can use these three experiments to find organic structures. I want to convince you that the long–range heteronuclear correlation is the key, and in order to do that it is interesting first to consider how far we can get without that experiment. Suppose as I suggested earlier that our carbon spectrum gives us, in the absence of symmetry, a single resonance for each distinct carbon site – a very reasonable assumption for most functionalised molecules with molecular weight below, say, 1000. The 1–bond hetero–correlation will let us associate each proton–bearing carbon with the ^{1}H shift(s) of its attached proton(s) – one or two shifts, since CH and CH_3 groups will certainly have only one associated ^{1}H shift, while CH_2's may have two if the protons are diastereotopic. It will probably not be so obvious as in the ^{13}C case, but in fact nearly all these proton shifts will be distinguishable provided we obtain the 2D spectrum with adequate resolution, since *exact* degeneracy of ^{1}H shifts (as opposed to partial overlap of multiplets) is rather rare, and the dispersion *via* the ^{13}C shifts will solve any problem of multiplet overlap. For clarity, let us assume that all the proton shifts are in fact different. Then we can work with the COSY spectrum to try to connect up fragments of the molecule through *vicinal* couplings, and our task is aided by the fact that we can identify and exclude *geminal* couplings by means of the offending partners' connection to the same carbon atom. Unfortunately, we soon come upon the problem that paths of H–H coupling are interrupted, for instance by heteroatoms or quaternary carbons, so although we may be able to build up some fragments – and this is very useful in itself – we will certainly not be able to assemble the whole molecular skeleton.

The potential solution to this problem lies in the long–range hetero–correlation, because this can identify many pertinent relationships, for instance couplings to carbonyls (e.g. $\underline{H}$–C–$\underline{C}$=O and $\underline{H}$–C–C–$\underline{C}$=O pathways) and couplings bridging heteroatoms (e.g. $\underline{H}$–C–O–$\underline{C}$, $\underline{H}$–N–$\underline{C}$, etc.). The difficulties are that there are very many peaks obtained in this experiment, so that one can feel rather overwhelmed with information, and that there is no general way to distinguish *geminal* from *vicinal* relationships. A practical approach is to work up the fragments available from H–H coupling as described above, and then tabulate the connections exhibited from both protons and carbons at the fragment ends to other atoms *outside* the fragment. Identification of atoms in common between the outside correlations of different fragments then assists in pairing them up. Narrowing down the search in this way makes it more approachable, and usually allows the unambiguous assembly of fragments into structures.

In conclusion, it is instructive to compare the approach to structure determination that I have just described with the traditional approach discussed earlier. Using the "hard" information available from shift correlation experiments, and taking the clearly resolved ^{13}C spectrum as our starting point, we can assemble the major part of the carbon skeleton of an organic molecule rationally and without recourse to empirical information. Of course, in realistic cases some of the assumptions mentioned previously will not be satisified, and we will be left with some degree of ambiguity as a result, so the element of fun will not be completely absent from this new approach. Finally, I note that since the nature of the determinations to be made in this method is well–defined, it lends itself well to implementation by computer, and it is tantalising to speculate that such implementations will be forthcoming shortly.

REFERENCES

1 A. A. Bothner–By, R. L. Stephens, J–M. Lee, C. D. Warren and R. W. Jeanloz, *J. Am. Chem. Soc.* 1984, **106**, 811; A. Bax and D. G. Davis, *J. Magn. Reson.* 1985, **63**, 207.

2 J. Jeener, *Ampere International Summer School*, Basko Polje 1971; R. R. Ernst, G. Bodenhausen and A. Wokaun, *Principles of Nuclear Magnetic Resonance in One and Two Dimensions*, Clarendon Press, Oxford 1987.

3 G. Bodenhausen and R. Freeman, *J. Magn. Reson.* 1977, **28**, 471.

4 H. Kessler, C. Griesinger, J. Zarbock and H. R. Loosli, *J. Magn. Reson.* 1984, **57**, 331.

COMPUTER ASSISTED CHEMICAL STRUCTURE ELUCIDATION

K. L. Mannock and J. M. Phalp
DEPARTMENT OF COMPUTER SCIENCE, BIRKBECK COLLEGE, UNIVERSITY OF LONDON
A. W. Payne
ANALYTICAL RESEARCH DIVISION, KODAK UK LTD

1 Introduction

In this paper we present the architecture for a prototype system (QuEST) to perform the task of chemical structure elucidation from spectral information using a computer as an assistant. This work differs from other approaches to this problem (most notably [1]) by utilising information presented by the chemist together with their sample. This information may include base materials, reactant, expected structure, etc.

Section 2 outlines Quantum, a system that handles the spectral management, structure editing and sample logging aspects of the system. Section 3 presents a brief overview of expert systems methodology and techniques. Section 4 describes the architecture for QuEST, the Quantum Expert System Tool that acts as the intelligent assistant in analyzing the spectral data. Section 5 discusses briefly some of the relevant related research to our work. Section 6 presents our conclusions (thus far) and our future plans.

2 Quantum

The Quantum system [2] has been developed over the last several years to provide a sample-management system and integrated reference data manager for Kodak specific samples (although it does include commercially available reference databases). The system is essential for effectively obtaining and testing empirical structure-spectrum correlations and for obtaining model compounds. In addition Quantum can inform the spectroscopist about the existence of a reference spectrum for the same or a similar structure.

Quantum comprises three integrated elements:

- a spectral data manager (SDM),
- an interactive visual chemical structure editing package (CHEMwriter), and
- a sample management system (SOFTLOG).

These three components are provided as a computer networked resource available at the various Kodak sites across the world (e.g., Harrow, England; Rochester, New York, USA; Kingsport, Tennessee, USA). This enables samples run at other laboratories in the Kodak organisation to be viewed at a site and comparisons made. The Quantum reference database holds infrared, mass and nuclear magnetic resonance spectra.

The operation of the system is as follows. The user enters the chemical structure with the program CHEMwriter at a graphics terminal. After structure entry, a data-entry menu appears. The user positions the graphics cursor over an atom and then *accesses* the atom by pressing the mouse button. Next the chemical shift is entered by accessing various menu items.

The spectral data manager (SDM) has several commands to search the database of stored spectra. These commands can be broken down into five groups (1) full structure, (2) substructure, (3) full spectrum, (4) partial spectrum, and (5) boolean expressions which combine substructure and partial spectrum search results [3].

3 Expert Systems

An expert system is a computer program designed to encapsulate the knowledge of a human *expert* in a well-defined domain.

For the purpose of this paper only rule-based expert systems will be considered. While this is a popular method for representing knowledge in expert systems other knowledge representations are described in [4]. In a rule-based expert system the domain knowledge (e.g., the knowledge of how molecules behave in the mass spectrometer) is held as coded rules in a knowledge base.

The components of an expert system are (typically) as follows:

- A User Interface.
  This is the mechanism by which the expert system and the user communicate.

- An Explanation Facility.
  This is a very important component of an expert system. An explanation facility is often used to distinguish an expert system from conventional computer programs. When the expert system has made a decision the user must be able to check the reasoning process which led to that advice.

- A Working Memory.
  The working memory contains facts which are currently available to satisfy rule constraints and lead to rule activation.

- An Inference Engine.
  This is the component of the expert system which decides in any given circumstance which rules will be activated. It may be the case that several rules can be satisfied by the facts held in working memory. In such circumstances the inference engine must be able to select a rule for activation from the set of rules which could be activated.

- A Knowledge Acquisition Facility.
  This is a facility to allow the user to automatically enter knowledge into the expert system (rather than the knowledge needing to be encoded by a knowledge engineer).

- A Knowledge Base.
  This part of the expert system contains the domain knowledge in the form of rules.

To illustrate the concept of rules consider two production rules which describe the process of crossing the road at a pedestrian crossing. The condition on the left hand side of the arrow in the rules below will precipitate the action on the right hand side.

$$\text{the light is red} \Rightarrow \text{stop} \quad \text{(rule 1)}$$
$$\text{the light is green} \Rightarrow \text{go} \quad \text{(rule 2)}$$

The two production rules may now be expressed in an equivalent pseudo-code IF THEN format.

```
Rule: Red_Light
      IF The light is red        (Rule 1)
      THEN Stop

Rule: Green_Light
      IF The light is green      (Rule 2)
      THEN Go
```

Each rule is identified by a name. Following the name is the IF part of the rule. The condition between the IF and THEN parts is known by various names (typically the *situation* part). The part of the rule following the THEN is the *action* part and gives the action(s) to be performed if the rule is activated. It is the task of the inference engine to decide whether the situation part of the rule is matched by facts present in the working memory.

The inference engine also determines the method of inferencing. This is the way in which rules are applied to deduce the solution to a problem. Several methods of inferencing exist; forward and backward chaining are common inferencing methods. DENDRAL uses a *generate-test* inferencing method. This involves a generator which *generates* all possible molecular structures which can be devised for a given empirical formula. Rules are then applied to *test* the structures generated to find those candidate structures which are compatible with the mass spectral data. There is also a rule driven planning phase which is required to restrict the large number of possible structures that can be generated for a given empirical formula. Consequently the inferencing method of DENDRAL is termed a *plan-generate-test* paradigm.

Generate-test is therefore a process of activating those rules which match the facts in working memory. The action of the rules may be to generate more facts to which the rules may be applied. The inferencing chain resulting from this process may lead to a solution being found (i.e., in the case of the DENDRAL system the correct structure may be found).

An additional inferencing method (which is used in QuEST) is *constraint satisfaction*. This method contrasts with the *plan-generate-test* method in that it may not involve an exhaustive generating phase. It is essentially the location of a solution within defined constraints.

4 QuEST

The Quantum Expert System Tool (QuEST) is an expert system module which acts as an add-on to the Quantum system. The major component of QuEST is a production rule interpreter QueRI (Quest Rule Interpreter), which operates on rules of a similar structure to those described previously. The environment in which the system is to function is as follows. The chemist logs the sample into Quantum and hands the sample on to the spectroscopists. The various tests are then run on the sample (IR, NMR and MS). For the purposes of this research mass spectroscopy is the only technique under consideration. In future developments QuEST will incorporate information from the other techniques. The sample is entered into a queue of samples which are then run via a robot sample loading facility. This enables the samples to be run as a batch without any human intervention. The various tests are then run on the sample (e.g., EI, CI, GC-MS, etc). The QuEST system then interrogates the data sets to determine whether the sample run has produced *good* data (i.e., data suitable for analysis).

The QuEST system then selects appropriate spectra to interpret given the extra information that was not considered in systems such as DENDRAL. This information typically includes the base components of the sample, the reactant, the expected structure, etc. This additional information then enforces constraints on the search of QuESTs knowledge base and the reference databases of SDM.

If an appropriate correlation can be made (utilising a peak assignment knowledge base) it is presented to the user for confirmation. The user then has the capability to ask *why* the system has come to that conclusion. A dialogue is created to present to the user the reasoning process of the system. A *how* facility is also available so that the user can investigate alternative solutions to the problem and to modify the constraints the system has applied. If the solution QuEST has arrived at is acceptable to the user then a report is produced and fed back to the chemist (via electronic mail). If the system has not produced an acceptable solution the user has two options:

1. enter into a dialogue with QuEST to further explore the available information, or

2. exit the QuEST dialogue and proceed to carry out a manual interpretation of the data with the aid of Quantum.

5 Related Work

The obvious comparison for this work is the DENDRAL project. In this short paper we therefore concentrate on an analysis of that work and its relation to our work.

5.1 The DENDRAL System

DENDRAL is a set of programs which can be used alone or in conjunction to solve the problems of chemical structure elucidation. The DENDRAL system incorporates two different approaches to this problem.

- Heuristic DENDRAL uses specific knowledge of chemistry and Mass Spectrometry. The programs associated with heuristic DENDRAL accept a Mass Spectrum and other data as input. From this information a set of chemical structures are generated. The structures so enumerated may be constrained by the user to include those which conform with user given specifications. Testing of the possible candidates can then be carried out to determine the structure which gave rise to the spectrum. This plan-generate-test paradigm forms the basis of both heuristic and meta DENDRAL.

- Meta DENDRAL differs from heuristic DENDRAL as it is a learning rather than a performance system. It accepts known structure-spectrum pairs as input and uses these pairs to infer specific knowledge about classes of compounds and their behaviour in the Mass Spectrometer. This knowledge (possibly in the form of rules) could then be used by heuristic DENDRAL to propose structures for experimental spectra.

5.2 A Consideration of the DENDRAL Project

Gray [5] argues that the process of chemical structure elucidation from spectroscopic data is essentially a 3 stage process:

1. Use of chemical and spectroscopic data to obtain constraints regarding the number and location of substructures present.
2. Enumeration of all possible candidate structures consistent with the constraints.
3. Evaluation of the candidate structures and planning additional discriminatory experiments to distinguish between the likely ones.

The DENDRAL programs are designed to assist with these 3 stages.

It is stated by Gray that these processes are basically algorithmic and other parts of the system are largely concerned only with representation of knowledge. As such Gray suggests that DENDRAL does not represent an artificial intelligence system. This point will not be developed in this paper as it is irrelevant to the practical applications of DENDRAL.

The other area in which Gray criticises the DENDRAL project is the applicability of the system to real world problems encountered by chemists. DENDRAL was developed to assist the chemist in the process of structural elucidation from mass spectral data. The need for this facility stemmed from the NASA Viking mission to Mars. The premise was to include a mass spectrometer in the Viking probe, the mass spectra obtained from sampling the Martian surface would then be transmitted back to Earth and analyzed. The aim was to carry out structure elucidation on the spectra to see if any structures such as amino acids, alcohols, ketones or amines could be detected. The presence of such compounds would lend strength to the contention that life had existed (or still existed) on the planet.

Gray states that this problem is not typical of those normally encountered by organic chemists. Rather than seeking to identify the detailed molecular composition of a simple compound Gray stated that chemists were mainly concerned with identifying molecules such as pollutants by their class. In short the task of many chemists is felt by Gray to be one of isolating and re-identifying known compounds about which information was available.

Criticism was also levelled by Gray at some of the tests carried out by workers on the DENDRAL project. These tests involved comparisons of the relative success of the DENDRAL system and graduate chemists to carry out structure elucidation given only a mass spectrum for the compound. Gray stated that such tests were misleading and cited the fact that the chemist would have other information available to assist in the interpretation in any real life situation. Such criticisms bring into question the actual achievements of the DENDRAL project.

Meta-DENDRAL was criticised by Gray on the grounds that it failed to adequately consider the effects of functional groups on fragmentation patterns. Although the Meta-DENDRAL approach enjoyed some success when applied to the steroids considered earlier in this report Gray doubted whether similar success could be achieved in other cases given the generality of the rules produced.

Gray believes the actual achievements of the DENDRAL project relate to its use of graph representations (vertex graphs) for chemical structures and its development of algorithms for enumeration of structural isomers. Also cited as an achievement of the project was the way in which it demonstrated that real world problems could be approached by computer systems given that sufficient domain specific knowledge was available.

In reply to Gray's comments workers on the DENDRAL project accuse Gray of failing to understand the basic goals of the DENDRAL project. With respect to the criticisms of the practical application of the DENDRAL programs they state that DENDRAL was not devised with the belief that a Mass Spectrum was the only source of information available to the chemist when attempting to carry out structural elucidation. The purpose of the research was to treat the mass spectrum as an exemplary source of information about the structure, other sources of information had also been considered.

6 Conclusions

In this paper we have briefly described the QuEST system and its relation to Quantum and DENDRAL. The main aim of our research is to address the problem of whether an experimental spectrum is consistent with a chemical structure. The structure has been suggested by a chemist to have given rise to the spectrum. The problem is therefore one of assessing whether a structure and a spectrum are consistent. This is a somewhat different problem from the problem of full structure elucidation addressed by the DENDRAL project.

Despite the differences between the aims of the DENDRAL project and our research we can identify several areas of the DENDRAL project which may be of use in our work.

- Planning and Testing Phases.
  Both these phases carry out ***break analysis*** on candidate structures. The fragments produced by the breaks will produce spectral features. Experimental spectra may be checked for these features.

- Location of Molecular Ions.
  One of the interests of our research has been to carry out analyses on the whole data set (over 100 spectra) collected for a mass spectrometry run. The purpose of this is to try to solve problems arising due to the presence of impurities in the sample prior to analyzing the peaks in an individual spectrum. Subject to our finding a suitable method for locating these components and obtaining relatively pure spectra for them, then the MOLION program would be useful in determining possible molecular weights for these components. Proposed molecular ion candidates could then be compared with reported starting materials for the reaction carried out to attempt to locate the source of the impurity.

- Representation of Chemical Structures and Spectra.
  The ability to represent structures in terms of classes of compounds by using structural skeletons will be of use. It is not envisaged that structure generation will initially be a feature in our system. Similar methods to those used by DENDRAL to assess and represent significant features in mass spectra may be used in our work.

- Representing Knowledge as Rules.
  The production rule paradigm used by DENDRAL appears to be a successful

means of representing chemical knowledge. A rule based approach is of current interest in this project. It is considered more appropriate than alternative representations such as object oriented systems. Efficient pattern matching in rule-based systems is now common place allowing such systems to operate efficiently.

- The Experimental Approach.
  DENDRAL used comparisons of the performance of the system with chemists. As such it was criticised [5] for making unrealistic comparisons as it was felt that the task was not one generally encountered by chemists. It is also our intention to carry out comparisons of the systems performance with that of chemists. We justify this approach as the task of correlating a structure and a spectrum to decide whether they are consistent is known to be a process carried out by chemists. In addition the system will be tested for its ability to confirm that paired structures and spectra stored in a reference database are consistent. The important feature of both these tests is that they will both be carried out on real data for compounds of interest to Kodak.

Acknowledgments

This work is funded by a grant from Eastman Kodak.

Disclaimer

The views presented in this paper are those of the authors and are not attributable to Kodak UK Ltd.

References

[1] R.K. Lindsay, B.G. Buchanan, E.A. Feigenbaum, and J. Lederberg. *Applications of artificial intelligence for organic chemistry. The DENDRAL project.* McGraw-Hill Book Company, 1980.

[2] C.A. Shelley. Problems that prevent computer-assisted structure elucidation from becoming a practical tool. In *Computer-supported Spectroscopic Databases*, Eliis Horwood Limited, 1986.

[3] C.A. Shelley and M.E. Munk. *Analytical Chemistry*, 54, 1982.

[4] J.C. Giarratano and G. Riley. *Expert Systems: Principles and Programming.* PWS-KENT Publishing Company, Boston, 1989.

[5] N.A.B. Gray. Dendral and meta-dendral- the myth and the reality. *Chemometrics and Intelligent Laboratory Systems*, 5:11–22, 1988.

AUTOMATED SPECTROSCOPY AND CHEMOMETRICS

B. Davies

ANALYTICAL SYSTEMS DEVELOPMENT DEPARTMENT, GLAXO OPERATIONS UK LTD, BARNARD CASTLE, CO. DURHAM, DL12 8DT

1 INTRODUCTION

Increasingly we are finding an emerging role for the analytical chemist focusing on providing rapid and relevant analytical measurements for production control purposes. This may be used for final product testing prior to release or in-process testing for control purposes. The use of highly controlled automated analysis systems placed in production areas and run by production based staff is seen as key to meeting these objectives.

The major analytical sensors in these process analysis systems are spectroscopic, using UV, NIR and FTIR technologies. Chemometric data analysis techniques are used to calibrate the often complex data sets needed to model process systems and to predict the properties of interest in the samples. This paper will discusss the role of automation in sample preparation and the role of automated data reduction techniques in the data to information transform required to provide analytical science answers to production control problems. Examples will be drawn from experiences in pharmaceutical applications but the approaches are readily modified to suit other environments.

2 WHY AUTOMATE ?

A key question to consider is what are the benefits of automating an analytical process ?

Increased efficiency and optimisation of the use of what may be high cost analytical equipment is often given as the major benefit. Replacing or releasing analytical staff usually comes next on the list. Technology transfer is not usually on the list but automation has a major part to play in controlling the analytical process and therefore the efficient transfer of analytical methodology between laboratories and from laboratories into production areas.

The pressures on production areas are increasingly to find alternative means of improving the production of products at optimum costs and consistent high quality timed to meet their customer requirements. Analytical science has a major part to play in assisting production areas to meet these objectives. In essence this will impact in two areas

- by allowing the release of product onto the market immediately after processing.

- by impacting on the level of inventory required to sustain any given operation.

To accomplish this the traditional route of the sample to a laboratory needs to be reviewed. The time overheads of traditional laboratory scheduling and testing are unacceptable in this scenario. Analysis must be performed at or near to the process area. And to successfully advance analytical science into the production environment requires technology transfer to occur. In analytical chemistry, technology transfer is the moving of analytical testing from a central laboratory function into a process area or factory. It does not necessarily mean simply moving equipment and personnel out of the lab and into the factory. Many analytical systems are simply not robust enough in both hardware and analytical terms to perform adequately on the factory floor.

Technology transfer is therefore dynamic providing an opportunity to reappraise both the information required to control the production process and the most effective analytical technique applied to the problem. Is the analysis system there to meet some pre-shipping final test requirement which requires a precise quantitative assay or will an objective qualitative measurement of product quality or conformity suffice ?

It is also an opportunity to give ownership of the measurement technique to the process staff responsible for making products by providing almost immediate feedback on the quality of their products. This ownership of analytical systems by process people poses the problem of what level of expertise is required by its users ? Ideally process operators should be able to run analytical systems with little or no analytical chemistry skills. Detailed technical or scientific support of the system being the responsibility of a highly skilled group of support analysts. The ideal analytical system for this type of application is therefore one that requires little or no sample preparation, can be integrated into a computer controlled system and is analytically robust enough to meet the information demand. These requirements lead logically to the use of spectroscopic sensors.

3 WHY SPECTROSCOPY ?

Modern spectrometers coupled with the latest in sample handling technologies amply meet the basic requirements set down above. They require little or no sample preparation for many types of materials to be analysed . Diffuse reflectance measurements in the NIR often simply require a sample to poured from its container into a cup and the spectrum measured using an integrating sphere. In the FTIR region CIR flowcells or their equivalent represent a simple sample interface for liquids of solutions and in the UV flowcells are now firmly established as the most robust means of introducing samples to the spectrometers.

Spectrometers are readily controlled through external computer systems which can play the role of orchestrating the analysis. Routine users are not required to have any knowledge of setting up the instrument scan or data collection options or the 'trading rules' of the technique as the optimum conditions are programmed from the control computer through the interface to the spectrometer. At the user level the spectrometer can be thought of as a 'black box' or sensor. It is in the development phase of the analytical system that the detailed understanding of the spectroscopy is vital and this together with software safeguards and fail safes must be programmed into the final control method for the analysis.

The control of the sensor system is vital in guaranteeing the robustness of the total analytical process. If the system is not to be run by analytical chemists then the users cannot and should not be required to make decisions on the system performance. This should be performed by the control computer through its software method. Instrument specifications can be programmed into the method and tested in a totally objective way by use of suitable standards. The operator can be prompted to introduce the samples to the spectrometers or an automated sampling system can be used. The system then obtains the standard's response and evaluates this against the expected response in accordance with the instrument specification. If the results do not comply with the specification then the system is shut down, fails safe and the user is requested to call for technical support. If the operator fails to present the sample properly or introduces an incorrect standard the system will refuse to assay materials. The process of system suitability testing is removed from the hands of the routine operator and a very high degree of robustness is introduced into the system.

4 AUTOMATED SAMPLE PREPARATION

Some samples may not be in a form or concentration range suitable for direct introduction into a spectroscopic system, some degree of sample work-up or preparation must be performed before a spectrum can be obtained. UV measurements in particular often require a sample to be dissolved and/or diluted before measurements can be made. Again this is an area where automation can be used to aid the routine user by its control of the sample preparation. Integrating a flexible, software programmed automation system with a spectrometer and its flowcell can dissolve, dilute and pump samples with a minimum of operator interaction. The chore of accurate weighing, dilution and quantitative transfer of materials is removed from the human and placed under very precise automated control.

Method Creation Language

Control of these individual hardware modules is through the use of a software language developed in our laboratories. The Method Creation Language (MCL).

MCL is a high level analytical applications orientated language designed to allow automated systems to be configured by the non computer specialist analytical scientist. Once the application has been developed and tested it can be secured and used in non laboratory areas by non specialist staff. A diagram of such a system is shown in Fig 1. Here an automation system is involved in weighing out the test sample, dissolving the solid and diluting the solution to a concentration suitable for measurement in a UV flow cell. All the routine functions of priming, calibration and testing of the hardware prior to running a sample must be completed successfully before a sample can be measured. The system has complete control of the analysis through to the generation of the final report.

Sample Interfaces

Clearly the best sample preparation method is no sample preparation and for many solid materials spectroscopy in the Near Infrared region can allow this. Samples can be poured from their containers directly into a sample cup and placed into an integrating sphere where a diffuse reflectance or transmission measurement can be made.

Spectroscopy in the Mid-IR is somewhat complicated for solid materials as in most cases solid samples must be diluted in KBr or Nujol before presentation to the spectrometer. Both techniques pose little problems to an accomplished analyst but are not readily automated. Diffuse reflectance in the Mid-IR also generally requires some dilution of the sample material and for precise quantitative analyses is not readily automated. Liquids and semi-solids can easily be integrated into an automated system. For liquids, transmission flow cells are readily available over a wide range of pathlengths and wavelength ranges. For liquids requiring extremely short pathlength cells the alternative approach of using ATR and Circular ATR cells is now firmly established. These approaches are of great use in dealing with aqueous solutions. Semi-solid, pastes, ointments and creams can be sampled again using ATR, this time in a horizontally mounted Contact Sampler. Putting the samples in such accessories as these can be done simply by a non-spectroscopist and do not require automation.

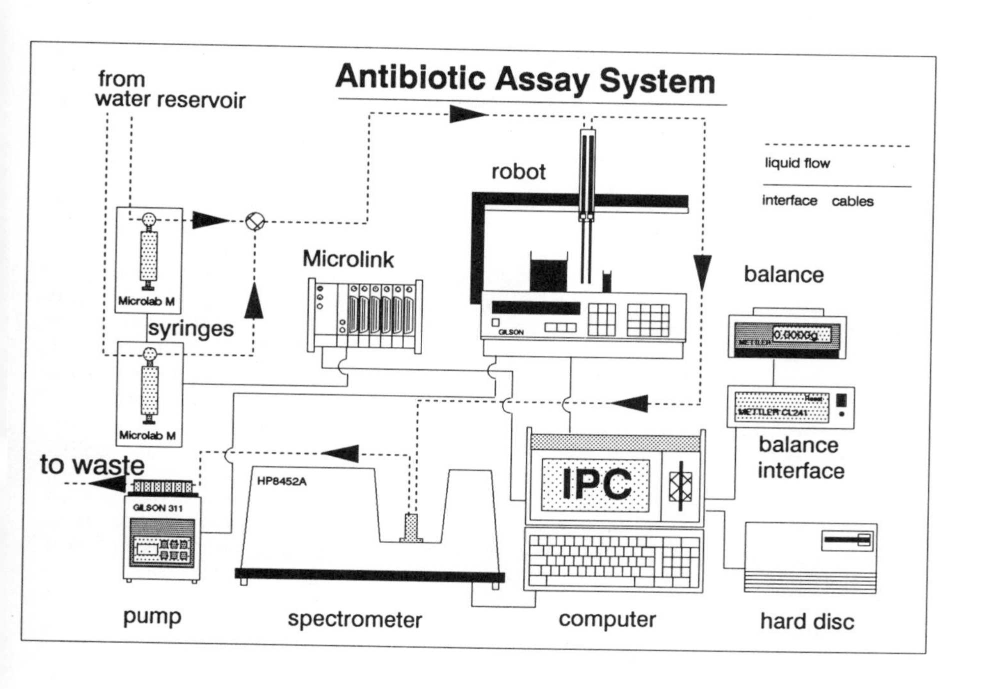

FIGURE 1 Diagram of an Automated Sample Preparation System for UV Process Spectroscopy

Within the pharmaceutical industry UV is now becoming established as an alternative technique to that of HPLC for many non stability indicating assays.

An increasing number of UV assays are being registered with the licensing authorities as part of the batch release tests necessary before a product can be delivered to a customer. Invariably, as most samples are solids some sort of sample preparation must be performed on the material before its spectrum can be obtained. As discussed above this is usually no more than dissolution and subsequent dilution of the material to match its solution concentration and a matching of the cell pathlength to optimise the spectroscopic measurement. These types of sample handling tasks are readily automated.

5 CHEMOMETRIC DATA REDUCTION

Automated spectroscopy does not stop at the sample preparation and presentations stage. The role of automation continues into the data reduction phase of an assay. Properly implemented automation of sample preparation coupled with controlled and calibrated instrumentation will provide a highly valid source of spectroscopic data. This must then be converted into relevant information before it can be successfully used in a production environment.

One potential drawback of using little or no sample preparation is that of the complexity of the resultant spectrum. Not only is the spectrum of the analyte of interest obtained but also any spectral signature of the sample matrix. In many cases the spectrum of the matrix material may swamp the analyte spectrum or make it difficult to identify analytical features which correlate with analyte concentration. Almost certainly the spectrum of the analyte and matrix will be so overlapped, especially in the Mid-IR, that traditional quantitative techniques relying on peak height or area are rendered useless.

Running parallel with the developments in automation and computer assisted spectroscopic development have been the developments in multi-variate statistical techniques for exploring large sets of data. Mathematical, statistical and expert systems approaches have been brought together to solve a variety of problems in analytical science. One of the problem areas is the analytical spectroscopy of complex mixtures of the type discussed above.

Spectra and Multivariate Methods

Spectra represent a rich source of multivariate data with each wavelength element or digitised data point representing a variable. The information resident in a spectrum is spread throughout these variables. Depending on the technique and sampling methods used on a sample this information can be both of a qualitative and quantitative nature.

The chemometric techniques of data reduction are used to investigate spectral data sets and isolate areas, patterns or latent variables in the data which can be correlated with the information of interest. If a correlation is established then a mathematical model linking the spectral data to the information is produced and this model then used to predict the properties of unknown test materials from their spectra.

Chemometric methods fall broadly into two categories, those using discriminant analysis and those using multivariate calibration techniques. Discriminant analysis gives qualitative information from samples and multivariate calibration qualitative and quantitative information. Both approaches have been used successfully within Glaxo to tackle production based spectroscopic problems.

Calibration of Chemometric Techniques

Chemometric techniques require a calibration or training set of spectra to build their model of the system under investigation. It should be realised that this model will represent the total knowledge the mathematics has of the system it will ultimately predict the chemical properties of. If the training set of spectra do not include all the relevant sources of variation likely to be found in the test samples then the model cannot be expected to model them with any degree of precision. It may then be seriously compromised in predicting the properties of unknown materials using this limited model. These sources of variation are especially important in developing models of analytes to be measured in product matrices. Here the variances seen in the calibration spectra should not only represent those associated with different analyte concentrations but should extend out to include sample matrix effects and physical characteristics of the material. In quantitative analysis applications the calibration samples must represent the working range of analyte concentrations of the material and be chosen to include any interaction effects between components or components and the matrix. Ideally these calibration samples should therefore contain all of the sources of chemical and matrix variances likely to be seen in the final test samples. With such a training set the chemometric technique can then construct a robust model with good predictive ability. Considerable thought must be put into the generation of such a training set. Often the analytical scientists must work closely with their production colleagues to identify the critical sources of variances and the optimum method of introducing them into a learning set.

Traditional approaches to producing such a training set may involve the preparation of many tens of samples if all the identified sources of variation are to be emulated in the calibration materials. To minimise this workload experimental design techniques can be used to suggest a scheme for the calibration set which minimise the number of discrete samples that must be prepared while maximising the information content of the samples. These techniques are widely available in text books and as computer programs and can represent a major saving of laboratory resources.

Often a set of realistic training materials cannot be produced in a laboratory because of the complexity of the system under investigation. It is very difficult or impossible to duplicate fully industrial processing in a laboratory even with access to pilot plant facilities. The approach of calibration using precisely controlled laboratory produced materials must be replaced by turning to the 'real' production materials as a source of calibration.

Here actual production samples are assayed using some other analytical technique. These assay values are then used with the spectra of the samples to establish the model. The potential weak link in this exercise is the analytical technique used to analyse the production samples. If the technique has a poor accuracy or precision then the assay values it produces will compromise the spectroscopic assay it is being used to calibrate. Considerable care and effort must be put into this type of secondary calibration of a spectroscopic assay.

Modelling Techniques

Discriminant Analysis. In qualitative analysis applications in the NIR discriminant analysis, cluster analysis and pattern recognition techniques are used. Here groups of spectra representing different chemical materials or classes are distinguished from each other using mathematical distance metrics derived from their spectra. These distance metrics are then used to classify an unknown material's spectrum as belonging to or not belonging to a set of spectra in a database. The application would be that of a classical library search or material identification. The basic approach is that of a supervised learning, where a set of spectra of materials of validated chemical composition are used as a training set. They are used to develop a computational decision rule for evaluating similarity or dissimilarity between spectra. These techniques are highly objective in their comparison of spectra and are preferable to the traditional methods of comparing unknown spectra with those in a library by eye.

Within Glaxo the discrimination decision rules used for raw material identification applications have been developed using Mahalanobis metric calculations.

<u>Quantitative Techniques.</u> The most well developed of the chemometric techniques are those used to develop multicomponent analyses. These are used in the Mid-IR and NIR and are now extending into the UV region of the spectrum. The statistical tools used for these applications within Glaxo are generally those of Principal Component Regression (PCR) and Partial Least Squares (PLS).

PCR uses factor analysis coupled with multiple linear regression to establish correlations between spectra and the properties of interest in the samples. Factor analysis is a technique for reducing the number of dimensions in a large set of data by extracting from them a set of factors which represent the number of independently varying contributions to the variance in the data. These factors present the underlying pattern of variance in an easily interpreted form. When the technique is applied to spectra it can represent a spectrum by a smaller set of numbers than in the original set of wavelengths. This is of particular use when a Mid-IR spectrum can typically consist of over 1000 data points and a calibration set may include over 50 of such spectra.

These factors when combined linearly enable the original spectra to be reconstructed. The contributions of the factors necessary to reconstruct an original spectrum are known as factor loadings. Each calibration spectrum is represented as a set of these loadings which correspond to the contributions of each factor to that spectrum. A spectrum of typically 1000 data points may therefore be represented by only several of these loading coefficients.

Multiple Linear Regression (MLR) is used to establish any correlations between the factor loadings and the composition of the calibration samples. Once established this correlation can be used in the analysis of unknown samples. The factor loadings of the unknown material's spectrum are multiplied by the appropriate coefficients from the regression to yield the composition of the unknown sample.

Partial Least Squares is a similar approach to PCR but here Principal Components are extracted from both the Spectral data and the composition values for the calibration set of spectra. These are then used through iteration to build a model of the system. The chemical composition data is used to identify patterns in the spectroscopic data that correlate with them. This has the effect of ensuring that the estimated regression factors have direct relevance to the chemical values. In PCR the regression stage effectively screens its factors for relevance to the chemical properties of interest. In principle PLS should be able to model a Spectra/Composition system with fewer and more relevant dimensions than PCR.

Currently the programs of use within Glaxo are CIRCOM which uses PCR and UNSCRAMBLER which uses PLS modelling. Both programs are capable of handling UV, NIR and Mid-IR data sets.

6 APPLICATIONS

UV

Applications of UV spectroscopy in process areas have mainly been in the automation and control of assays registered as release tests for some products. Here the UV sensor is front ended by automated sample preparation/presentation systems and forms part of an integrated analytical system.

Near Infrared

NIR is principally used within Glaxo for the identification of some incoming raw materials. Instruments have been placed in or adjacent to stores receipt areas and are run routinely by non-spectroscopists. A large multi-national database of NIR spectra is now established and using a set of spectral transfer standards data can be transferred between instruments.

Some quantitative applications have been established for monitoring powder blends, solvent mixtures and whole tablet core analysis.

Mid-IR

The Mid-IR is used in applications requiring minimal sample preparation and where the concentration ranges of the materials of interest are too low to be identified by NIR. Current applications are in the analysis of ointments, injections, and solvent recovery monitoring.

7 SUMMARY

The process for developing analytical methodologies which can be transferred to production areas is now well established and continuing to gain momentum. The benefits of placing these systems in such areas will continue to drive the search for other means of providing rapid and robust measurement systems. Molecular spectroscopy, in particular UV, NIR and Mid-IR are ideally placed to act as sensors in such systems.

The use of advanced statistical methods, chemometrics, as calibration tools is enabling assays to be developed on multicomponent systems previously regarded as to complex for spectroscopic measurements and requiring the use of separative techniques. These tools are seen as a supplement to the skills of the spectroscopist and analytical scientist rather than their replacement.

Within this decade we shall see an increasing use of these spectroscopic tools as they are extended in applications of parametric release and non destructive testing.

ACKNOWLEDGEMENTS

Mr S L Boucher
Mr C M Harland
Miss C Harris
Dr C J P Scott

of the Analytical Systems Development Department, Glaxo Operations UK Ltd., Barnard Castle

Subject Index

Author Index